Lawrence C. Paulson

Isabelle

A Generic Theorem Prover

With Contributions by Tobias Nipkow

Springer-Verlag
Berlin Heidelberg New York
London Paris Tokyo
Hong Kong Barcelona
Budapest

Series Editors

Gerhard Goos
Universität Karlsruhe
Postfach 69 80
Vincenz-Priessnitz-Straße 1
D-76131 Karlsruhe, Germany

Juris Hartmanis
Cornell University
Department of Computer Science
4130 Upson Hall
Ithaca, NY 14853, USA

Author

Lawrence C. Paulson
Computer Laboratory, University of Cambridge
Pembroke Street, Cambridge CB2 3QG, United Kingdom

CR Subject Classification (1991): F.4.1, F.4.3, F.3.1, D.2.4, I.2.3

ISBN 3-540-58244-4 Springer-Verlag Berlin Heidelberg New York
ISBN 0-387-58244-4 Springer-Verlag New York Berlin Heidelberg

CIP data applied for

Printed in Germany

Typesetting: Camera-ready by author
SPIN: 10472615 45/3140-543210 - Printed on acid-free paper

To Nathan and Sarah

You can only find truth with logic
if you have already found truth without it.

G.K. Chesterton, *The Man who was Orthodox*

Preface

Most theorem provers support a fixed logic, such as first-order or equational logic. They bring sophisticated proof procedures to bear upon the conjectured formula. The resolution prover Otter [60] is an impressive example.

ALF [30], Coq [14] and Nuprl [9] each support a fixed logic too. These are higher-order type theories, explicitly concerned with computation and capable of expressing developments in constructive mathematics. They are far removed from classical first-order logic.

A diverse collection of logics — type theories, process calculi, λ-calculi — may be found in the Computer Science literature. Such logics require proof support. Few proof procedures are known for them, but the theorem prover can at least automate routine steps.

A **generic** theorem prover is one that supports a variety of logics. Some generic provers are noteworthy for their user interfaces [11, 29, 55]. Most of them work by implementing a syntactic framework that can express typical inference rules. Isabelle's distinctive feature is its representation of logics within a fragment of higher-order logic, called the meta-logic. The proof theory of higher-order logic may be used to demonstrate that the representation is correct [43]. The approach has much in common with the Edinburgh Logical Framework [24] and with Felty's [18] use of λProlog to implement logics.

An inference rule in Isabelle is a generalized Horn clause. Rules are joined to make proofs by resolving such clauses. Logical variables in goals can be instantiated incrementally. But Isabelle is not a resolution theorem prover like Otter. Isabelle's clauses are drawn from a richer language and a fully automatic search would be impractical. Isabelle does not resolve clauses automatically, but under user direction. You can conduct single-step proofs, use Isabelle's built-in proof procedures, or develop new proof procedures using tactics and tacticals.

Isabelle's meta-logic is higher-order, based on the simply typed λ-calculus. So resolution cannot use ordinary unification, but higher-order unification [27]. This complicated procedure gives Isabelle strong support for many logical formalisms involving variable binding.

The diagram below illustrates some of the logics distributed with Isabelle. These include first-order logic (intuitionistic and classical), the sequent calculus, higher-order logic, Zermelo-Fraenkel set theory [56], a version of Constructive Type Theory [39], several modal logics, and a Logic for Computable Functions [42]. Several experimental logics are being developed, such as linear logic.

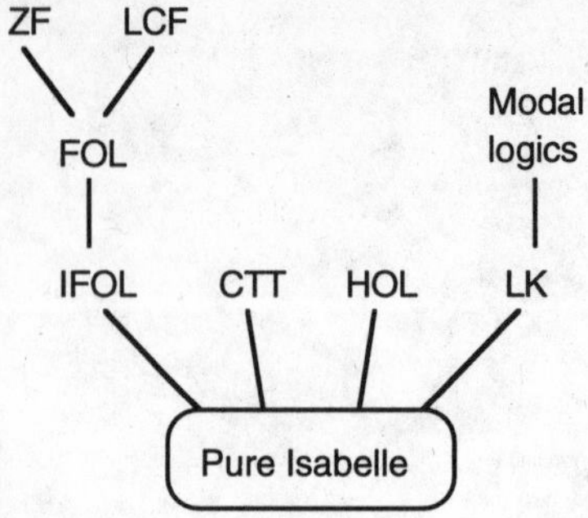

How to read this book

Isabelle is a complex system, but beginners can get by with a few commands and a basic knowledge of how Isabelle works. Some knowledge of Standard ML is essential because ML is Isabelle's user interface. Advanced Isabelle theorem proving can involve writing ML code, possibly with Isabelle's sources at hand. My book on ML [46] covers much material connected with Isabelle, including a simple theorem prover.

The Isabelle documentation is divided into three parts, which serve distinct purposes:

- *Introduction to Isabelle* describes the basic features of Isabelle. This part is intended to be read through. If you are impatient to get started, you might skip the first chapter, which describes Isabelle's meta-logic in some detail. The other chapters present on-line sessions of increasing difficulty. It also explains how to derive rules define theories, and concludes with an extended example: a Prolog interpreter.

- *The Isabelle Reference Manual* provides detailed information about Isabelle's facilities, excluding the object-logics. This part would make boring reading, though browsing might be useful. Mostly you should use it to locate facts quickly.

- *Isabelle's Object-Logics* describes the various logics distributed with Isabelle. The chapters are intended for reference only; they overlap somewhat so that each chapter can be read in isolation.

This book should not be read from start to finish. Instead you might read a couple of chapters from *Introduction to Isabelle*, then try some examples referring to the other parts, return to the *Introduction*, and so forth. Starred sections discuss obscure matters and may be skipped on a first reading.

Releases of Isabelle

Isabelle was first distributed in 1986. The 1987 version introduced a higher-order meta-logic with an improved treatment of quantifiers. The 1988 version added limited polymorphism and support for natural deduction. The 1989 version included a parser and pretty printer generator. The 1992 version introduced type classes, to support many-sorted and higher-order logics. The 1993 version provides greater support for theories and is much faster.

Isabelle is still under development. Projects under consideration include better support for inductive definitions, some means of recording proofs, a graphical user interface, and developments in the standard object-logics. I hope but cannot promise to maintain upwards compatibility.

Isabelle is available by anonymous ftp:

- University of Cambridge
  host `ftp.cl.cam.ac.uk`
  directory `ml`

- Technical University of Munich
  host `ftp.informatik.tu-muenchen.de`
  directory `local/lehrstuhl/nipkow`

The electronic distribution list `isabelle-users@cl.cam.ac.uk` provides a forum for discussing problems and applications involving Isabelle. To join, send me a message via `lcp@cl.cam.ac.uk`. Please notify me of any errors you find in this book.

Acknowledgements

Tobias Nipkow has made immense contributions to Isabelle, including the parser generator, type classes, the simplifier, and several object-logics. He also arranged for several of his students to help. Carsten Clasohm implemented the theory database; Markus Wenzel implemented macros; Sonia Mahjoub and Karin Nimmermann also contributed.

Nipkow and his students wrote much of the documentation underlying this book. Nipkow wrote the first versions of Sect. 3.2, Sect. 9.1, Chap. 10, Chap. 13 and App. A. Carsten Clasohm contributed to Chap. 9. Markus Wenzel contributed to Chap. 11. Nipkow also provided the quotation at the front.

David Aspinall, Sara Kalvala, Ina Kraan, Chris Owens, Zhenyu Qian, Norbert Völker and Markus Wenzel suggested changes and corrections to the documentation.

Martin Coen, Rajeev Goré, Philippe de Groote and Philippe Noël helped to develop Isabelle's standard object-logics. David Aspinall performed some useful research into theories and implemented an Isabelle Emacs mode. Isabelle was developed using Dave Matthews's Standard ML compiler, Poly/ML.

The research has been funded by numerous SERC grants dating from the Alvey programme (grants GR/E0355.7, GR/G53279, GR/H40570) and by ESPRIT (projects 3245: Logical Frameworks and 6453: Types).

Table of Contents

I Introduction to Isabelle 1

II The Isabelle Reference Manual 63

List of Figures

Part I

Introduction to Isabelle

1. Foundations

The following sections discuss Isabelle's logical foundations in detail: representing logical syntax in the typed λ-calculus; expressing inference rules in Isabelle's meta-logic; combining rules by resolution.

If you wish to use Isabelle immediately, please turn to page 25. You can always read about foundations later, either by returning to this point or by looking up particular items in the index.

1.1 Formalizing logical syntax in Isabelle

Figure 1.1 presents intuitionistic first-order logic, including equality. Let us see how to formalize this logic in Isabelle, illustrating the main features of Isabelle's polymorphic meta-logic.

Isabelle represents syntax using the simply typed λ-calculus. We declare a type for each syntactic category of the logic. We declare a constant for each symbol of the logic, giving each n-place operation an n-argument curried function type. Most importantly, λ-abstraction represents variable binding in quantifiers.

Isabelle has ML-style polymorphic types such as $(\alpha)list$, where *list* is a type constructor and α is a type variable; for example, $(bool)list$ is the type of lists of booleans. Function types have the form $(\sigma, \tau)fun$ or $\sigma \Rightarrow \tau$, where σ and τ are types. Curried function types may be abbreviated:

$$\sigma_1 \Rightarrow (\cdots \sigma_n \Rightarrow \tau \cdots) \quad \text{as} \quad [\sigma_1, \ldots, \sigma_n] \Rightarrow \tau$$

The syntax for terms is summarised below. Note that function application is written $t(u)$ rather than the usual $t\,u$.

$t :: \tau$	type constraint, on a term or bound variable
$\lambda x \,.\, t$	abstraction
$\lambda x_1 \ldots x_n \,.\, t$	curried abstraction, $\lambda x_1 \,.\, \ldots \lambda x_n \,.\, t$
$t(u)$	application
$t(u_1, \ldots, u_n)$	curried application, $t(u_1) \ldots (u_n)$

1.1.1 Simple types and constants

The syntactic categories of our logic (Fig. 1.1) are **formulae** and **terms**. Formulae denote truth values, so (following tradition) let us call their type o. To

$$\neg P \quad \text{abbreviates} \quad P \to \bot$$
$$P \leftrightarrow Q \quad \text{abbreviates} \quad (P \to Q) \land (Q \to P)$$

$$\frac{P \quad Q}{P \land Q}\,(\land I) \qquad \frac{P \land Q}{P}\,(\land E1) \qquad \frac{P \land Q}{Q}\,(\land E2)$$

$$\frac{P}{P \lor Q}\,(\lor I1) \qquad \frac{Q}{P \lor Q}\,(\lor I2) \qquad \frac{P \lor Q \quad \begin{array}{c}[P]\\ \vdots\\ R\end{array} \quad \begin{array}{c}[Q]\\ \vdots\\ R\end{array}}{R}\,(\lor E)$$

$$\frac{\begin{array}{c}[P]\\ \vdots\\ Q\end{array}}{P \to Q}\,(\to I) \qquad \frac{P \to Q \quad P}{Q}\,(\to E)$$

$$\frac{\bot}{P}\,(\bot E)$$

$$\frac{P}{\forall x \,.\, P}\,(\forall I)* \qquad \frac{\forall x \,.\, P}{P[t/x]}\,(\forall E)$$

$$\frac{P[t/x]}{\exists x \,.\, P}\,(\exists I) \qquad \frac{\exists x \,.\, P \quad \begin{array}{c}[P]\\ \vdots\\ Q\end{array}}{Q}\,(\exists E)*$$

$$t = t\ (\mathit{refl}) \qquad \frac{t = u \quad P[t/x]}{P[u/x]}\,(\mathit{subst})$$

**Eigenvariable conditions*:
$\forall I$: provided x is not free in the assumptions
$\exists E$: provided x is not free in Q or any assumption except P

Fig. 1.1. Intuitionistic first-order logic

allow 0 and $Suc(t)$ as terms, let us declare a type *nat* of natural numbers. Later, we shall see how to admit terms of other types.

After declaring the types o and *nat*, we may declare constants for the symbols of our logic. Since $\bot$ denotes a truth value (falsity) and 0 denotes a number, we put

$$\begin{aligned} \bot &:: o \\ 0 &:: nat. \end{aligned}$$

If a symbol requires operands, the corresponding constant must have a function type. In our logic, the successor function (*Suc*) is from natural numbers to natural numbers, negation ($\neg$) is a function from truth values to truth values, and the binary connectives are curried functions taking two truth values as arguments:

$$\begin{aligned} Suc &:: nat \Rightarrow nat \\ \neg &:: o \Rightarrow o \\ \wedge, \vee, \rightarrow, \leftrightarrow &:: [o, o] \Rightarrow o \end{aligned}$$

The binary connectives can be declared as infixes, with appropriate precedences, so that we write $P \wedge Q \vee R$ instead of $\vee(\wedge(P, Q), R)$.

Section 3.2 below describes the syntax of Isabelle theory files and illustrates it by extending our logic with mathematical induction.

1.1.2 Polymorphic types and constants

Which type should we assign to the equality symbol? If we tried $[nat, nat] \Rightarrow o$, then equality would be restricted to the natural numbers; we should have to declare different equality symbols for each type. Isabelle's type system is polymorphic, so we could declare

$$= \;::\; [\alpha, \alpha] \Rightarrow o,$$

where the type variable α ranges over all types. But this is also wrong. The declaration is too polymorphic; α includes types like o and $nat \Rightarrow nat$. Thus, it admits $\bot = \neg(\bot)$ and $Suc = Suc$ as formulae, which is acceptable in higher-order logic but not in first-order logic.

Isabelle's **type classes** control polymorphism [37]. Each type variable belongs to a class, which denotes a set of types. Classes are partially ordered by the subclass relation, which is essentially the subset relation on the sets of types. They closely resemble the classes of the functional language Haskell [25, 26].

Isabelle provides the built-in class *logic*, which consists of the logical types: the ones we want to reason about. Let us declare a class *term*, to consist of all legal types of terms in our logic. The subclass structure is now $term \leq logic$.

We put *nat* in class *term* by declaring *nat::term*. We declare the equality constant by

$$= \;::\; [\alpha{::}term, \alpha] \Rightarrow o$$

where α::*term* constrains the type variable α to class *term*. Such type variables resemble Standard ML's equality type variables.

We give o and function types the class *logic* rather than *term*, since they are not legal types for terms. We may introduce new types of class *term* — for instance, type *string* or *real* — at any time. We can even declare type constructors such as *list*, and state that type (τ)*list* belongs to class *term* provided τ does; equality applies to lists of natural numbers but not to lists of formulae. We may summarize this paragraph by a set of **arity declarations** for type constructors:

$$\begin{aligned} o &:: \mathit{logic} \\ \mathit{fun} &:: (\mathit{logic}, \mathit{logic})\mathit{logic} \\ \mathit{nat}, \mathit{string}, \mathit{real} &:: \mathit{term} \\ \mathit{list} &:: (\mathit{term})\mathit{term} \end{aligned}$$

(Recall that *fun* is the type constructor for function types.) In higher-order logic, equality does apply to truth values and functions; this requires the arity declarations $o :: \mathit{term}$ and $\mathit{fun} :: (\mathit{term}, \mathit{term})\mathit{term}$. The class system can also handle overloading. We could declare *arith* to be the subclass of *term* consisting of the 'arithmetic' types, such as *nat*. Then we could declare the operators

$$+, -, \times, / :: [\alpha{::}\mathit{arith}, \alpha] \Rightarrow \alpha$$

If we declare new types *real* and *complex* of class *arith*, then we in effect have three sets of operators:

$$\begin{aligned} +, -, \times, / &:: [\mathit{nat}, \mathit{nat}] \Rightarrow \mathit{nat} \\ +, -, \times, / &:: [\mathit{real}, \mathit{real}] \Rightarrow \mathit{real} \\ +, -, \times, / &:: [\mathit{complex}, \mathit{complex}] \Rightarrow \mathit{complex} \end{aligned}$$

Isabelle will regard these as distinct constants, each of which can be defined separately. We could even introduce the type (α)*vector* and declare its arity as $(\mathit{arith})\mathit{arith}$. Then we could declare the constant

$$+ :: [(\alpha)\mathit{vector}, (\alpha)\mathit{vector}] \Rightarrow (\alpha)\mathit{vector}$$

and specify it in terms of $+ :: [\alpha, \alpha] \Rightarrow \alpha$.

A type variable may belong to any finite number of classes. Suppose that we had declared yet another class $\mathit{ord} \leq \mathit{term}$, the class of all 'ordered' types, and a constant

$$\leq :: [\alpha{::}\mathit{ord}, \alpha] \Rightarrow o.$$

In this context the variable x in $x \leq (x+x)$ will be assigned type $\alpha{::}\{\mathit{arith}, \mathit{ord}\}$, which means α belongs to both *arith* and *ord*. Semantically the set $\{\mathit{arith}, \mathit{ord}\}$ should be understood as the intersection of the sets of types represented by *arith* and *ord*. Such intersections of classes are called **sorts**. The empty intersection of classes, $\{\}$, contains all types and is thus the **universal sort**.

Even with overloading, each term has a unique, most general type. For this to be possible, the class and type declarations must satisfy certain technical constraints; see Sect. 9.1.

1.1.3 Higher types and quantifiers

Quantifiers are regarded as operations upon functions. Ignoring polymorphism for the moment, consider the formula $\forall x \,.\, P(x)$, where x ranges over type *nat*. This is true if $P(x)$ is true for all x. Abstracting $P(x)$ into a function, this is the same as saying that $\lambda x \,.\, P(x)$ returns true for all arguments. Thus, the universal quantifier can be represented by a constant

$$\forall \;::\; (nat \Rightarrow o) \Rightarrow o,$$

which is essentially an infinitary truth table. The representation of $\forall x \,.\, P(x)$ is $\forall(\lambda x \,.\, P(x))$.

The existential quantifier is treated in the same way. Other binding operators are also easily handled; for instance, the summation operator $\Sigma_{k=i}^{j} f(k)$ can be represented as $\Sigma(i, j, \lambda k \,.\, f(k))$, where

$$\Sigma \;::\; [nat, nat, nat \Rightarrow nat] \Rightarrow nat.$$

Quantifiers may be polymorphic. We may define $\forall$ and $\exists$ over all legal types of terms, not just the natural numbers, and allow summations over all arithmetic types:

$$\begin{aligned} \forall, \exists \;&::\; (\alpha{::}term \Rightarrow o) \Rightarrow o \\ \Sigma \;&::\; [nat, nat, nat \Rightarrow \alpha{::}arith] \Rightarrow \alpha \end{aligned}$$

Observe that the index variables still have type *nat*, while the values being summed may belong to any arithmetic type.

1.2 Formalizing logical rules in Isabelle

Object-logics are formalized by extending Isabelle's meta-logic [43], which is intuitionistic higher-order logic. The meta-level connectives are **implication**, the **universal quantifier**, and **equality**.

- The implication $\phi \Longrightarrow \psi$ means 'ϕ implies ψ', and expresses logical **entailment**.

- The quantification $\bigwedge x \,.\, \phi$ means 'ϕ is true for all x', and expresses **generality** in rules and axiom schemes.

- The equality $a \equiv b$ means 'a equals b', for expressing **definitions** (see Sect. 3.1.2). Equalities left over from the unification process, so called **flex-flex constraints**, are written $a \stackrel{?}{\equiv} b$. The two equality symbols have the same logical meaning.

The syntax of the meta-logic is formalized in the same manner as object-logics, using the simply typed λ-calculus. Analogous to type o above, there is a built-in type *prop* of meta-level truth values. Meta-level formulae will have this type. Type *prop* belongs to class *logic*; also, $\sigma \Rightarrow \tau$ belongs to *logic* provided σ and τ do. Here are the types of the built-in connectives:

$$\begin{aligned} \Longrightarrow &:: [prop, prop] \Rightarrow prop \\ \bigwedge &:: (\alpha{::}logic \Rightarrow prop) \Rightarrow prop \\ \equiv &:: [\alpha{::}\{\}, \alpha] \Rightarrow prop \\ \stackrel{?}{\equiv} &:: [\alpha{::}\{\}, \alpha] \Rightarrow prop \end{aligned}$$

The polymorphism in $\bigwedge$ is restricted to class *logic* to exclude certain types, those used just for parsing. The type variable $\alpha{::}\{\}$ ranges over the universal sort.

In our formalization of first-order logic, we declared a type o of object-level truth values, rather than using *prop* for this purpose. If we declared the object-level connectives to have types such as $\neg :: prop \Rightarrow prop$, then these connectives would be applicable to meta-level formulae. Keeping *prop* and o as separate types maintains the distinction between the meta-level and the object-level. To formalize the inference rules, we shall need to relate the two levels; accordingly, we declare the constant

$$Trueprop \;\; :: \;\; o \Rightarrow prop.$$

We may regard *Trueprop* as a meta-level predicate, reading *Trueprop*(P) as 'P is true at the object-level.' Put another way, *Trueprop* is a coercion from o to *prop*.

1.2.1 Expressing propositional rules

We shall illustrate the use of the meta-logic by formalizing the rules of Fig. 1.1. Each object-level rule is expressed as a meta-level axiom.

One of the simplest rules is $(\wedge E1)$. Making everything explicit, its formalization in the meta-logic is

$$\bigwedge P\ Q\ .\ Trueprop(P \wedge Q) \Longrightarrow Trueprop(P). \qquad (\wedge E1)$$

This may look formidable, but it has an obvious reading: for all object-level truth values P and Q, if $P \wedge Q$ is true then so is P. The reading is correct because the meta-logic has simple models, where types denote sets and $\bigwedge$ really means 'for all.'

Isabelle adopts notational conventions to ease the writing of rules. We may hide the occurrences of *Trueprop* by making it an implicit coercion. Outer universal quantifiers may be dropped. Finally, the nested implication

$$\phi_1 \Longrightarrow (\cdots \phi_n \Longrightarrow \psi \cdots)$$

may be abbreviated as $[\![\phi_1; \ldots ; \phi_n]\!] \Longrightarrow \psi$, which formalizes a rule of n premises.

Using these conventions, the conjunction rules become the following axioms. These fully specify the properties of $\wedge$:

$$[\![P; Q]\!] \Longrightarrow P \wedge Q \qquad (\wedge I)$$

$$P \wedge Q \Longrightarrow P \qquad P \wedge Q \Longrightarrow Q \qquad (\wedge E1, 2)$$

Next, consider the disjunction rules. The discharge of assumption in $(\vee E)$ is expressed using $\Longrightarrow$:

$$P \Longrightarrow P \vee Q \qquad Q \Longrightarrow P \vee Q \qquad (\vee I1, 2)$$

$$[\![P \vee Q; P \Longrightarrow R; Q \Longrightarrow R]\!] \Longrightarrow R \qquad (\vee E)$$

To understand this treatment of assumptions in natural deduction, look at implication. The rule $(\rightarrow I)$ is the classic example of natural deduction: to prove that $P \rightarrow Q$ is true, assume P is true and show that Q must then be true. More concisely, if P implies Q (at the meta-level), then $P \rightarrow Q$ is true (at the object-level). Showing the coercion explicitly, this is formalized as

$$(\mathit{Trueprop}(P) \Longrightarrow \mathit{Trueprop}(Q)) \Longrightarrow \mathit{Trueprop}(P \rightarrow Q).$$

The rule $(\rightarrow E)$ is straightforward; hiding *Trueprop*, the axioms to specify $\rightarrow$ are

$$(P \Longrightarrow Q) \Longrightarrow P \rightarrow Q \qquad (\rightarrow I)$$

$$[\![P \rightarrow Q; P]\!] \Longrightarrow Q. \qquad (\rightarrow E)$$

Finally, the intuitionistic contradiction rule is formalized as the axiom

$$\bot \Longrightarrow P. \qquad (\bot E)$$

! Earlier versions of Isabelle, and certain papers [43, 45], use $[\![P]\!]$ to mean *Trueprop*(P).

1.2.2 Quantifier rules and substitution

Isabelle expresses variable binding using λ-abstraction; for instance, $\forall x \,.\, P$ is formalized as $\forall(\lambda x \,.\, P)$. Recall that $F(t)$ is Isabelle's syntax for application of the function F to the argument t; it is not a meta-notation for substitution. On the other hand, a substitution will take place if F has the form $\lambda x \,.\, P$; Isabelle transforms $(\lambda x . P)(t)$ to $P[t/x]$ by β-conversion. Thus, we can express inference rules that involve substitution for bound variables.

A logic may attach provisos to certain of its rules, especially quantifier rules. We cannot hope to formalize arbitrary provisos. Fortunately, those typical of quantifier rules always have the same form, namely 'x not free in ... *(some set of formulae),*' where x is a variable (called a **parameter** or **eigenvariable**) in some premise. Isabelle treats provisos using $\bigwedge$, its inbuilt notion of 'for all'.

The purpose of the proviso 'x not free in ...' is to ensure that the premise may not make assumptions about the value of x, and therefore holds for all x. We formalize $(\forall I)$ by

$$\left(\bigwedge x \,.\, \mathit{Trueprop}(P(x))\right) \Longrightarrow \mathit{Trueprop}(\forall x \,.\, P(x)).$$

This means, 'if $P(x)$ is true for all x, then $\forall x \,.\, P(x)$ is true.' The $\forall E$ rule exploits β-conversion. Hiding *Trueprop*, the $\forall$ axioms are

$$\left(\bigwedge x \,.\, P(x)\right) \Longrightarrow \forall x \,.\, P(x) \qquad (\forall I)$$

$$(\forall x \,.\, P(x)) \Longrightarrow P(t). \qquad (\forall E)$$

We have defined the object-level universal quantifier ($\forall$) using $\bigwedge$. But we do not require meta-level counterparts of all the connectives of the object-logic! Consider the existential quantifier:

$$P(t) \Longrightarrow \exists x \,.\, P(x) \qquad (\exists I)$$

$$[\![\exists x \,.\, P(x);\ \bigwedge x \,.\, P(x) \Longrightarrow Q]\!] \Longrightarrow Q \qquad (\exists E)$$

Let us verify $(\exists E)$ semantically. Suppose that the premises hold; since $\exists x \,.\, P(x)$ is true, we may choose an a such that $P(a)$ is true. Instantiating $\bigwedge x.P(x) \Longrightarrow Q$ with a yields $P(a) \Longrightarrow Q$, and we obtain the desired conclusion, Q.

The treatment of substitution deserves mention. The rule

$$\frac{t = u \quad P}{P[u/t]}$$

would be hard to formalize in Isabelle. It calls for replacing t by u throughout P, which cannot be expressed using β-conversion. Our rule (*subst*) uses P as a template for substitution, inferring $P[u/x]$ from $P[t/x]$. When we formalize this as an axiom, the template becomes a function variable:

$$[\![t = u; P(t)]\!] \Longrightarrow P(u). \qquad (\mathit{subst})$$

1.2.3 Signatures and theories

A **signature** contains the information necessary for type checking, parsing and pretty printing a term. It specifies classes and their relationships, types and their arities, constants and their types, etc. It also contains syntax rules, specified using mixfix declarations.

Two signatures can be merged provided their specifications are compatible — they must not, for example, assign different types to the same constant. Under similar conditions, a signature can be extended. Signatures are managed internally by Isabelle; users seldom encounter them.

A **theory** consists of a signature plus a collection of axioms. The **pure** theory contains only the meta-logic. Theories can be combined provided their signatures are compatible. A theory definition extends an existing theory with further signature specifications — classes, types, constants and mixfix declarations — plus a list of axioms, expressed as strings to be parsed. A theory can formalize a small piece of mathematics, such as lists and their operations, or an entire logic. A mathematical development typically involves many theories in a hierarchy. For example, the pure theory could be extended to form a theory for Fig. 1.1; this could be extended in two separate ways to form a theory for natural numbers and a theory for lists; the union of these two could be extended into a theory defining the length of a list:

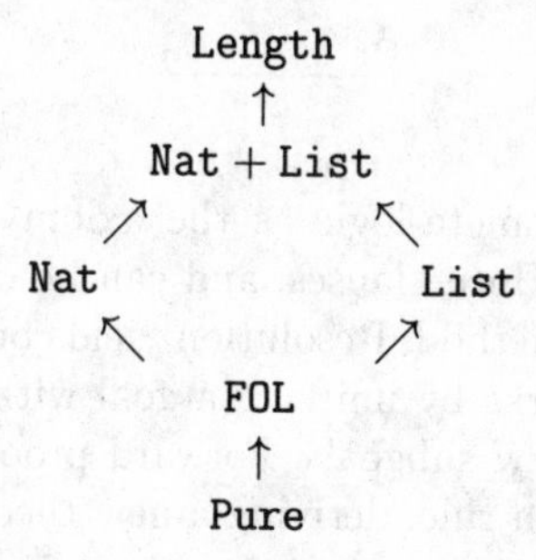

Each Isabelle proof typically works within a single theory, which is associated with the proof state. However, many different theories may coexist at the same time, and you may work in each of these during a single session.

! Confusing problems arise if you work in the wrong theory. Each theory defines its own syntax. An identifier may be regarded in one theory as a constant and in another as a variable.

1.3 Proof construction in Isabelle

I have elsewhere described the meta-logic and demonstrated it by formalizing first-order logic [43]. There is a one-to-one correspondence between meta-level proofs and object-level proofs. To each use of a meta-level axiom, such as $(\forall I)$, there is a use of the corresponding object-level rule. Object-level assumptions and parameters have meta-level counterparts. The meta-level formalization is **faithful**, admitting no incorrect object-level inferences, and **adequate**, admitting all correct object-level inferences. These properties must be demonstrated separately for each object-logic.

The meta-logic is defined by a collection of inference rules, including equational rules for the λ-calculus and logical rules. The rules for $\Longrightarrow$ and $\bigwedge$ resemble those for $\rightarrow$ and $\forall$ in Fig. 1.1. Proofs performed using the primitive meta-rules would be lengthy; Isabelle proofs normally use certain derived rules. **Resolution**, in particular, is convenient for backward proof.

Unification is central to theorem proving. It supports quantifier reasoning by allowing certain 'unknown' terms to be instantiated later, possibly in stages. When proving that the time required to sort n integers is proportional to n^2, we need not state the constant of proportionality; when proving that a hardware adder will deliver the sum of its inputs, we need not state how many clock ticks will be required. Such quantities often emerge from the proof.

Isabelle provides **schematic variables**, or **unknowns**, for unification. Logically, unknowns are free variables. But while ordinary variables remain fixed, unification may instantiate unknowns. Unknowns are written with a ? prefix and are frequently subscripted: $?a$, $?a_1$, $?a_2$, ..., $?P$, $?P_1$, ...

Recall that an inference rule of the form

$$\frac{\phi_1 \quad \ldots \quad \phi_n}{\phi}$$

is formalized in Isabelle's meta-logic as the axiom $[\![\phi_1; \ldots; \phi_n]\!] \Longrightarrow \phi$. Such axioms resemble Prolog's Horn clauses, and can be combined by resolution — Isabelle's principal proof method. Resolution yields both forward and backward proof. Backward proof works by unifying a goal with the conclusion of a rule, whose premises become new subgoals. Forward proof works by unifying theorems with the premises of a rule, deriving a new theorem.

Isabelle formulae require an extended notion of resolution. They differ from Horn clauses in two major respects:

- They are written in the typed λ-calculus, and therefore must be resolved using higher-order unification.
- The constituents of a clause need not be atomic formulae. Any formula of the form $Trueprop(\cdots)$ is atomic, but axioms such as $\to I$ and $\forall I$ contain non-atomic formulae.

Isabelle has little in common with classical resolution theorem provers such as Otter [60]. At the meta-level, Isabelle proves theorems in their positive form, not by refutation. However, an object-logic that includes a contradiction rule may employ a refutation proof procedure.

1.3.1 Higher-order unification

Unification is equation solving. The solution of $f(?x, c) \stackrel{?}{\equiv} f(d, ?y)$ is $?x \equiv d$ and $?y \equiv c$. **Higher-order unification** is equation solving for typed λ-terms. To handle β-conversion, it must reduce $(\lambda x \,.\, t)u$ to $t[u/x]$. That is easy — in the typed λ-calculus, all reduction sequences terminate at a normal form. But it must guess the unknown function $?f$ in order to solve the equation

$$?f(t) \stackrel{?}{\equiv} g(u_1, \ldots, u_k). \tag{1.1}$$

Huet's [27] search procedure solves equations by imitation and projection. **Imitation** makes $?f$ apply the leading symbol (if a constant) of the right-hand side. To solve equation (1.1), it guesses

$$?f \equiv \lambda x \,.\, g(?h_1(x), \ldots, ?h_k(x)),$$

where $?h_1, \ldots, ?h_k$ are new unknowns. Assuming there are no other occurrences of $?f$, equation (1.1) simplifies to the set of equations

$$?h_1(t) \stackrel{?}{\equiv} u_1 \quad \ldots \quad ?h_k(t) \stackrel{?}{\equiv} u_k.$$

If the procedure solves these equations, instantiating $?h_1, \ldots, ?h_k$, then it yields an instantiation for $?f$.

Projection makes $?f$ apply one of its arguments. To solve equation (1.1), if t expects m arguments and delivers a result of suitable type, it guesses

$$?f \equiv \lambda x \,.\, x(?h_1(x), \ldots, ?h_m(x)),$$

where $?h_1, \ldots, ?h_m$ are new unknowns. Assuming there are no other occurrences of $?f$, equation (1.1) simplifies to the equation

$$t(?h_1(t), \ldots, ?h_m(t)) \stackrel{?}{\equiv} g(u_1, \ldots, u_k).$$

! Huet's unification procedure is complete. Isabelle's polymorphic version, which solves for type unknowns as well as for term unknowns, is incomplete. The problem is that projection requires type information. In equation (1.1), if the type of t is unknown, then projections are possible for all $m \geq 0$, and the types of the $?h_i$ will be similarly unconstrained. Therefore, Isabelle never attempts such projections, and may fail to find unifiers where a type unknown turns out to be a function type.

Given $?f(t_1, \ldots, t_n) \stackrel{?}{\equiv} u$, Huet's procedure could make up to $n+1$ guesses. The search tree and set of unifiers may be infinite. But higher-order unification can work effectively, provided you are careful with **function unknowns**:

- Equations with no function unknowns are solved using first-order unification, extended to treat bound variables. For example, $\lambda x \,.\, x \stackrel{?}{\equiv} \lambda x \,.\, ?y$ has no solution because $?y \equiv x$ would capture the free variable x.

- An occurrence of the term $?f(x, y, z)$, where the arguments are distinct bound variables, causes no difficulties. Its projections can only match the corresponding variables.

- Even an equation such as $?f(a) \stackrel{?}{\equiv} a + a$ is all right. It has four solutions, but Isabelle evaluates them lazily, trying projection before imitation. The first solution is usually the one desired:

  $$?f \equiv \lambda x \,.\, x + x \quad ?f \equiv \lambda x \,.\, a + x \quad ?f \equiv \lambda x \,.\, x + a \quad ?f \equiv \lambda x \,.\, a + a$$

- Equations such as $?f(?x, ?y) \stackrel{?}{\equiv} t$ and $?f(?g(x)) \stackrel{?}{\equiv} t$ admit vast numbers of unifiers, and must be avoided.

In problematic cases, you may have to instantiate some unknowns before invoking unification.

1.3.2 Joining rules by resolution

Let $[\![\psi_1; \ldots; \psi_m]\!] \Longrightarrow \psi$ and $[\![\phi_1; \ldots; \phi_n]\!] \Longrightarrow \phi$ be two Isabelle theorems, representing object-level rules. Choosing some i from 1 to n, suppose that ψ and ϕ_i have a higher-order unifier. Writing Xs for the application of substitution s to expression X, this means there is some s such that $\psi s \equiv \phi_i s$. By resolution, we may conclude

$$([\![\phi_1; \ldots; \phi_{i-1}; \psi_1; \ldots; \psi_m; \phi_{i+1}; \ldots; \phi_n]\!] \Longrightarrow \phi)s.$$

The substitution s may instantiate unknowns in both rules. In short, resolution is the following rule:

$$\frac{[\![\psi_1; \ldots; \psi_m]\!] \Longrightarrow \psi \qquad [\![\phi_1; \ldots; \phi_n]\!] \Longrightarrow \phi}{([\![\phi_1; \ldots; \phi_{i-1}; \psi_1; \ldots; \psi_m; \phi_{i+1}; \ldots; \phi_n]\!] \Longrightarrow \phi)s} \ (\psi s \equiv \phi_i s)$$

It operates at the meta-level, on Isabelle theorems, and is justified by the properties of $\Longrightarrow$ and $\bigwedge$. It takes the number i (for $1 \leq i \leq n$) as a parameter and may yield infinitely many conclusions, one for each unifier of ψ with ϕ_i. Isabelle returns these conclusions as a sequence (lazy list).

Resolution expects the rules to have no outer quantifiers ($\bigwedge$). It may rename or instantiate any schematic variables, but leaves free variables unchanged. When constructing a theory, Isabelle puts the rules into a standard form containing no free variables; for instance, $(\rightarrow E)$ becomes

$$[\![?P \rightarrow ?Q; ?P]\!] \Longrightarrow ?Q.$$

When resolving two rules, the unknowns in the first rule are renamed, by subscripting, to make them distinct from the unknowns in the second rule. To resolve $(\rightarrow E)$ with itself, the first copy of the rule becomes

$$[\![?P_1 \rightarrow ?Q_1; ?P_1]\!] \Longrightarrow ?Q_1.$$

Resolving this with $(\rightarrow E)$ in the first premise, unifying $?Q_1$ with $?P \rightarrow ?Q$, is the meta-level inference

$$\frac{[\![?P_1 \rightarrow ?Q_1; ?P_1]\!] \Longrightarrow ?Q_1 \qquad [\![?P \rightarrow ?Q; ?P]\!] \Longrightarrow ?Q}{[\![?P_1 \rightarrow (?P \rightarrow ?Q); ?P_1; ?P]\!] \Longrightarrow ?Q.}$$

Renaming the unknowns in the resolvent, we have derived the object-level rule

$$\frac{R \rightarrow (P \rightarrow Q) \quad R \quad P}{Q.}$$

Joining rules in this fashion is a simple way of proving theorems. The derived rules are conservative extensions of the object-logic, and may permit simpler proofs. Let us consider another example. Suppose we have the axiom

$$\forall x\, y\,.\, Suc(x) = Suc(y) \rightarrow x = y. \qquad (inject)$$

The standard form of $(\forall E)$ is $\forall x . ?P(x) \Longrightarrow ?P(?t)$. Resolving (*inject*) with $(\forall E)$ replaces $?P$ by $\lambda x . \forall y . Suc(x) = Suc(y) \rightarrow x = y$ and leaves $?t$ unchanged, yielding

$$\forall y . Suc(?t) = Suc(y) \rightarrow ?t = y.$$

Resolving this with $(\forall E)$ puts a subscript on $?t$ and yields

$$Suc(?t_1) = Suc(?t) \rightarrow ?t_1 = ?t.$$

Resolving this with $(\rightarrow E)$ increases the subscripts and yields

$$Suc(?t_2) = Suc(?t_1) \Longrightarrow ?t_2 = ?t_1.$$

We have derived the rule

$$\frac{Suc(m) = Suc(n)}{m = n,}$$

which goes directly from $Suc(m) = Suc(n)$ to $m = n$. It is handy for simplifying an equation like $Suc(Suc(Suc(m))) = Suc(Suc(Suc(0)))$.

1.4 Lifting a rule into a context

The rules $(\rightarrow I)$ and $(\forall I)$ may seem unsuitable for resolution. They have non-atomic premises, namely $P \Longrightarrow Q$ and $\bigwedge x . P(x)$, while the conclusions of all the rules are atomic (they have the form $Trueprop(\cdots)$). Isabelle gets round the problem through a meta-inference called **lifting**. Let us consider how to construct proofs such as

$$\cfrac{\cfrac{\begin{matrix}[P, Q]\\ \vdots \\ R\end{matrix}}{Q \rightarrow R}\,(\rightarrow I)}{P \rightarrow (Q \rightarrow R)}\,(\rightarrow I) \qquad \cfrac{\cfrac{P(x, y)}{\forall y . P(x, y)}\,(\forall I)}{\forall x\, y . P(x, y)}\,(\forall I)$$

1.4.1 Lifting over assumptions

Lifting over $\theta \Longrightarrow$ is the following meta-inference rule:

$$\frac{[\![\phi_1; \ldots; \phi_n]\!] \Longrightarrow \phi}{[\![\theta \Longrightarrow \phi_1; \ldots; \theta \Longrightarrow \phi_n]\!] \Longrightarrow (\theta \Longrightarrow \phi)}$$

This is clearly sound: if $[\![\phi_1; \ldots; \phi_n]\!] \Longrightarrow \phi$ is true and $\theta \Longrightarrow \phi_1$, ..., $\theta \Longrightarrow \phi_n$ and θ are all true then ϕ must be true. Iterated lifting over a series of meta-formulae θ_k, ..., θ_1 yields an object-rule whose conclusion is $[\![\theta_1; \ldots; \theta_k]\!] \Longrightarrow \phi$. Typically the θ_i are the assumptions in a natural deduction proof; lifting copies them into a rule's premises and conclusion.

When resolving two rules, Isabelle lifts the first one if necessary. The standard form of $(\rightarrow I)$ is

$$(?P \Longrightarrow ?Q) \Longrightarrow ?P \rightarrow ?Q.$$

To resolve this rule with itself, Isabelle modifies one copy as follows: it renames the unknowns to $?P_1$ and $?Q_1$, then lifts the rule over $?P \Longrightarrow$ to obtain

$$(?P \Longrightarrow (?P_1 \Longrightarrow ?Q_1)) \Longrightarrow (?P \Longrightarrow (?P_1 \rightarrow ?Q_1)).$$

Using the $[\![\cdots]\!]$ abbreviation, this can be written as

$$[\![[\![?P; ?P_1]\!] \Longrightarrow ?Q_1; ?P]\!] \Longrightarrow ?P_1 \rightarrow ?Q_1.$$

Unifying $?P \Longrightarrow ?P_1 \rightarrow ?Q_1$ with $?P \Longrightarrow ?Q$ instantiates $?Q$ to $?P_1 \rightarrow ?Q_1$. Resolution yields

$$([\![?P; ?P_1]\!] \Longrightarrow ?Q_1) \Longrightarrow ?P \rightarrow (?P_1 \rightarrow ?Q_1).$$

This represents the derived rule

$$\frac{\begin{matrix}[P, Q] \\ \vdots \\ R\end{matrix}}{P \rightarrow (Q \rightarrow R).}$$

1.4.2 Lifting over parameters

An analogous form of lifting handles premises of the form $\bigwedge x \ldots$. Here, lifting prefixes an object-rule's premises and conclusion with $\bigwedge x$. At the same time, lifting introduces a dependence upon x. It replaces each unknown $?a$ in the rule by $?a'(x)$, where $?a'$ is a new unknown (by subscripting) of suitable type — necessarily a function type. In short, lifting is the meta-inference

$$\frac{[\![\phi_1; \ldots; \phi_n]\!] \Longrightarrow \phi}{[\![\bigwedge x \,.\, \phi_1^x; \ldots; \bigwedge x \,.\, \phi_n^x]\!] \Longrightarrow \bigwedge x \,.\, \phi^x,}$$

where ϕ^x stands for the result of lifting unknowns over x in ϕ. It is not hard to verify that this meta-inference is sound. If $\phi \Longrightarrow \psi$ then $\phi^x \Longrightarrow \psi^x$ for all x; so if ϕ^x is true for all x then so is ψ^x. Thus, from $\phi \Longrightarrow \psi$ we conclude $(\bigwedge x \,.\, \phi^x) \Longrightarrow (\bigwedge x \,.\, \psi^x)$.

For example, $(\vee I)$ might be lifted to

$$(\bigwedge x \,.\, ?P_1(x)) \Longrightarrow (\bigwedge x \,.\, ?P_1(x) \vee ?Q_1(x))$$

and $(\forall I)$ to

$$(\bigwedge x\, y \,.\, ?P_1(x, y)) \Longrightarrow (\bigwedge x \,.\, \forall y \,.\, ?P_1(x, y)).$$

Isabelle has renamed a bound variable in $(\forall I)$ from x to y, avoiding a clash. Resolving the above with $(\forall I)$ is the meta-inference

$$\frac{(\bigwedge x\, y\, .\, ?P_1(x,y)) \Longrightarrow (\bigwedge x\, .\, \forall y\, .\, ?P_1(x,y)) \quad (\bigwedge x\, .\, ?P(x)) \Longrightarrow (\forall x\, .\, ?P(x))}{\bigwedge x\, y\, .\, ?P_1(x,y)) \Longrightarrow \forall x\, y\, .\, ?P_1(x,y))}$$

Here, $?P$ is replaced by $\lambda x\, .\, \forall y\, .\, ?P_1(x,y)$; the resolvent expresses the derived rule

$$\frac{Q(x,y)}{\forall x\, y\, .\, Q(x,y)} \quad \text{provided } x,\ y \text{ not free in the assumptions}$$

I discuss lifting and parameters at length elsewhere [43]. Miller goes into even greater detail [35].

1.5 Backward proof by resolution

Resolution is convenient for deriving simple rules and for reasoning forward from facts. It can also support backward proof, where we start with a goal and refine it to progressively simpler subgoals until all have been solved. LCF and its descendants HOL and Nuprl provide tactics and tacticals, which constitute a high-level language for expressing proof searches. **Tactics** refine subgoals while **tacticals** combine tactics.

Isabelle's tactics and tacticals work differently from LCF's. An Isabelle rule is bidirectional: there is no distinction between inputs and outputs. LCF has a separate tactic for each rule; Isabelle performs refinement by any rule in a uniform fashion, using resolution.

Isabelle works with meta-level theorems of the form $[\![\phi_1; \ldots; \phi_n]\!] \Longrightarrow \phi$. We have viewed this as the **rule** with premises $\phi_1, \ldots, \phi_n$ and conclusion ϕ. It can also be viewed as the **proof state** with subgoals $\phi_1, \ldots, \phi_n$ and main goal ϕ.

To prove the formula ϕ, take $\phi \Longrightarrow \phi$ as the initial proof state. This assertion is, trivially, a theorem. At a later stage in the backward proof, a typical proof state is $[\![\phi_1; \ldots; \phi_n]\!] \Longrightarrow \phi$. This proof state is a theorem, ensuring that the subgoals $\phi_1, \ldots, \phi_n$ imply ϕ. If $n = 0$ then we have proved ϕ outright. If ϕ contains unknowns, they may become instantiated during the proof; a proof state may be $[\![\phi_1; \ldots; \phi_n]\!] \Longrightarrow \phi'$, where ϕ' is an instance of ϕ.

1.5.1 Refinement by resolution

To refine subgoal i of a proof state by a rule, perform the following resolution:

$$\frac{\text{rule} \qquad \text{proof state}}{\text{new proof state}}$$

Suppose the rule is $[\![\psi'_1; \ldots; \psi'_m]\!] \Longrightarrow \psi'$ after lifting over subgoal i's assumptions and parameters. If the proof state is $[\![\phi_1; \ldots; \phi_n]\!] \Longrightarrow \phi$, then the new proof state is (for $1 \leq i \leq n$)

$$([\![\phi_1; \ldots; \phi_{i-1}; \psi'_1; \ldots; \psi'_m; \phi_{i+1}; \ldots; \phi_n]\!] \Longrightarrow \phi)s.$$

Substitution s unifies ψ' with ϕ_i. In the proof state, subgoal i is replaced by m new subgoals, the rule's instantiated premises. If some of the rule's unknowns are left un-instantiated, they become new unknowns in the proof state. Refinement by $(\exists I)$, namely

$$?P(?t) \Longrightarrow \exists x \,.\, ?P(x),$$

inserts a new unknown derived from $?t$ by subscripting and lifting. We do not have to specify an 'existential witness' when applying $(\exists I)$. Further resolutions may instantiate unknowns in the proof state.

1.5.2 Proof by assumption

In the course of a natural deduction proof, parameters $x_1, \ldots, x_l$ and assumptions $\theta_1, \ldots, \theta_k$ accumulate, forming a context for each subgoal. Repeated lifting steps can lift a rule into any context. To aid readability, Isabelle puts contexts into a normal form, gathering the parameters at the front:

$$\bigwedge x_1 \ldots x_l \,.\, [\![\theta_1; \ldots; \theta_k]\!] \Longrightarrow \theta. \tag{1.2}$$

Under the usual reading of the connectives, this expresses that θ follows from $\theta_1, \ldots \theta_k$ for arbitrary $x_1, \ldots, x_l$. It is trivially true if θ equals any of $\theta_1, \ldots \theta_k$, or is unifiable with any of them. This models proof by assumption in natural deduction.

Isabelle automates the meta-inference for proof by assumption. Its arguments are the meta-theorem $[\![\phi_1; \ldots; \phi_n]\!] \Longrightarrow \phi$, and some i from 1 to n, where ϕ_i has the form (1.2). Its results are meta-theorems of the form

$$([\![\phi_1; \ldots; \phi_{i-1}; \phi_{i+1}; \phi_n]\!] \Longrightarrow \phi)s$$

for each s and j such that s unifies $\lambda x_1 \ldots x_l . \theta_j$ with $\lambda x_1 \ldots x_l . \theta$. Isabelle supplies the parameters $x_1, \ldots, x_l$ to higher-order unification as bound variables, which regards them as unique constants with a limited scope — this enforces parameter provisos [43].

The premise represents a proof state with n subgoals, of which the ith is to be solved by assumption. Isabelle searches the subgoal's context for an assumption θ_j that can solve it. For each unifier, the meta-inference returns an instantiated proof state from which the ith subgoal has been removed. Isabelle searches for a unifying assumption; for readability and robustness, proofs do not refer to assumptions by number.

Consider the proof state

$$([\![P(a); P(b)]\!] \Longrightarrow P(?x)) \Longrightarrow Q(?x).$$

Proof by assumption (with $i = 1$, the only possibility) yields two results:

- $Q(a)$, instantiating $?x \equiv a$
- $Q(b)$, instantiating $?x \equiv b$

Here, proof by assumption affects the main goal. It could also affect other subgoals; if we also had the subgoal $[\![P(b); P(c)]\!] \Longrightarrow P(?x)$, then $?x \equiv a$ would transform it to $[\![P(b); P(c)]\!] \Longrightarrow P(a)$, which might be unprovable.

1.5.3 A propositional proof

Our first example avoids quantifiers. Given the main goal $P \vee P \to P$, Isabelle creates the initial state

$$(P \vee P \to P) \Longrightarrow (P \vee P \to P).$$

Bear in mind that every proof state we derive will be a meta-theorem, expressing that the subgoals imply the main goal. Our aim is to reach the state $P \vee P \to P$; this meta-theorem is the desired result.

The first step is to refine subgoal 1 by $(\to I)$, creating a new state where $P \vee P$ is an assumption:

$$(P \vee P \Longrightarrow P) \Longrightarrow (P \vee P \to P)$$

The next step is $(\vee E)$, which replaces subgoal 1 by three new subgoals. Because of lifting, each subgoal contains a copy of the context — the assumption $P \vee P$. (In fact, this assumption is now redundant; we shall shortly see how to get rid of it!) The new proof state is the following meta-theorem, laid out for clarity:

$$\begin{array}{ll} [\![\, P \vee P \Longrightarrow ?P_1 \vee ?Q_1; & \text{(subgoal 1)} \\ \quad [\![P \vee P; ?P_1]\!] \Longrightarrow P; & \text{(subgoal 2)} \\ \quad [\![P \vee P; ?Q_1]\!] \Longrightarrow P & \text{(subgoal 3)} \\ ]\!] \Longrightarrow (P \vee P \to P) & \text{(main goal)} \end{array}$$

Notice the unknowns in the proof state. Because we have applied $(\vee E)$, we must prove some disjunction, $?P_1 \vee ?Q_1$. Of course, subgoal 1 is provable by assumption. This instantiates both $?P_1$ and $?Q_1$ to P throughout the proof state:

$$\begin{array}{ll} [\![\, [\![P \vee P; P]\!] \Longrightarrow P; & \text{(subgoal 1)} \\ \quad [\![P \vee P; P]\!] \Longrightarrow P & \text{(subgoal 2)} \\ ]\!] \Longrightarrow (P \vee P \to P) & \text{(main goal)} \end{array}$$

Both of the remaining subgoals can be proved by assumption. After two such steps, the proof state is $P \vee P \to P$.

1.5.4 A quantifier proof

To illustrate quantifiers and $\bigwedge$-lifting, let us prove $(\exists x \,.\, P(f(x))) \to (\exists x \,.\, P(x))$. The initial proof state is the trivial meta-theorem

$$(\exists x \,.\, P(f(x))) \to (\exists x \,.\, P(x)) \Longrightarrow (\exists x \,.\, P(f(x))) \to (\exists x \,.\, P(x)).$$

As above, the first step is refinement by $(\rightarrow I)$:

$$(\exists x\,.\,P(f(x)) \Longrightarrow \exists x\,.\,P(x)) \Longrightarrow (\exists x\,.\,P(f(x))) \rightarrow (\exists x\,.\,P(x))$$

The next step is $(\exists E)$, which replaces subgoal 1 by two new subgoals. Both have the assumption $\exists x\,.\,P(f(x))$. The new proof state is the meta-theorem

$$\begin{array}{ll} [\![\, \exists x\,.\,P(f(x)) \Longrightarrow \exists x\,.\,?P_1(x); & \text{(subgoal 1)} \\ \quad \bigwedge x\,.\,[\![\exists x\,.\,P(f(x)); ?P_1(x)]\!] \Longrightarrow \exists x\,.\,P(x) & \text{(subgoal 2)} \\ ]\!] \Longrightarrow (\exists x\,.\,P(f(x))) \rightarrow (\exists x\,.\,P(x)) & \text{(main goal)} \end{array}$$

The unknown $?P_1$ appears in both subgoals. Because we have applied $(\exists E)$, we must prove $\exists x\,.\,?P_1(x)$, where $?P_1(x)$ may become any formula possibly containing x. Proving subgoal 1 by assumption instantiates $?P_1$ to $\lambda x\,.\,P(f(x))$:

$$\left(\bigwedge x\,.\,[\![\exists x\,.\,P(f(x)); P(f(x))]\!] \Longrightarrow \exists x\,.\,P(x)\right) \Longrightarrow (\exists x\,.\,P(f(x))) \rightarrow (\exists x\,.\,P(x))$$

The next step is refinement by $(\exists I)$. The rule is lifted into the context of the parameter x and the assumption $P(f(x))$. This copies the context to the subgoal and allows the existential witness to depend upon x:

$$\left(\bigwedge x\,.\,[\![\exists x\,.\,P(f(x)); P(f(x))]\!] \Longrightarrow P(?x_2(x))\right) \Longrightarrow (\exists x\,.\,P(f(x))) \rightarrow (\exists x\,.\,P(x))$$

The existential witness, $?x_2(x)$, consists of an unknown applied to a parameter. Proof by assumption unifies $\lambda x\,.\,P(f(x))$ with $\lambda x\,.\,P(?x_2(x))$, instantiating $?x_2$ to f. The final proof state contains no subgoals: $(\exists x\,.\,P(f(x))) \rightarrow (\exists x\,.\,P(x))$.

1.5.5 Tactics and tacticals

Tactics perform backward proof. Isabelle tactics differ from those of LCF, HOL and Nuprl by operating on entire proof states, rather than on individual subgoals. An Isabelle tactic is a function that takes a proof state and returns a sequence (lazy list) of possible successor states. Lazy lists are coded in ML as functions, a standard technique [46]. Isabelle represents proof states by theorems.

Basic tactics execute the meta-rules described above, operating on a given subgoal. The **resolution tactics** take a list of rules and return next states for each combination of rule and unifier. The **assumption tactic** examines the subgoal's assumptions and returns next states for each combination of assumption and unifier. Lazy lists are essential because higher-order resolution may return infinitely many unifiers. If there are no matching rules or assumptions then no next states are generated; a tactic application that returns an empty list is said to **fail**.

Sequences realize their full potential with **tacticals** — operators for combining tactics. Depth-first search, breadth-first search and best-first search (where a heuristic function selects the best state to explore) return their outcomes as a sequence. Isabelle provides such procedures in the form of tacticals. Simpler procedures can be expressed directly using the basic tacticals `THEN`, `ORELSE` and `REPEAT`:

*tac*1 `THEN` *tac*2 is a tactic for sequential composition. Applied to a proof state, it returns all states reachable in two steps by applying *tac*1 followed by *tac*2.

*tac*1 `ORELSE` *tac*2 is a choice tactic. Applied to a state, it tries *tac*1 and returns the result if non-empty; otherwise, it uses *tac*2.

`REPEAT` *tac* is a repetition tactic. Applied to a state, it returns all states reachable by applying *tac* as long as possible — until it would fail.

For instance, this tactic repeatedly applies *tac*1 and *tac*2, giving *tac*1 priority:

`REPEAT(`*tac*1 `ORELSE` *tac*2`)`

1.6 Variations on resolution

In principle, resolution and proof by assumption suffice to prove all theorems. However, specialized forms of resolution are helpful for working with elimination rules. Elim-resolution applies an elimination rule to an assumption; destruct-resolution is similar, but applies a rule in a forward style.

The last part of the section shows how the techniques for proving theorems can also serve to derive rules.

1.6.1 Elim-resolution

Consider proving the theorem $((R \vee R) \vee R) \vee R \rightarrow R$. By $(\rightarrow I)$, we prove R from the assumption $((R \vee R) \vee R) \vee R$. Applying $(\vee E)$ to this assumption yields two subgoals, one that assumes R (and is therefore trivial) and one that assumes $(R \vee R) \vee R$. This subgoal admits another application of $(\vee E)$. Since natural deduction never discards assumptions, we eventually generate a subgoal containing much that is redundant:

$$[\![((R \vee R) \vee R) \vee R; (R \vee R) \vee R; R \vee R; R]\!] \Longrightarrow R.$$

In general, using $(\vee E)$ on the assumption $P \vee Q$ creates two new subgoals with the additional assumption P or Q. In these subgoals, $P \vee Q$ is redundant. Other elimination rules behave similarly. In first-order logic, only universally quantified assumptions are sometimes needed more than once — say, to prove $P(f(f(a)))$ from the assumptions $\forall x \,.\, P(x) \rightarrow P(f(x))$ and $P(a)$.

Many logics can be formulated as sequent calculi that delete redundant assumptions after use. The rule $(\vee E)$ might become

$$\frac{\Gamma, P, \Delta \vdash \Theta \qquad \Gamma, Q, \Delta \vdash \Theta}{\Gamma, P \vee Q, \Delta \vdash \Theta} \ \vee\text{-left}$$

In backward proof, a goal containing $P \vee Q$ on the left of the $\vdash$ (that is, as an assumption) splits into two subgoals, replacing $P \vee Q$ by P or Q. But the sequent calculus, with its explicit handling of assumptions, can be tiresome to use.

Elim-resolution is Isabelle's way of getting sequent calculus behaviour from natural deduction rules. It lets an elimination rule consume an assumption. Elim-resolution combines two meta-theorems:

- a rule $[\![\psi_1; \ldots; \psi_m]\!] \Longrightarrow \psi$
- a proof state $[\![\phi_1; \ldots; \phi_n]\!] \Longrightarrow \phi$

The rule must have at least one premise, thus $m > 0$. Write the rule's lifted form as $[\![\psi'_1; \ldots; \psi'_m]\!] \Longrightarrow \psi'$. Suppose we wish to change subgoal number i.

Ordinary resolution would attempt to reduce ϕ_i, replacing subgoal i by m new ones. Elim-resolution tries simultaneously to reduce ϕ_i and to solve ψ'_1 by assumption; it returns a sequence of next states. Each of these replaces subgoal i by instances of $\psi'_2, \ldots, \psi'_m$ from which the selected assumption has been deleted. Suppose ϕ_i has the parameter x and assumptions $\theta_1, \ldots, \theta_k$. Then ψ'_1, the rule's first premise after lifting, will be $\bigwedge x \,.\, [\![\theta_1; \ldots; \theta_k]\!] \Longrightarrow \psi_1^x$. Elim-resolution tries to unify $\psi' \stackrel{?}{\equiv} \phi_i$ and $\lambda x \,.\, \theta_j \stackrel{?}{\equiv} \lambda x \,.\, \psi_1^x$ simultaneously, for $j = 1, \ldots, k$.

Let us redo the example from Sect. 1.5.3. The elimination rule is $(\vee E)$,

$$[\![?P \vee ?Q;\ ?P \Longrightarrow ?R;\ ?Q \Longrightarrow ?R]\!] \Longrightarrow ?R$$

and the proof state is $(P \vee P \Longrightarrow P) \Longrightarrow (P \vee P \to P)$. The lifted rule is

$$\begin{array}{l}[\![\ P \vee P \Longrightarrow ?P_1 \vee ?Q_1; \\ \quad [\![P \vee P;\ ?P_1]\!] \Longrightarrow ?R_1; \\ \quad [\![P \vee P;\ ?Q_1]\!] \Longrightarrow ?R_1 \\ ]\!] \Longrightarrow ?R_1\end{array}$$

Unification takes the simultaneous equations $P \vee P \stackrel{?}{\equiv} ?P_1 \vee ?Q_1$ and $?R_1 \stackrel{?}{\equiv} P$, yielding $?P_1 \equiv ?Q_1 \equiv ?R_1 \equiv P$. The new proof state is simply

$$[\![P \Longrightarrow P;\ P \Longrightarrow P]\!] \Longrightarrow (P \vee P \to P).$$

Elim-resolution's simultaneous unification gives better control than ordinary resolution. Recall the substitution rule:

$$[\![?t = ?u;\ ?P(?t)]\!] \Longrightarrow ?P(?u) \qquad (subst)$$

Unsuitable for ordinary resolution because $?P(?u)$ admits many unifiers, (*subst*) works well with elim-resolution. It deletes some assumption of the form $x = y$ and replaces every y by x in the subgoal formula. The simultaneous unification instantiates $?u$ to y; if y is not an unknown, then $?P(y)$ can easily be unified with another formula.

In logical parlance, the premise containing the connective to be eliminated is called the **major premise**. Elim-resolution expects the major premise to come first. The order of the premises is significant in Isabelle.

1.6.2 Destruction rules

Looking back to Fig. 1.1, notice that there are two kinds of elimination rule. The rules $(\wedge E1)$, $(\wedge E2)$, $(\rightarrow E)$ and $(\forall E)$ extract the conclusion from the major premise. In Isabelle parlance, such rules are called **destruction rules**; they are readable and easy to use in forward proof. The rules $(\vee E)$, $(\bot E)$ and $(\exists E)$ work by discharging assumptions; they support backward proof in a style reminiscent of the sequent calculus.

The latter style is the most general form of elimination rule. In natural deduction, there is no way to recast $(\vee E)$, $(\bot E)$ or $(\exists E)$ as destruction rules. But we can write general elimination rules for $\wedge$, $\rightarrow$ and $\forall$:

$$\frac{P \wedge Q \quad \begin{matrix}[P, Q] \\ \vdots \\ R\end{matrix}}{R} \qquad \frac{P \rightarrow Q \quad P \quad \begin{matrix}[Q] \\ \vdots \\ R\end{matrix}}{R} \qquad \frac{\forall x \,.\, P \quad \begin{matrix}[P[t/x]] \\ \vdots \\ Q\end{matrix}}{Q}$$

Because they are concise, destruction rules are simpler to derive than the corresponding elimination rules. To facilitate their use in backward proof, Isabelle provides a means of transforming a destruction rule such as

$$\frac{P_1 \quad \ldots \quad P_m}{Q} \quad \text{to the elimination rule} \quad \frac{P_1 \quad \ldots \quad P_m \quad \begin{matrix}[Q] \\ \vdots \\ R\end{matrix}}{R.}$$

Destruct-resolution combines this transformation with elim-resolution. It applies a destruction rule to some assumption of a subgoal. Given the rule above, it replaces the assumption P_1 by Q, with new subgoals of showing instances of P_2, ..., P_m. Destruct-resolution works forward from a subgoal's assumptions. Ordinary resolution performs forward reasoning from theorems, as illustrated in Sect. 1.3.2.

1.6.3 Deriving rules by resolution

The meta-logic, itself a form of the predicate calculus, is defined by a system of natural deduction rules. Each theorem may depend upon meta-assumptions. The theorem that ϕ follows from the assumptions $\phi_1, \ldots, \phi_n$ is written

$$\phi \quad [\phi_1, \ldots, \phi_n].$$

A more conventional notation might be $\phi_1, \ldots, \phi_n \vdash \phi$, but Isabelle's notation is more readable with large formulae.

Meta-level natural deduction provides a convenient mechanism for deriving new object-level rules. To derive the rule

$$\frac{\theta_1 \quad \ldots \quad \theta_k}{\phi,}$$

assume the premises $\theta_1, \ldots, \theta_k$ at the meta-level. Then prove ϕ, possibly using these assumptions. Starting with a proof state $\phi \Longrightarrow \phi$, assumptions may accumulate, reaching a final state such as

$$\phi \quad [\theta_1, \ldots, \theta_k].$$

The meta-rule for $\Longrightarrow$ introduction discharges an assumption. Discharging them in the order $\theta_k, \ldots, \theta_1$ yields the meta-theorem $[\![\theta_1; \ldots; \theta_k]\!] \Longrightarrow \phi$, with no assumptions. This represents the desired rule. Let us derive the general $\wedge$ elimination rule:

$$\frac{P \wedge Q \qquad \begin{matrix}[P, Q] \\ \vdots \\ R\end{matrix}}{R} \quad (\wedge E)$$

We assume $P \wedge Q$ and $[\![P; Q]\!] \Longrightarrow R$, and commence backward proof in the state $R \Longrightarrow R$. Resolving this with the second assumption yields the state

$$[\![P; Q]\!] \Longrightarrow R \quad [\,[\![P; Q]\!] \Longrightarrow R\,].$$

Resolving subgoals 1 and 2 with $(\wedge E1)$ and $(\wedge E2)$, respectively, yields the state

$$[\![P \wedge ?Q_1;\ ?P_2 \wedge Q]\!] \Longrightarrow R \quad [\,[\![P; Q]\!] \Longrightarrow R\,].$$

The unknowns $?Q_1$ and $?P_2$ arise from unconstrained subformulae in the premises of $(\wedge E1)$ and $(\wedge E2)$. Resolving both subgoals with the assumption $P \wedge Q$ instantiates the unknowns to yield

$$R \quad [\,[\![P; Q]\!] \Longrightarrow R, P \wedge Q\,].$$

The proof may use the meta-assumptions in any order, and as often as necessary; when finished, we discharge them in the correct order to obtain the desired form:

$$[\![P \wedge Q;\ [\![P; Q]\!] \Longrightarrow R]\!] \Longrightarrow R$$

We have derived the rule using free variables, which prevents their premature instantiation during the proof; we may now replace them by schematic variables.

! Schematic variables are not allowed in meta-assumptions, for a variety of reasons. Meta-assumptions remain fixed throughout a proof.

2. Getting Started with Isabelle

Let us consider how to perform simple proofs using Isabelle. At present, Isabelle's user interface is ML. Proofs are conducted by applying certain ML functions, which update a stored proof state. All syntax must be expressed using ASCII characters. Menu-driven graphical interfaces are under construction, but Isabelle users will always need to know some ML, at least to use tacticals.

Object-logics are built upon Pure Isabelle, which implements the meta-logic and provides certain fundamental data structures: types, terms, signatures, theorems and theories, tactics and tacticals. These data structures have the corresponding ML types `typ`, `term`, `Sign.sg`, `thm`, `theory` and `tactic`; tacticals have function types such as `tactic->tactic`. Isabelle users can operate on these data structures by writing ML programs.

2.1 Forward proof

This section describes the concrete syntax for types, terms and theorems, and demonstrates forward proof.

2.1.1 Lexical matters

An **identifier** is a string of letters, digits, underscores (_) and single quotes ('), beginning with a letter. Single quotes are regarded as primes; for instance `x'` is read as x'. Identifiers are separated by white space and special characters. **Reserved words** are identifiers that appear in Isabelle syntax definitions.

An Isabelle theory can declare symbols composed of special characters, such as `=`, `==`, `=>` and `==>`. (The latter three are part of the syntax of the meta-logic.) Such symbols may be run together; thus if `}` and `{` are used for set brackets then `{{a},{a,b}}` is valid notation for a set of sets — but only if `}}` and `{{` have not been declared as symbols! The parser resolves any ambiguity by taking the longest possible symbol that has been declared. Thus the string `==>` is read as a single symbol. But `= =>` is read as two symbols.

Identifiers that are not reserved words may serve as free variables or constants. A **type identifier** consists of an identifier prefixed by a prime, for example `'a` and `'hello`. Type identifiers stand for (free) type variables, which remain fixed during a proof.

An **unknown** (or type unknown) consists of a question mark, an identifier (or type identifier), and a subscript. The subscript, a non-negative integer, allows the renaming of unknowns prior to unification.[1]

2.1.2 Syntax of types and terms

Classes are denoted by identifiers; the built-in class `logic` contains the 'logical' types. Sorts are lists of classes enclosed in braces } and {; singleton sorts may be abbreviated by dropping the braces.

Types are written with a syntax like ML's. The built-in type `prop` is the type of propositions. Type variables can be constrained to particular classes or sorts, for example `'a::term` and `?'b::{ord,arith}`.

ASCII Notation for Types

`α::C`	$\alpha :: C$	class constraint
`α::{C1,...,Cn}`	$\alpha :: \{C_1, \ldots, C_n\}$	sort constraint
`σ => τ`	$\sigma \Rightarrow \tau$	function type
`[σ1,...,σn] => τ`	$[\sigma_1, \ldots, \sigma_n] \Rightarrow \tau$	curried function type
`(τ1,...,τn) tycon`	$(\tau_1, \ldots, \tau_n) tycon$	type construction

Terms are those of the typed λ-calculus.

ASCII Notation for Terms

`t::σ`	$t :: \sigma$	type constraint
`%x.t`	$\lambda x \,.\, t$	abstraction
`%x1...xn.t`	$\lambda x_1 \ldots x_n \,.\, t$	curried abstraction
`t(u1,...,un)`	$t(u_1, \ldots, u_n)$	curried application

The theorems and rules of an object-logic are represented by theorems in the meta-logic, which are expressed using meta-formulae. Since the meta-logic is higher-order, meta-formulae ϕ, ψ, θ, ... are just terms of type `prop`.

ASCII Notation for Meta-Formulae

`a == b`	$a \equiv b$	meta-equality
`a =?= b`	$a \stackrel{?}{\equiv} b$	flex-flex constraint
`φ ==> ψ`	$\phi \Longrightarrow \psi$	meta-implication
`[|φ1;...;φn|] ==> ψ`	$[\![\phi_1; \ldots; \phi_n]\!] \Longrightarrow \psi$	nested implication
`!!x.φ`	$\bigwedge x \,.\, \phi$	meta-quantification
`!!x1...xn.φ`	$\bigwedge x_1 \ldots \bigwedge x_n \,.\, \phi$	nested quantification

Flex-flex constraints are meta-equalities arising from unification; they require special treatment. See Sect. 2.1.4.

[1] The subscript may appear after the identifier, separated by a dot; this prevents ambiguity when the identifier ends with a digit. Thus `?z6.0` has identifier `"z6"` and subscript 0, while `?a0.5` has identifier `"a0"` and subscript 5. If the identifier does not end with a digit, then no dot appears and a subscript of 0 is omitted; for example, `?hello` has identifier `"hello"` and subscript zero, while `?z6` has identifier `"z"` and subscript 6. The same conventions apply to type unknowns. The question mark is *not* part of the identifier!

Most logics define the implicit coercion *Trueprop* from object-formulae to propositions. This could cause an ambiguity: in $P \Longrightarrow Q$, do the variables P and Q stand for meta-formulae or object-formulae? If the latter, $P \Longrightarrow Q$ really abbreviates $Trueprop(P) \Longrightarrow Trueprop(Q)$. To prevent such ambiguities, Isabelle's syntax does not allow a meta-formula to consist of a variable. Variables of type `prop` are seldom useful, but you can make a variable stand for a meta-formula by prefixing it with the symbol `PROP`:

```
PROP ?psi ==> PROP ?theta
```

Symbols of object-logics also must be rendered into ASCII, typically as follows:

`True`	$\top$	true
`False`	$\bot$	false
`P & Q`	$P \wedge Q$	conjunction
`P | Q`	$P \vee Q$	disjunction
`~ P`	$\neg P$	negation
`P --> Q`	$P \to Q$	implication
`P <-> Q`	$P \leftrightarrow Q$	bi-implication
`ALL x y z . P`	$\forall x\, y\, z\,.\, P$	for all
`EX x y z . P`	$\exists x\, y\, z\,.\, P$	there exists

To illustrate the notation, consider two axioms for first-order logic:

$$[\![P; Q]\!] \Longrightarrow P \wedge Q \qquad (\wedge I)$$

$$[\![\exists x\,.\, P(x); \textstyle\bigwedge x\,.\, P(x) \to Q]\!] \Longrightarrow Q \qquad (\exists E)$$

Using the `[|...|]` shorthand, $(\wedge I)$ translates into ASCII characters as

```
[| ?P; ?Q |] ==> ?P & ?Q
```

The schematic variables let unification instantiate the rule. To avoid cluttering logic definitions with question marks, Isabelle converts any free variables in a rule to schematic variables; we normally declare $(\wedge I)$ as

```
[| P; Q |] ==> P & Q
```

This variables convention agrees with the treatment of variables in goals. Free variables in a goal remain fixed throughout the proof. After the proof is finished, Isabelle converts them to scheme variables in the resulting theorem. Scheme variables in a goal may be replaced by terms during the proof, supporting answer extraction, program synthesis, and so forth.

For a final example, the rule $(\exists E)$ is rendered in ASCII as

```
[| EX x.P(x);  !!x. P(x) ==> Q |] ==> Q
```

2.1.3 Basic operations on theorems

Meta-level theorems have the ML type `thm`. They represent the theorems and inference rules of object-logics. Isabelle's meta-logic is implemented using the

LCF approach: each meta-level inference rule is represented by a function from theorems to theorems. Object-level rules are taken as axioms.

The main theorem printing commands are `prth`, `prths` and `prthq`. Of the other operations on theorems, most useful are `RS` and `RSN`, which perform resolution.

`prth` *thm*; pretty-prints *thm* at the terminal.

`prths` *thms*; pretty-prints *thms*, a list of theorems.

`prthq` *thmq*; pretty-prints *thmq*, a sequence of theorems; this is useful for inspecting the output of a tactic.

*thm*1 `RS` *thm*2 resolves the conclusion of *thm*1 with the first premise of *thm*2.

*thm*1 `RSN` $(i, thm2)$ resolves the conclusion of *thm*1 with the ith premise of *thm*2.

`standard` *thm* puts *thm* into a standard format. It also renames schematic variables to have subscript zero, improving readability and reducing subscript growth.

The rules of a theory are normally bound to ML identifiers. Suppose we are running an Isabelle session containing theory `FOL`, natural deduction first-order logic.[2] Let us try an example given in Sect. 1.3.2. We first print `mp`, which is the rule $(\rightarrow E)$, then resolve it with itself.

```
prth mp;
  [| ?P --> ?Q; ?P |] ==> ?Q
  val it = "[| ?P --> ?Q; ?P |] ==> ?Q" : thm
prth (mp RS mp);
  [| ?P1 --> ?P --> ?Q; ?P1; ?P |] ==> ?Q
  val it = "[| ?P1 --> ?P --> ?Q; ?P1; ?P |] ==> ?Q" : thm
```

User input appears in `typewriter characters`, and output appears in *`slanted typewriter characters`*. ML's response *`val`* ... is compiler-dependent and will sometimes be suppressed. This session illustrates two formats for the display of theorems. Isabelle's top-level displays theorems as ML values, enclosed in quotes. Printing commands like `prth` omit the quotes and the surrounding `val` ... : `thm`. Ignoring their side-effects, the commands are identity functions.

To contrast `RS` with `RSN`, we resolve `conjunct1`, which stands for $(\wedge E1)$, with `mp`.

```
conjunct1 RS mp;
  val it = "[| (?P --> ?Q) & ?Q1; ?P |] ==> ?Q" : thm
conjunct1 RSN (2,mp);
  val it = "[| ?P --> ?Q; ?P & ?Q1 |] ==> ?Q" : thm
```

[2]For a listing of the `FOL` rules and their ML names, turn to page 188.

These correspond to the following proofs:

$$\frac{\dfrac{(P \to Q) \land Q_1}{P \to Q}(\land E1) \quad P}{Q}(\to E) \qquad \frac{P \to Q \quad \dfrac{P \land Q_1}{P}(\land E1)}{Q}(\to E)$$

Rules can be derived by pasting other rules together. Let us join `spec`, which stands for $(\forall E)$, with `mp` and `conjunct1`. In ML, the identifier `it` denotes the value just printed.

```
spec;
  val it = "ALL x. ?P(x) ==> ?P(?x)" : thm
it RS mp;
  val it = "[| ALL x. ?P3(x) --> ?Q2(x); ?P3(?x1) |] ==>
            ?Q2(?x1)" : thm
it RS conjunct1;
  val it = "[| ALL x. ?P4(x) --> ?P6(x) & ?Q5(x); ?P4(?x2) |] ==>
            ?P6(?x2)" : thm
standard it;
  val it = "[| ALL x. ?P(x) --> ?Pa(x) & ?Q(x); ?P(?x) |] ==>
            ?Pa(?x)" : thm
```

By resolving $(\forall E)$ with $(\to E)$ and $(\land E1)$, we have derived a destruction rule for formulae of the form $\forall x . P(x) \to (Q(x) \land R(x))$. Used with destruct-resolution, such specialized rules provide a way of referring to particular assumptions.

2.1.4 *Flex-flex constraints

In higher-order unification, **flex-flex** equations are those where both sides begin with a function unknown, such as $?f(0) \stackrel{?}{\equiv} ?g(0)$. They admit a trivial unifier, here $?f \equiv \lambda x . ?a$ and $?g \equiv \lambda y . ?a$, where $?a$ is a new unknown. They admit many other unifiers, such as $?f \equiv \lambda x . ?g(0)$ and $\{?f \equiv \lambda x . x,\ ?g \equiv \lambda x . 0\}$. Huet's procedure does not enumerate the unifiers; instead, it retains flex-flex equations as constraints on future unifications. Flex-flex constraints occasionally become attached to a proof state; more frequently, they appear during use of `RS` and `RSN`:

```
refl;
  val it = "?a = ?a" : thm
exI;
  val it = "?P(?x) ==> EX x. ?P(x)" : thm
refl RS exI;
  val it = "?a3(?x) =?= ?a2(?x) ==> EX x. ?a3(x) = ?a2(x)" : thm
```

Renaming variables, this is $\exists x . ?f(x) = ?g(x)$ with the constraint $?f(?u) \stackrel{?}{\equiv} ?g(?u)$. Instances satisfying the constraint include $\exists x . ?f(x) = ?f(x)$ and $\exists x . x = ?u$. Calling `flexflex_rule` removes all constraints by applying the trivial unifier:

```
prthq (flexflex_rule it);
  EX x. ?a4 = ?a4
```

Isabelle simplifies flex-flex equations to eliminate redundant bound variables. In $\lambda x\, y . ?f(k(y), x) \stackrel{?}{\equiv} \lambda x\, y . ?g(y)$, there is no bound occurrence of x on the right side; thus, there will be none on the left in a common instance of these terms.

Choosing a new variable $?h$, Isabelle assigns $?f \equiv \lambda u\, v.?h(u)$, simplifying the left side to $\lambda x\, y\, .\, ?h(k(y))$. Dropping x from the equation leaves $\lambda y\, .\, ?h(k(y)) \stackrel{?}{\equiv} \lambda y\, .\, ?g(y)$. By η-conversion, this simplifies to the assignment $?g \equiv \lambda y.?h(k(y))$.

! RS and RSN fail (by raising exception THM) unless the resolution delivers **exactly one** resolvent. For multiple results, use RL and RLN, which operate on theorem lists. The following example uses `read_instantiate` to create an instance of `refl` containing no schematic variables.

```
val reflk = read_instantiate [("a","k")] refl;
  val reflk = "k = k" : thm
```

A flex-flex constraint is no longer possible; resolution does not find a unique unifier:

```
reflk RS exI;
  uncaught exception THM
```

Using RL this time, we discover that there are four unifiers, and four resolvents:

```
[reflk] RL [exI];
  val it = ["EX x. x = x", "EX x. k = x",
            "EX x. x = k", "EX x. k = k"] : thm list
```

2.2 Backward proof

Although RS and RSN are fine for simple forward reasoning, large proofs require tactics. Isabelle provides a suite of commands for conducting a backward proof using tactics.

2.2.1 The basic tactics

The tactics `assume_tac`, `resolve_tac`, `eresolve_tac`, and `dresolve_tac` suffice for most single-step proofs. Although `eresolve_tac` and `dresolve_tac` are not strictly necessary, they simplify proofs involving elimination and destruction rules. All the tactics act on a subgoal designated by a positive integer i, failing if i is out of range. The resolution tactics try their list of theorems in left-to-right order.

`assume_tac` i is the tactic that attempts to solve subgoal i by assumption. Proof by assumption is not a trivial step; it can falsify other subgoals by instantiating shared variables. There may be several ways of solving the subgoal by assumption.

`resolve_tac` *thms* i is the basic resolution tactic, used for most proof steps. The *thms* represent object-rules, which are resolved against subgoal i of the proof state. For each rule, resolution forms next states by unifying the conclusion with the subgoal and inserting instantiated premises in its place. A rule can admit many higher-order unifiers. The tactic fails if none of the rules generates next states.

`eresolve_tac` *thms i* performs elim-resolution. Like `resolve_tac` *thms i* followed by `assume_tac` *i*, it applies a rule then solves its first premise by assumption. But `eresolve_tac` additionally deletes that assumption from any subgoals arising from the resolution.

`dresolve_tac` *thms i* performs destruct-resolution with the *thms*, as described in Sect. 1.6.2. It is useful for forward reasoning from the assumptions.

2.2.2 Commands for backward proof

Tactics are normally applied using the subgoal module, which maintains a proof state and manages the proof construction. It allows interactive backtracking through the proof space, going away to prove lemmas, etc.; of its many commands, most important are the following:

`goal` *theory formula*; begins a new proof, where *theory* is usually an ML identifier and the *formula* is written as an ML string.

`by` *tactic*; applies the *tactic* to the current proof state, raising an exception if the tactic fails.

`undo();` reverts to the previous proof state. Undo can be repeated but cannot be undone. Do not omit the parentheses; typing `undo;` merely causes ML to echo the value of that function.

`result()` returns the theorem just proved, in a standard format. It fails if unproved subgoals are left, etc.

The commands and tactics given above are cumbersome for interactive use. Although our examples will use the full commands, you may prefer Isabelle's shortcuts:

`ba` *i*;	abbreviates	`by (assume_tac` *i*`);`
`br` *thm i*;	abbreviates	`by (resolve_tac [`*thm*`]` *i*`);`
`be` *thm i*;	abbreviates	`by (eresolve_tac [`*thm*`]` *i*`);`
`bd` *thm i*;	abbreviates	`by (dresolve_tac [`*thm*`]` *i*`);`

2.2.3 A trivial example in propositional logic

Directory `FOL` of the Isabelle distribution defines the theory of first-order logic. Let us try the example from Sect. 1.5.3, entering the goal $P \vee P \rightarrow P$ in that theory.[3]

```
goal FOL.thy "P|P --> P";
  Level 0
  P | P --> P
  1. P | P --> P
```

[3]To run these examples, see the file `FOL/ex/intro.ML`. The files `README` and `Makefile` on the directories `Pure` and `FOL` explain how to build first-order logic.

Isabelle responds by printing the initial proof state, which has $P \vee P \rightarrow P$ as the main goal and the only subgoal. The **level** of the state is the number of by commands that have been applied to reach it. We now use `resolve_tac` to apply the rule `impI`, or $(\rightarrow I)$, to subgoal 1:

```
by (resolve_tac [impI] 1);
  Level 1
  P | P --> P
  1. P | P ==> P
```

In the new proof state, subgoal 1 is P under the assumption $P \vee P$. (The meta-implication ==> indicates assumptions.) We apply `disjE`, or $(\vee E)$, to that subgoal:

```
by (resolve_tac [disjE] 1);
  Level 2
  P | P --> P
  1. P | P ==> ?P1 | ?Q1
  2. [| P | P; ?P1 |] ==> P
  3. [| P | P; ?Q1 |] ==> P
```

At Level 2 there are three subgoals, each provable by assumption. We deviate from Sect. 1.5.3 by tackling subgoal 3 first, using `assume_tac`. This affects subgoal 1, updating `?Q1` to `P`.

```
by (assume_tac 3);
  Level 3
  P | P --> P
  1. P | P ==> ?P1 | P
  2. [| P | P; ?P1 |] ==> P
```

Next we tackle subgoal 2, instantiating `?P1` to `P` in subgoal 1.

```
by (assume_tac 2);
  Level 4
  P | P --> P
  1. P | P ==> P | P
```

Lastly we prove the remaining subgoal by assumption:

```
by (assume_tac 1);
  Level 5
  P | P --> P
  No subgoals!
```

Isabelle tells us that there are no longer any subgoals: the proof is complete. Calling `result` returns the theorem.

```
val mythm = result();
  val mythm = "?P | ?P --> ?P" : thm
```

Isabelle has replaced the free variable `P` by the scheme variable `?P`. Free variables in the proof state remain fixed throughout the proof. Isabelle finally converts them to scheme variables so that the resulting theorem can be instantiated with any formula.

As an exercise, try doing the proof as in Sect. 1.5.3, observing how instantiations affect the proof state.

2.2.4 Part of a distributive law

To demonstrate the tactics `eresolve_tac`, `dresolve_tac` and the tactical `REPEAT`, let us prove part of the distributive law

$$(P \wedge Q) \vee R \leftrightarrow (P \vee R) \wedge (Q \vee R).$$

We begin by stating the goal to Isabelle and applying $(\rightarrow I)$ to it:

```
goal FOL.thy "(P & Q) | R  --> (P | R)";
  Level 0
  P & Q | R --> P | R
   1. P & Q | R --> P | R
by (resolve_tac [impI] 1);
  Level 1
  P & Q | R --> P | R
   1. P & Q | R ==> P | R
```

Previously we applied $(\vee E)$ using `resolve_tac`, but `eresolve_tac` deletes the assumption after use. The resulting proof state is simpler.

```
by (eresolve_tac [disjE] 1);
  Level 2
  P & Q | R --> P | R
   1. P & Q ==> P | R
   2. R ==> P | R
```

Using `dresolve_tac`, we can apply $(\wedge E1)$ to subgoal 1, replacing the assumption $P \wedge Q$ by P. Normally we should apply the rule $(\wedge E)$, given in Sect. 1.6.2. That is an elimination rule and requires `eresolve_tac`; it would replace $P \wedge Q$ by the two assumptions P and Q. Because the present example does not need Q, we may try out `dresolve_tac`.

```
by (dresolve_tac [conjunct1] 1);
  Level 3
  P & Q | R --> P | R
   1. P ==> P | R
   2. R ==> P | R
```

The next two steps apply $(\vee I1)$ and $(\vee I2)$ in an obvious manner.

```
by (resolve_tac [disjI1] 1);
  Level 4
  P & Q | R --> P | R
   1. P ==> P
   2. R ==> P | R
by (resolve_tac [disjI2] 2);
  Level 5
  P & Q | R --> P | R
   1. P ==> P
   2. R ==> R
```

Two calls of `assume_tac` can finish the proof. The tactical `REPEAT` here expresses a tactic that calls `assume_tac 1` as many times as possible. We can restrict attention to subgoal 1 because the other subgoals move up after subgoal 1 disappears.

```
by (REPEAT (assume_tac 1));
  Level 6
  P & Q | R --> P | R
  No subgoals!
```

2.3 Quantifier reasoning

This section illustrates how Isabelle enforces quantifier provisos. Suppose that x, y and z are parameters of a subgoal. Quantifier rules create terms such as $?f(x,z)$, where $?f$ is a function unknown. Instantiating $?f$ to $\lambda x\, z\, .\, t$ has the effect of replacing $?f(x,z)$ by t, where the term t may contain free occurrences of x and z. On the other hand, no instantiation of $?f$ can replace $?f(x,z)$ by a term containing free occurrences of y, since parameters are bound variables.

2.3.1 Two quantifier proofs: a success and a failure

Let us contrast a proof of the theorem $\forall x\, .\, \exists y\, .\, x = y$ with an attempted proof of the non-theorem $\exists y\, .\, \forall x\, .\, x = y$. The former proof succeeds, and the latter fails, because of the scope of quantified variables [43]. Unification helps even in these trivial proofs. In $\forall x\, .\, \exists y\, .\, x = y$ the y that 'exists' is simply x, but we need never say so. This choice is forced by the reflexive law for equality, and happens automatically.

The successful proof. The proof of $\forall x\, .\, \exists y\, .\, x = y$ demonstrates the introduction rules $(\forall I)$ and $(\exists I)$. We state the goal and apply $(\forall I)$:

```
goal FOL.thy "ALL x. EX y. x=y";
  Level 0
  ALL x. EX y. x = y
   1. ALL x. EX y. x = y
by (resolve_tac [allI] 1);
  Level 1
  ALL x. EX y. x = y
   1. !!x. EX y. x = y
```

The variable x is no longer universally quantified, but is a parameter in the subgoal; thus, it is universally quantified at the meta-level. The subgoal must be proved for all possible values of x.

To remove the existential quantifier, we apply the rule $(\exists I)$:

```
by (resolve_tac [exI] 1);
  Level 2
  ALL x. EX y. x = y
   1. !!x. x = ?y1(x)
```

The bound variable y has become ?y1(x). This term consists of the function unknown ?y1 applied to the parameter x. Instances of ?y1(x) may or may not contain x. We resolve the subgoal with the reflexivity axiom.

```
by (resolve_tac [refl] 1);
  Level 3
  ALL x. EX y. x = y
  No subgoals!
```

Let us consider what has happened in detail. The reflexivity axiom is lifted over x to become $\bigwedge x\,.\,?f(x) = ?f(x)$, which is unified with $\bigwedge x\,.\,x = ?y_1(x)$. The function unknowns $?f$ and $?y_1$ are both instantiated to the identity function, and $x = ?y_1(x)$ collapses to $x = x$ by β-reduction.

The unsuccessful proof. We state the goal $\exists y.\forall x.x = y$, which is not a theorem, and try $(\exists I)$:

```
goal FOL.thy "EX y. ALL x. x=y";
  Level 0
  EX y. ALL x. x = y
   1. EX y. ALL x. x = y
by (resolve_tac [exI] 1);
  Level 1
  EX y. ALL x. x = y
   1. ALL x. x = ?y
```

The unknown ?y may be replaced by any term, but this can never introduce another bound occurrence of x. We now apply $(\forall I)$:

```
by (resolve_tac [allI] 1);
  Level 2
  EX y. ALL x. x = y
   1. !!x. x = ?y
```

Compare our position with the previous Level 2. Instead of ?y1(x) we have ?y, whose instances may not contain the bound variable x. The reflexivity axiom does not unify with subgoal 1.

```
by (resolve_tac [refl] 1);
  by: tactic returned no results
```

There can be no proof of $\exists y.\forall x.x = y$ by the soundness of first-order logic. I have elsewhere proved the faithfulness of Isabelle's encoding of first-order logic [43]; there could, of course, be faults in the implementation.

2.3.2 Nested quantifiers

Multiple quantifiers create complex terms. Proving

$$(\forall x\, y\,.\,P(x, y)) \rightarrow (\forall z\, w\,.\,P(w, z))$$

will demonstrate how parameters and unknowns develop. If they appear in the wrong order, the proof will fail.

This section concludes with a demonstration of REPEAT and ORELSE.

```
goal FOL.thy "(ALL x y.P(x,y))  -->  (ALL z w.P(w,z))";
  Level 0
  (ALL x y. P(x,y)) --> (ALL z w. P(w,z))
   1. (ALL x y. P(x,y)) --> (ALL z w. P(w,z))
by (resolve_tac [impI] 1);
  Level 1
  (ALL x y. P(x,y)) --> (ALL z w. P(w,z))
   1. ALL x y. P(x,y) ==> ALL z w. P(w,z)
```

The wrong approach. Using dresolve_tac, we apply the rule $(\forall E)$, bound to the ML identifier spec. Then we apply $(\forall I)$.

```
by (dresolve_tac [spec] 1);
  Level 2
  (ALL x y. P(x,y)) --> (ALL z w. P(w,z))
   1. ALL y. P(?x1,y) ==> ALL z w. P(w,z)
by (resolve_tac [allI] 1);
  Level 3
  (ALL x y. P(x,y)) --> (ALL z w. P(w,z))
   1. !!z. ALL y. P(?x1,y) ==> ALL w. P(w,z)
```

The unknown ?x1 and the parameter z have appeared. We again apply $(\forall E)$ and $(\forall I)$.

```
by (dresolve_tac [spec] 1);
  Level 4
  (ALL x y. P(x,y)) --> (ALL z w. P(w,z))
   1. !!z. P(?x1,?y3(z)) ==> ALL w. P(w,z)
by (resolve_tac [allI] 1);
  Level 5
  (ALL x y. P(x,y)) --> (ALL z w. P(w,z))
   1. !!z w. P(?x1,?y3(z)) ==> P(w,z)
```

The unknown ?y3 and the parameter w have appeared. Each unknown is applied to the parameters existing at the time of its creation; instances of ?x1 cannot contain z or w, while instances of ?y3(z) can only contain z. Due to the restriction on ?x1, proof by assumption will fail.

```
by (assume_tac 1);
  by: tactic returned no results
  uncaught exception ERROR
```

The right approach. To do this proof, the rules must be applied in the correct order. Parameters should be created before unknowns. The choplev command returns to an earlier stage of the proof; let us return to the result of applying $(\to I)$:

```
choplev 1;
  Level 1
  (ALL x y. P(x,y)) --> (ALL z w. P(w,z))
   1. ALL x y. P(x,y) ==> ALL z w. P(w,z)
```

Previously we made the mistake of applying $(\forall E)$ before $(\forall I)$.

```
by (resolve_tac [allI] 1);
  Level 2
  (ALL x y. P(x,y)) --> (ALL z w. P(w,z))
   1. !!z. ALL x y. P(x,y) ==> ALL w. P(w,z)
by (resolve_tac [allI] 1);
  Level 3
  (ALL x y. P(x,y)) --> (ALL z w. P(w,z))
   1. !!z w. ALL x y. P(x,y) ==> P(w,z)
```

The parameters `z` and `w` have appeared. We now create the unknowns:

```
by (dresolve_tac [spec] 1);
  Level 4
  (ALL x y. P(x,y)) --> (ALL z w. P(w,z))
   1. !!z w. ALL y. P(?x3(z,w),y) ==> P(w,z)
by (dresolve_tac [spec] 1);
  Level 5
  (ALL x y. P(x,y)) --> (ALL z w. P(w,z))
   1. !!z w. P(?x3(z,w),?y4(z,w)) ==> P(w,z)
```

Both `?x3(z,w)` and `?y4(z,w)` could become any terms containing `z` and `w`:

```
by (assume_tac 1);
  Level 6
  (ALL x y. P(x,y)) --> (ALL z w. P(w,z))
  No subgoals!
```

A one-step proof using tacticals. Repeated application of rules can be effective, but the rules should be attempted in the correct order. Let us return to the original goal using `choplev`:

```
choplev 0;
  Level 0
  (ALL x y. P(x,y)) --> (ALL z w. P(w,z))
   1. (ALL x y. P(x,y)) --> (ALL z w. P(w,z))
```

As we have just seen, `allI` should be attempted before `spec`, while `assume_tac` generally can be attempted first. Such priorities can easily be expressed using `ORELSE`, and repeated using `REPEAT`.

```
by (REPEAT (assume_tac 1 ORELSE resolve_tac [impI,allI] 1
      ORELSE dresolve_tac [spec] 1));
  Level 1
  (ALL x y. P(x,y)) --> (ALL z w. P(w,z))
  No subgoals!
```

2.3.3 A realistic quantifier proof

To see the practical use of parameters and unknowns, let us prove half of the equivalence

$$(\forall x \,.\, P(x) \to Q) \leftrightarrow ((\exists x \,.\, P(x)) \to Q).$$

We state the left-to-right half to Isabelle in the normal way. Since $\rightarrow$ is nested to the right, $(\rightarrow I)$ can be applied twice; we use REPEAT:

```
goal FOL.thy "(ALL x.P(x) --> Q) --> (EX x.P(x)) --> Q";
  Level 0
  (ALL x. P(x) --> Q) --> (EX x. P(x)) --> Q
   1. (ALL x. P(x) --> Q) --> (EX x. P(x)) --> Q
by (REPEAT (resolve_tac [impI] 1));
  Level 1
  (ALL x. P(x) --> Q) --> (EX x. P(x)) --> Q
   1. [| ALL x. P(x) --> Q; EX x. P(x) |] ==> Q
```

We can eliminate the universal or the existential quantifier. The existential quantifier should be eliminated first, since this creates a parameter. The rule $(\exists E)$ is bound to the identifier exE.

```
by (eresolve_tac [exE] 1);
  Level 2
  (ALL x. P(x) --> Q) --> (EX x. P(x)) --> Q
   1. !!x. [| ALL x. P(x) --> Q; P(x) |] ==> Q
```

The only possibility now is $(\forall E)$, a destruction rule. We use dresolve_tac, which discards the quantified assumption; it is only needed once.

```
by (dresolve_tac [spec] 1);
  Level 3
  (ALL x. P(x) --> Q) --> (EX x. P(x)) --> Q
   1. !!x. [| P(x); P(?x3(x)) --> Q |] ==> Q
```

Because we applied $(\exists E)$ before $(\forall E)$, the unknown term ?x3(x) may depend upon the parameter x.

Although $(\rightarrow E)$ is a destruction rule, it works with eresolve_tac to perform backward chaining. This technique is frequently useful.

```
by (eresolve_tac [mp] 1);
  Level 4
  (ALL x. P(x) --> Q) --> (EX x. P(x)) --> Q
   1. !!x. P(x) ==> P(?x3(x))
```

The tactic has reduced Q to P(?x3(x)), deleting the implication. The final step is trivial, thanks to the occurrence of x.

```
by (assume_tac 1);
  Level 5
  (ALL x. P(x) --> Q) --> (EX x. P(x)) --> Q
  No subgoals!
```

2.3.4 The classical reasoner

Although Isabelle cannot compete with fully automatic theorem provers, it provides enough automation to tackle substantial examples. The classical reasoner can be set up for any classical natural deduction logic; see Chap. 14.

Rules are packaged into **classical sets**. The classical reasoner provides several tactics, which apply rules using naive algorithms. Unification handles quantifiers as shown above. The most useful tactic is `fast_tac`.

Let us solve problems 40 and 60 of Pelletier [52]. (The backslashes \...\ are an ML string escape sequence, to break the long string over two lines.)

```
goal FOL.thy "(EX y. ALL x. J(y,x) <-> ~J(x,x))  \
\       -->  ~ (ALL x. EX y. ALL z. J(z,y) <-> ~ J(z,x))";
  Level 0
  (EX y. ALL x. J(y,x) <-> ~J(x,x)) -->
  ~(ALL x. EX y. ALL z. J(z,y) <-> ~J(z,x))
   1. (EX y. ALL x. J(y,x) <-> ~J(x,x)) -->
      ~(ALL x. EX y. ALL z. J(z,y) <-> ~J(z,x))
```

The rules of classical logic are bundled as `FOL_cs`. We may solve subgoal 1 at a stroke, using `fast_tac`.

```
by (fast_tac FOL_cs 1);
  Level 1
  (EX y. ALL x. J(y,x) <-> ~J(x,x)) -->
  ~(ALL x. EX y. ALL z. J(z,y) <-> ~J(z,x))
  No subgoals!
```

Sceptics may examine the proof by calling the package's single-step tactics, such as `step_tac`. This would take up much space, however, so let us proceed to the next example:

```
goal FOL.thy "ALL x. P(x,f(x)) <-> \
\       (EX y. (ALL z. P(z,y) --> P(z,f(x))) & P(x,y))";
  Level 0
  ALL x. P(x,f(x)) <-> (EX y. (ALL z. P(z,y) --> P(z,f(x))) & P(x,y))
   1. ALL x. P(x,f(x)) <->
      (EX y. (ALL z. P(z,y) --> P(z,f(x))) & P(x,y))
```

Again, subgoal 1 succumbs immediately.

```
by (fast_tac FOL_cs 1);
  Level 1
  ALL x. P(x,f(x)) <-> (EX y. (ALL z. P(z,y) --> P(z,f(x))) & P(x,y))
  No subgoals!
```

The classical reasoner is not restricted to the usual logical connectives. The natural deduction rules for unions and intersections resemble those for disjunction and conjunction. The rules for infinite unions and intersections resemble those for quantifiers. Given such rules, the classical reasoner is effective for reasoning in set theory.

3. Advanced Methods

Before continuing, it might be wise to try some of your own examples in Isabelle, reinforcing your knowledge of the basic functions.

Look through *Isabelle's Object-Logics* and try proving some simple theorems. You probably should begin with first-order logic (`FOL` or `LK`). Try working some of the examples provided, and others from the literature. Set theory (`ZF`) and Constructive Type Theory (`CTT`) form a richer world for mathematical reasoning and, again, many examples are in the literature. Higher-order logic (`HOL`) is Isabelle's most sophisticated logic because its types and functions are identified with those of the meta-logic.

Choose a logic that you already understand. Isabelle is a proof tool, not a teaching tool; if you do not know how to do a particular proof on paper, then you certainly will not be able to do it on the machine. Even experienced users plan large proofs on paper.

We have covered only the bare essentials of Isabelle, but enough to perform substantial proofs. By occasionally dipping into the *Reference Manual*, you can learn additional tactics, subgoal commands and tacticals.

3.1 Deriving rules in Isabelle

A mathematical development goes through a progression of stages. Each stage defines some concepts and derives rules about them. We shall see how to derive rules, perhaps involving definitions, using Isabelle. The following section will explain how to declare types, constants, rules and definitions.

3.1.1 Deriving a rule using tactics and meta-level assumptions

The subgoal module supports the derivation of rules, as discussed in Sect. 1.6.3. The `goal` command, when supplied a goal of the form $[\![\theta_1; \ldots; \theta_k]\!] \Longrightarrow \phi$, creates $\phi \Longrightarrow \phi$ as the initial proof state and returns a list consisting of the theorems θ_i $[\theta_i]$, for $i = 1, \ldots, k$. These meta-assumptions are also recorded internally, allowing `result` to discharge them in the original order.

Let us derive $\wedge$ elimination using Isabelle. Until now, calling `goal` has returned an empty list, which we have thrown away. In this example, the list contains the two premises of the rule. We bind them to the ML identifiers major and minor:[1]

```
val [major,minor] = goal FOL.thy
    "[| P&Q;  [| P; Q |] ==> R |] ==> R";
  Level 0
  R
   1. R
  val major = "P & Q  [P & Q]" : thm
  val minor = "[| P; Q |] ==> R  [[| P; Q |] ==> R]" : thm
```

Look at the minor premise, recalling that meta-level assumptions are shown in brackets. Using minor, we reduce R to the subgoals P and Q:

```
by (resolve_tac [minor] 1);
  Level 1
  R
   1. P
   2. Q
```

Deviating from Sect. 1.6.3, we apply $(\wedge E1)$ forwards from the assumption $P \wedge Q$ to obtain the theorem P $[P \wedge Q]$.

```
major RS conjunct1;
  val it = "P  [P & Q]" : thm
by (resolve_tac [major RS conjunct1] 1);
  Level 2
  R
   1. Q
```

Similarly, we solve the subgoal involving Q.

```
major RS conjunct2;
  val it = "Q  [P & Q]" : thm
by (resolve_tac [major RS conjunct2] 1);
  Level 3
  R
  No subgoals!
```

Calling topthm returns the current proof state as a theorem. Note that it contains assumptions. Calling result discharges the assumptions — both occurrences of $P \wedge Q$ are discharged as one — and makes the variables schematic.

```
topthm();
  val it = "R  [P & Q, P & Q, [| P; Q |] ==> R]" : thm
val conjE = result();
  val conjE = "[| ?P & ?Q; [| ?P; ?Q |] ==> ?R |] ==> ?R" : thm
```

3.1.2 Definitions and derived rules

Definitions are expressed as meta-level equalities. Let us define negation and the if-and-only-if connective:

[1] Some ML compilers will print a message such as *binding not exhaustive.* This warns that goal must return a 2-element list. Otherwise, the pattern-match will fail; ML will raise exception Match.

$$\begin{aligned} \neg ?P &\equiv ?P \rightarrow \bot \\ ?P \leftrightarrow ?Q &\equiv (?P \rightarrow ?Q) \wedge (?Q \rightarrow ?P) \end{aligned}$$

Isabelle permits **meta-level rewriting** using definitions such as these. **Unfolding** replaces every instance of $\neg ?P$ by the corresponding instance of $?P \rightarrow \bot$. For example, $\forall x\, .\, \neg(P(x) \wedge \neg R(x,0))$ unfolds to

$$\forall x\, .\, (P(x) \wedge R(x,0) \rightarrow \bot) \rightarrow \bot.$$

Folding a definition replaces occurrences of the right-hand side by the left. The occurrences need not be free in the entire formula.

When you define new concepts, you should derive rules asserting their abstract properties, and then forget their definitions. This supports modularity: if you later change the definitions without affecting their abstract properties, then most of your proofs will carry through without change. Indiscriminate unfolding makes a subgoal grow exponentially, becoming unreadable.

Taking this point of view, Isabelle does not unfold definitions automatically during proofs. Rewriting must be explicit and selective. Isabelle provides tactics and meta-rules for rewriting, and a version of the `goal` command that unfolds the conclusion and premises of the rule being derived.

For example, the intuitionistic definition of negation given above may seem peculiar. Using Isabelle, we shall derive pleasanter negation rules:

$$\frac{\begin{array}{c}[P] \\ \vdots \\ \bot\end{array}}{\neg P}\,(\neg I) \qquad \frac{\neg P \quad P}{Q}\,(\neg E)$$

This requires proving the following meta-formulae:

$$(P \Longrightarrow \bot) \Longrightarrow \neg P \qquad (\neg I)$$

$$[\![\neg P; P]\!] \Longrightarrow Q. \qquad (\neg E)$$

3.1.3 Deriving the ¬ introduction rule

To derive $(\neg I)$, we may call `goal` with the appropriate formula. Again, `goal` returns a list consisting of the rule's premises. We bind this one-element list to the ML identifier `prems`.

```
val prems = goal FOL.thy "(P ==> False) ==> ~P";
  Level 0
  ~P
   1. ~P
  val prems = ["P ==> False  [P ==> False]"] : thm list
```

Calling `rewrite_goals_tac` with `not_def`, which is the definition of negation, unfolds that definition in the subgoals. It leaves the main goal alone.

```
not_def;
  val it = "~?P == ?P --> False" : thm
by (rewrite_goals_tac [not_def]);
  Level 1
  ~P
   1. P --> False
```

Using `impI` and the premise, we reduce subgoal 1 to a triviality:

```
by (resolve_tac [impI] 1);
  Level 2
  ~P
   1. P ==> False
by (resolve_tac prems 1);
  Level 3
  ~P
   1. P ==> P
```

The rest of the proof is routine. Note the form of the final result.

```
by (assume_tac 1);
  Level 4
  ~P
  No subgoals!
val notI = result();
  val notI = "(?P ==> False) ==> ~?P" : thm
```

There is a simpler way of conducting this proof. The `goalw` command starts a backward proof, as does `goal`, but it also unfolds definitions. Thus there is no need to call `rewrite_goals_tac`:

```
val prems = goalw FOL.thy [not_def]
    "(P ==> False) ==> ~P";
  Level 0
  ~P
   1. P --> False
  val prems = ["P ==> False  [P ==> False]"] : thm list
```

3.1.4 Deriving the ¬ elimination rule

Let us derive the rule ($\neg E$). The proof follows that of `conjE` above, with an additional step to unfold negation in the major premise. Although the `goalw` command is best for this, let us try `goal` to see another way of unfolding definitions. After binding the premises to ML identifiers, we apply `FalseE`:

```
val [major,minor] = goal FOL.thy "[| ~P;  P |] ==> R";
  Level 0
  R
   1. R
  val major = "~ P  [~ P]" : thm
  val minor = "P  [P]" : thm
```

```
by (resolve_tac [FalseE] 1);
  Level 1
  R
   1. False
```

Everything follows from falsity. And we can prove falsity using the premises and Modus Ponens:

```
by (resolve_tac [mp] 1);
  Level 2
  R
   1. ?P1 --> False
   2. ?P1
```

For subgoal 1, we transform the major premise from $\neg P$ to $P \to \bot$. The function `rewrite_rule`, given a list of definitions, unfolds them in a theorem. Rewriting does not affect the theorem's hypothesis, which remains $\neg P$:

```
rewrite_rule [not_def] major;
  val it = "P --> False  [~P]" : thm
by (resolve_tac [it] 1);
  Level 3
  R
   1. P
```

The subgoal `?P1` has been instantiated to `P`, which we can prove using the minor premise:

```
by (resolve_tac [minor] 1);
  Level 4
  R
  No subgoals!
val notE = result();
  val notE = "[| ~?P; ?P |] ==> ?R" : thm
```

Again, there is a simpler way of conducting this proof. Recall that the `goalw` command unfolds definitions the conclusion; it also unfolds definitions in the premises:

```
val [major,minor] = goalw FOL.thy [not_def]
    "[| ~P;  P |] ==> R";
  val major = "P --> False  [~ P]" : thm
  val minor = "P  [P]" : thm
```

Observe the difference in `major`; the premises are unfolded without calling `rewrite_rule`. Incidentally, the four calls to `resolve_tac` above can be collapsed to one, with the help of `RS`; this is a typical example of forward reasoning from a complex premise.

```
minor RS (major RS mp RS FalseE);
  val it = "?P  [P, ~P]" : thm
by (resolve_tac [it] 1);
  Level 1
  R
  No subgoals!
```

Finally, here is a trick that is sometimes useful. If the goal has an outermost meta-quantifier, then goal and goalw do not return the rule's premises in the list of theorems; instead, the premises become assumptions in subgoal 1.

```
goalw FOL.thy [not_def] "!!P R. [| ~P;  P |] ==> R";
  Level 0
  !!P R. [| ~ P; P |] ==> R
   1. !!P R. [| P --> False; P |] ==> R
val it = [] : thm list
```

The proof continues as before. But instead of referring to ML identifiers, we refer to assumptions using eresolve_tac or assume_tac:

```
by (resolve_tac [FalseE] 1);
  Level 1
  !!P R. [| ~ P; P |] ==> R
   1. !!P R. [| P --> False; P |] ==> False
by (eresolve_tac [mp] 1);
  Level 2
  !!P R. [| ~ P; P |] ==> R
   1. !!P R. P ==> P
by (assume_tac 1);
  Level 3
  !!P R. [| ~ P; P |] ==> R
  No subgoals!
```

Calling result strips the meta-quantifiers, so the resulting theorem is the same as before.

```
val notE = result();
  val notE = "[| ~?P; ?P |] ==> ?R" : thm
```

Do not use the !! trick if the premises contain meta-level connectives, because eresolve_tac and assume_tac would not be able to handle the resulting assumptions. The trick is not suitable for deriving the introduction rule $(\neg I)$.

3.2 Defining theories

Isabelle makes no distinction between simple extensions of a logic — like defining a type *bool* with constants *true* and *false* — and defining an entire logic. A theory definition has the form

```
T = S1 + ··· + Sn +
classes       class declarations
default       sort
types         type declarations and synonyms
arities       arity declarations
consts        constant declarations
rules         rule declarations
translations  translation declarations
end
ML            ML code
```

This declares the theory T to extend the existing theories $S_1, \ldots, S_n$. It may declare new classes, types, arities (overloadings of existing types), constants and rules; it can specify the default sort for type variables. A constant declaration can specify an associated concrete syntax. The translations section specifies rewrite rules on abstract syntax trees, for defining notations and abbreviations. The ML section contains code to perform arbitrary syntactic transformations. The main declaration forms are discussed below. The full syntax can be found in App. A.

All the declaration parts can be omitted. In the simplest case, T is just the union of $S_1, \ldots, S_n$. New theories always extend one or more other theories, inheriting their types, constants, syntax, etc. The theory `Pure` contains nothing but Isabelle's meta-logic.

Each theory definition must reside in a separate file, whose name is the theory's with `.thy` appended. For example, theory `ListFn` resides on a file named `ListFn.thy`. Isabelle uses this convention to locate the file containing a given theory; `use_thy` automatically loads a theory's parents before loading the theory itself.

Calling `use_thy "`T`"` reads a theory from the file T`.thy`, writes the corresponding ML code to the file `.`T`.thy.ML`, reads the latter file, and deletes it if no errors occurred. This declares the ML structure T, which contains a component `thy` denoting the new theory, a component for each rule, and everything declared in *ML code.*

Errors may arise during the translation to ML (say, a misspelled keyword) or during creation of the new theory (say, a type error in a rule). But if all goes well, `use_thy` will finally read the file T`.ML`, if it exists. This file typically begins with the ML declaration `open` T and contains proofs that refer to the components of T.

When a theory file is modified, many theories may have to be reloaded. Isabelle records the modification times and dependencies of theory files. See Sect. 9.3 for more details.

3.2.1 Declaring constants and rules

Most theories simply declare constants and rules. The **constant declaration part** has the form

```
consts  c1 :: "τ1"
        ⋮
        cn :: "τn"
```

where $c_1, \ldots, c_n$ are constants and $\tau_1, \ldots, \tau_n$ are types. Each type *must* be enclosed in quotation marks. Each constant must be enclosed in quotation marks unless it is a valid identifier. To declare $c_1, \ldots, c_n$ as constants of type τ, the n declarations may be abbreviated to a single line:

```
        c1, ..., cn :: "τ"
```

The **rule declaration part** has the form

rules id_1 "$rule_1$"
⋮
id_n "$rule_n$"

where id_1, ..., id_n are ML identifiers and $rule_1$, ..., $rule_n$ are expressions of type *prop*. Each rule *must* be enclosed in quotation marks.

Definitions are rules of the form $t \equiv u$. Normally definitions should be conservative, serving only as abbreviations. As of this writing, Isabelle does not provide a separate declaration part for definitions; it is your responsibility to ensure that your definitions are conservative. However, Isabelle's rewriting primitives will reject $t \equiv u$ unless all variables free in u are also free in t.

This theory extends first-order logic with two constants *nand* and *xor*, and declares rules to define them:

```
Gate = FOL +
consts  nand,xor :: "[o,o] => o"
rules   nand_def "nand(P,Q) == ~(P & Q)"
        xor_def  "xor(P,Q)  == P & ~Q | ~P & Q"
end
```

3.2.2 Declaring type constructors

Types are composed of type variables and **type constructors**. Each type constructor takes a fixed number of arguments. They are declared with an ML-like syntax. If *list* takes one type argument, *tree* takes two arguments and *nat* takes no arguments, then these type constructors can be declared by

```
types 'a list
      ('a,'b) tree
      nat
```

The **type declaration part** has the general form

types $tids_1$ id_1
⋮
$tids_1$ id_n

where id_1, ..., id_n are identifiers and $tids_1$, ..., $tids_n$ are type argument lists as shown in the example above. It declares each id_i as a type constructor with the specified number of argument places.

The **arity declaration part** has the form

arities $tycon_1$ :: $arity_1$
⋮
$tycon_n$ :: $arity_n$

where $tycon_1$, ..., $tycon_n$ are identifiers and $arity_1$, ..., $arity_n$ are arities. Arity declarations add arities to existing types; they do not declare the types themselves. In the simplest case, for an 0-place type constructor, an arity is simply the type's class. Let us declare a type *bool* of class *term*, with constants *tt* and *ff*. (In first-order logic, booleans are distinct from formulae, which have type *o* :: *logic*.)

```
Bool = FOL +
types   bool
arities bool    :: term
consts  tt,ff   :: "bool"
end
```

A k-place type constructor may have arities of the form $(s_1, \ldots, s_k)c$, where $s_1, \ldots, s_n$ are sorts and c is a class. Each sort specifies a type argument; it has the form $\{c_1, \ldots, c_m\}$, where c_1, ..., c_m are classes. Mostly we deal with singleton sorts, and may abbreviate them by dropping the braces. The arity $(term)term$ is short for $(\{term\})term$. Recall the discussion in Sect. 1.1.2.

A type constructor may be overloaded (subject to certain conditions) by appearing in several arity declarations. For instance, the function type constructor *fun* has the arity $(logic, logic)logic$; in higher-order logic, it is declared also to have arity $(term, term)term$.

Theory `List` declares the 1-place type constructor *list*, gives it arity $(term)term$, and declares constants *Nil* and *Cons* with polymorphic types:[2]

```
List = FOL +
types   'a list
arities list    :: (term)term
consts  Nil     :: "'a list"
        Cons    :: "['a, 'a list] => 'a list"
end
```

Multiple arity declarations may be abbreviated to a single line:

```
arities tycon1, ..., tyconn :: arity
```

! Arity declarations resemble constant declarations, but there are *no* quotation marks! Types and rules must be quoted because the theory translator passes them verbatim to the ML output file.

3.2.3 Type synonyms

Isabelle supports **type synonyms (abbreviations)** which are similar to those found in ML. Such synonyms are defined in the type declaration part and are fairly self explanatory:

```
types gate       = "[o,o] => o"
      'a pred    = "'a => o"
      ('a,'b)nuf = "'b => 'a"
```

[2]In the `consts` part, type variable `'a` has the default sort, which is `term`. See the *Reference Manual* (Sect. 9.1) for more information.

Type declarations and synonyms can be mixed arbitrarily:

```
types nat
      'a stream = "nat => 'a"
      signal    = "nat stream"
      'a list
```

A synonym is merely an abbreviation for some existing type expression. Hence synonyms may not be recursive! Internally all synonyms are fully expanded. As a consequence Isabelle output never contains synonyms. Their main purpose is to improve the readability of theories. Synonyms can be used just like any other type:

```
consts and,or :: "gate"
       negate :: "signal => signal"
```

3.2.4 Infix and mixfix operators

Infix or mixfix syntax may be attached to constants. Consider the following theory:

```
Gate2 = FOL +
consts  "~&"     :: "[o,o] => o"          (infixl 35)
        "#"      :: "[o,o] => o"          (infixl 30)
rules   nand_def "P ~& Q == ~(P & Q)"
        xor_def  "P # Q  == P & ~Q | ~P & Q"
end
```

The constant declaration part declares two left-associating infix operators with their priorities, or precedences; they are $\neg\&$ of priority 35 and # of priority 30. Hence $P \# Q \# R$ is parsed as $(P \# Q) \# R$ and $P \# Q \neg\& R$ as $P \# (Q \neg\& R)$. Note the quotation marks in `"~&"` and `"#"`.

The constants `op ~&` and `op #` are declared automatically, just as in ML. Hence you may write propositions like `op #(True) == op ~&(True)`, which asserts that the functions $\lambda Q\,.\,True \# Q$ and $\lambda Q\,.\,True \neg\& Q$ are identical.

Mixfix operators may have arbitrary context-free syntaxes. Let us add a line to the constant declaration part:

```
If :: "[o,o,o] => o"        ("if _ then _ else _")
```

This declares a constant *If* of type $[o, o, o] \Rightarrow o$ with concrete syntax `if` P `then` Q `else` R as well as `If(`P`,`Q`,`R`)`. Underscores denote argument positions.

The declaration above does not allow the `if-then-else` construct to be split across several lines, even if it is too long to fit on one line. Pretty-printing information can be added to specify the layout of mixfix operators. For details, see Chap. 10.

Mixfix declarations can be annotated with priorities, just like infixes. The example above is just a shorthand for

```
If :: "[o,o,o] => o"        ("if _ then _ else _" [0,0,0] 1000)
```

The numeric components determine priorities. The list of integers defines, for each argument position, the minimal priority an expression at that position must have. The final integer is the priority of the construct itself. In the example above, any argument expression is acceptable because priorities are non-negative, and conditionals may appear everywhere because 1000 is the highest priority. On the other hand, the declaration

```
If :: "[o,o,o] => o"        ("if _ then _ else _" [100,0,0] 99)
```

defines concrete syntax for a conditional whose first argument cannot have the form `if` P `then` Q `else` R because it must have a priority of at least 100. We may of course write

`if (if` P `then` Q `else` R`) then` S `else` T

because expressions in parentheses have maximal priority.

Binary type constructors, like products and sums, may also be declared as infixes. The type declaration below introduces a type constructor $*$ with infix notation $\alpha * \beta$, together with the mixfix notation <_,_> for pairs.

```
Prod = FOL +
types   ('a,'b) "*"                              (infixl 20)
arities "*"      :: (term,term)term
consts  fst      :: "'a * 'b => 'a"
        snd      :: "'a * 'b => 'b"
        Pair     :: "['a,'b] => 'a * 'b"         ("(1<_,/_>)")
rules   fst      "fst(<a,b>) = a"
        snd      "snd(<a,b>) = b"
end
```

! The name of the type constructor is `*` and not `op *`, as it would be in the case of an infix constant. Only infix type constructors can have symbolic names like `*`. There is no general mixfix syntax for types.

3.2.5 Overloading

The **class declaration part** has the form

`classes` id_1 `<` c_1
$\vdots$
id_n `<` c_n

where $id_1, \ldots, id_n$ are identifiers and $c_1, \ldots, c_n$ are existing classes. It declares each id_i as a new class, a subclass of c_i. In the general case, an identifier may be declared to be a subclass of k existing classes:

id `<` c_1 `, ... ,` c_k

Type classes allow constants to be overloaded. As suggested in Sect. 1.1.2, let us define the class *arith* of arithmetic types with the constants $+ :: [\alpha, \alpha] \Rightarrow \alpha$

and $0, 1{::}\alpha$, for $\alpha{::}arith$. We introduce *arith* as a subclass of *term* and add the three polymorphic constants of this class.

```
Arith = FOL +
classes arith < term
consts "0"     :: "'a::arith"                ("0")
       "1"     :: "'a::arith"                ("1")
       "+"     :: "['a::arith,'a] => 'a"     (infixl 60)
end
```

No rules are declared for these constants: we merely introduce their names without specifying properties. On the other hand, classes with rules make it possible to prove **generic** theorems. Such theorems hold for all instances, all types in that class.

We can now obtain distinct versions of the constants of *arith* by declaring certain types to be of class *arith*. For example, let us declare the 0-place type constructors *bool* and *nat*:

```
BoolNat = Arith +
types    bool,nat
arities  bool,nat      :: arith
consts   Suc           :: "nat=>nat"
rules    add0          "0 + n = n::nat"
         addS          "Suc(m)+n = Suc(m+n)"
         nat1          "1 = Suc(0)"
         or0l          "0 + x = x::bool"
         or0r          "x + 0 = x::bool"
         or1l          "1 + x = 1::bool"
         or1r          "x + 1 = 1::bool"
end
```

Because *nat* and *bool* have class *arith*, we can use 0, 1 and + at either type. The type constraints in the axioms are vital. Without constraints, the x in $1+x=x$ would have type $\alpha{::}arith$ and the axiom would hold for any type of class *arith*. This would collapse *nat* to a trivial type:

$$Suc(1) = Suc(0+1) = Suc(0)+1 = 1+1 = 1!$$

3.3 Theory example: the natural numbers

We shall now work through a small example of formalized mathematics demonstrating many of the theory extension features.

3.3.1 Extending first-order logic with the natural numbers

Section 1.1 has formalized a first-order logic, including a type *nat* and the constants $0 :: nat$ and $Suc :: nat \Rightarrow nat$. Let us introduce the Peano axioms for mathematical induction and the freeness of 0 and *Suc*:

$$\frac{\begin{matrix} & [P] \\ & \vdots \\ P[0/x] & P[Suc(x)/x] \end{matrix}}{P[n/x]}\ (induct) \quad \begin{array}{l}\text{provided } x \text{ is not free in} \\ \text{any assumption except } P\end{array}$$

$$\frac{Suc(m) = Suc(n)}{m = n}\ (Suc\_inject) \qquad \frac{Suc(m) = 0}{R}\ (Suc\_neq\_0)$$

Mathematical induction asserts that $P(n)$ is true, for any $n :: nat$, provided $P(0)$ holds and that $P(x)$ implies $P(Suc(x))$ for all x. Some authors express the induction step as $\forall x . P(x) \rightarrow P(Suc(x))$. To avoid making induction require the presence of other connectives, we formalize mathematical induction as

$$[\![P(0); \bigwedge x . P(x) \Longrightarrow P(Suc(x))]\!] \Longrightarrow P(n). \qquad (induct)$$

Similarly, to avoid expressing the other rules using $\forall$, $\rightarrow$ and $\neg$, we take advantage of the meta-logic;[3] (Suc_neq_0) is an elimination rule for $Suc(m) = 0$:

$$Suc(m) = Suc(n) \Longrightarrow m = n \qquad (Suc\_inject)$$

$$Suc(m) = 0 \Longrightarrow R \qquad (Suc\_neq\_0)$$

We shall also define a primitive recursion operator, *rec*. Traditionally, primitive recursion takes a natural number a and a 2-place function f, and obeys the equations

$$\begin{aligned} rec(0, a, f) &= a \\ rec(Suc(m), a, f) &= f(m, rec(m, a, f)) \end{aligned}$$

Addition, defined by $m + n \equiv rec(m, n, \lambda x\, y \,.\, Suc(y))$, should satisfy

$$\begin{aligned} 0 + n &= n \\ Suc(m) + n &= Suc(m + n) \end{aligned}$$

Primitive recursion appears to pose difficulties: first-order logic has no function-valued expressions. We again take advantage of the meta-logic, which does have functions. We also generalise primitive recursion to be polymorphic over any type of class *term*, and declare the addition function:

$$\begin{aligned} rec &:: [nat, \alpha{::}term, [nat, \alpha] \Rightarrow \alpha] \Rightarrow \alpha \\ + &:: [nat, nat] \Rightarrow nat \end{aligned}$$

[3]On the other hand, the axioms $Suc(m) = Suc(n) \leftrightarrow m = n$ and $\neg(Suc(m) = 0)$ are logically equivalent to those given, and work better with Isabelle's simplifier.

3.3.2 Declaring the theory to Isabelle

Let us create the theory Nat starting from theory FOL, which contains only classical logic with no natural numbers. We declare the 0-place type constructor *nat* and the associated constants. Note that the constant 0 requires a mixfix annotation because 0 is not a legal identifier, and could not otherwise be written in terms:

```
Nat = FOL +
types   nat
arities nat         :: term
consts  "0"         :: "nat"                              ("0")
        Suc         :: "nat=>nat"
        rec         :: "[nat, 'a, [nat,'a]=>'a] => 'a"
        "+"         :: "[nat, nat] => nat"                (infixl 60)
rules   Suc_inject  "Suc(m)=Suc(n) ==> m=n"
        Suc_neq_0   "Suc(m)=0      ==> R"
        induct      "[| P(0);  !!x. P(x) ==> P(Suc(x)) |]  ==> P(n)"
        rec_0       "rec(0,a,f) = a"
        rec_Suc     "rec(Suc(m), a, f) = f(m, rec(m,a,f))"
        add_def     "m+n == rec(m, n, %x y. Suc(y))"
end
```

In axiom add_def, recall that % stands for λ. Loading this theory file creates the ML structure Nat, which contains the theory and axioms. Opening structure Nat lets us write induct instead of Nat.induct, and so forth.

```
open Nat;
```

3.3.3 Proving some recursion equations

File FOL/ex/Nat.ML contains proofs involving this theory of the natural numbers. As a trivial example, let us derive recursion equations for +. Here is the zero case:

```
goalw Nat.thy [add_def] "0+n = n";
  Level 0
  0 + n = n
   1. rec(0,n,%x y. Suc(y)) = n
by (resolve_tac [rec_0] 1);
  Level 1
  0 + n = n
  No subgoals!
val add_0 = result();
```

And here is the successor case:

```
goalw Nat.thy [add_def] "Suc(m)+n = Suc(m+n)";
  Level 0
  Suc(m) + n = Suc(m + n)
   1. rec(Suc(m),n,%x y. Suc(y)) = Suc(rec(m,n,%x y. Suc(y)))
```

```
by (resolve_tac [rec_Suc] 1);
  Level 1
  Suc(m) + n = Suc(m + n)
  No subgoals!
val add_Suc = result();
```

The induction rule raises some complications, which are discussed next.

3.4 Refinement with explicit instantiation

In order to employ mathematical induction, we need to refine a subgoal by the rule (*induct*). The conclusion of this rule is $?P(?n)$, which is highly ambiguous in higher-order unification. It matches every way that a formula can be regarded as depending on a subterm of type *nat*. To get round this problem, we could make the induction rule conclude $\forall n \,.\, ?P(n)$ — but putting a subgoal into this form requires refinement by $(\forall E)$, which is equally hard!

The tactic `res_inst_tac`, like `resolve_tac`, refines a subgoal by a rule. But it also accepts explicit instantiations for the rule's schematic variables.

`res_inst_tac` *insts thm i* instantiates the rule *thm* with the instantiations *insts*, and then performs resolution on subgoal *i*.

`eres_inst_tac` and `dres_inst_tac` are similar, but perform elim-resolution and destruct-resolution, respectively.

The list *insts* consists of pairs $[(v_1, e_1), \ldots, (v_n, e_n)]$, where $v_1, \ldots, v_n$ are names of schematic variables in the rule — with no leading question marks! — and $e_1, \ldots, e_n$ are expressions giving their instantiations. The expressions are type-checked in the context of a particular subgoal: free variables receive the same types as they have in the subgoal, and parameters may appear. Type variable instantiations may appear in *insts*, but they are seldom required: `res_inst_tac` instantiates type variables automatically whenever the type of e_i is an instance of the type of $?v_i$.

3.4.1 A simple proof by induction

Let us prove that no natural number k equals its own successor. To use (*induct*), we instantiate $?n$ to k; Isabelle finds a good instantiation for $?P$.

```
goal Nat.thy "~ (Suc(k) = k)";
  Level 0
  ~Suc(k) = k
   1. ~Suc(k) = k
by (res_inst_tac [("n","k")] induct 1);
  Level 1
  ~Suc(k) = k
   1. ~Suc(0) = 0
   2. !!x. ~Suc(x) = x ==> ~Suc(Suc(x)) = Suc(x)
```

We should check that Isabelle has correctly applied induction. Subgoal 1 is the base case, with k replaced by 0. Subgoal 2 is the inductive step, with k replaced by $Suc(x)$ and with an induction hypothesis for x. The rest of the proof demonstrates notI, notE and the other rules of theory Nat. The base case holds by Suc_neq_0:

```
by (resolve_tac [notI] 1);
  Level 2
  ~Suc(k) = k
   1. Suc(0) = 0 ==> False
   2. !!x. ~Suc(x) = x ==> ~Suc(Suc(x)) = Suc(x)
by (eresolve_tac [Suc_neq_0] 1);
  Level 3
  ~Suc(k) = k
   1. !!x. ~Suc(x) = x ==> ~Suc(Suc(x)) = Suc(x)
```

The inductive step holds by the contrapositive of Suc_inject. Negation rules transform the subgoal into that of proving $Suc(x) = x$ from $Suc(Suc(x)) = Suc(x)$:

```
by (resolve_tac [notI] 1);
  Level 4
  ~Suc(k) = k
   1. !!x. [| ~Suc(x) = x; Suc(Suc(x)) = Suc(x) |] ==> False
by (eresolve_tac [notE] 1);
  Level 5
  ~Suc(k) = k
   1. !!x. Suc(Suc(x)) = Suc(x) ==> Suc(x) = x
by (eresolve_tac [Suc_inject] 1);
  Level 6
  ~Suc(k) = k
  No subgoals!
```

3.4.2 An example of ambiguity in resolve_tac

If you try the example above, you may observe that res_inst_tac is not actually needed. Almost by chance, resolve_tac finds the right instantiation for (*induct*) to yield the desired next state. With more complex formulae, our luck fails.

```
goal Nat.thy "(k+m)+n = k+(m+n)";
  Level 0
  k + m + n = k + (m + n)
   1. k + m + n = k + (m + n)
by (resolve_tac [induct] 1);
  Level 1
  k + m + n = k + (m + n)
   1. k + m + n = 0
   2. !!x. k + m + n = x ==> k + m + n = Suc(x)
```

This proof requires induction on k. The occurrence of 0 in subgoal 1 indicates that induction has been applied to the term $k+(m+n)$; this application is sound but will not lead to a proof here. Fortunately, Isabelle can (lazily!) generate all the valid applications of induction. The back command causes backtracking to an alternative outcome of the tactic.

```
back();
  Level 1
  k + m + n = k + (m + n)
   1. k + m + n = k + 0
   2. !!x. k + m + n = k + x ==> k + m + n = k + Suc(x)
```

Now induction has been applied to $m + n$. This is equally useless. Let us call back again.

```
back();
  Level 1
  k + m + n = k + (m + n)
   1. k + m + 0 = k + (m + 0)
   2. !!x. k + m + x = k + (m + x) ==>
              k + m + Suc(x) = k + (m + Suc(x))
```

Now induction has been applied to n. What is the next alternative?

```
back();
  Level 1
  k + m + n = k + (m + n)
   1. k + m + n = k + (m + 0)
   2. !!x. k + m + n = k + (m + x) ==> k + m + n = k + (m + Suc(x))
```

Inspecting subgoal 1 reveals that induction has been applied to just the second occurrence of n. This perfectly legitimate induction is useless here.

The main goal admits fourteen different applications of induction. The number is exponential in the size of the formula.

3.4.3 Proving that addition is associative

Let us invoke the induction rule properly, using res_inst_tac. At the same time, we shall have a glimpse at Isabelle's simplification tactics, which are described in Chap. 13.

Isabelle's simplification tactics repeatedly apply equations to a subgoal, perhaps proving it. For efficiency, the rewrite rules must be packaged into a **simplification set**, or **simpset**. We take the standard simpset for first-order logic and insert the equations proved in the previous section, namely $0 + n = n$ and $\mathtt{Suc}(m) + n = \mathtt{Suc}(m + n)$:

```
val add_ss = FOL_ss addrews [add_0, add_Suc];
```

We state the goal for associativity of addition, and use res_inst_tac to invoke induction on k:

```
goal Nat.thy "(k+m)+n = k+(m+n)";
  Level 0
  k + m + n = k + (m + n)
   1. k + m + n = k + (m + n)
```

```
by (res_inst_tac [("n","k")] induct 1);
  Level 1
  k + m + n = k + (m + n)
   1. 0 + m + n = 0 + (m + n)
   2. !!x. x + m + n = x + (m + n) ==>
               Suc(x) + m + n = Suc(x) + (m + n)
```

The base case holds easily; both sides reduce to $m + n$. The tactic `simp_tac` rewrites with respect to the given simplification set, applying the rewrite rules for addition:

```
by (simp_tac add_ss 1);
  Level 2
  k + m + n = k + (m + n)
   1. !!x. x + m + n = x + (m + n) ==>
               Suc(x) + m + n = Suc(x) + (m + n)
```

The inductive step requires rewriting by the equations for addition together the induction hypothesis, which is also an equation. The tactic `asm_simp_tac` rewrites using a simplification set and any useful assumptions:

```
by (asm_simp_tac add_ss 1);
  Level 3
  k + m + n = k + (m + n)
  No subgoals!
```

3.5 A Prolog interpreter

To demonstrate the power of tacticals, let us construct a Prolog interpreter and execute programs involving lists.[4] The Prolog program consists of a theory. We declare a type constructor for lists, with an arity declaration to say that $(\tau)list$ is of class *term* provided τ is:

$$list :: (term)term$$

We declare four constants: the empty list *Nil*; the infix list constructor :; the list concatenation predicate *app*; the list reverse predicate *rev*. (In Prolog, functions on lists are expressed as predicates.)

$$\begin{aligned} Nil &:: \alpha list \\ : &:: [\alpha, \alpha list] \Rightarrow \alpha list \\ app &:: [\alpha list, \alpha list, \alpha list] \Rightarrow o \\ rev &:: [\alpha list, \alpha list] \Rightarrow o \end{aligned}$$

The predicate *app* should satisfy the Prolog-style rules

$$app(Nil, ys, ys) \qquad \frac{app(xs, ys, zs)}{app(x : xs, ys, x : zs)}$$

[4]To run these examples, see the file `FOL/ex/Prolog.ML`.

We define the naive version of *rev*, which calls *app*:

$$rev(Nil, Nil) \qquad \frac{rev(xs, ys) \quad app(ys, x : Nil, zs)}{rev(x : xs, zs)}$$

Theory `Prolog` extends first-order logic in order to make use of the class *term* and the type *o*. The interpreter does not use the rules of `FOL`.

```
Prolog = FOL +
types    'a list
arities list    :: (term)term
consts  Nil     :: "'a list"
        ":"     :: "['a, 'a list]=> 'a list"            (infixr 60)
        app     :: "['a list, 'a list, 'a list] => o"
        rev     :: "['a list, 'a list] => o"
rules   appNil  "app(Nil,ys,ys)"
        appCons "app(xs,ys,zs) ==> app(x:xs, ys, x:zs)"
        revNil  "rev(Nil,Nil)"
        revCons "[| rev(xs,ys); app(ys,x:Nil,zs) |] ==> rev(x:xs,zs)"
end
```

3.5.1 Simple executions

Repeated application of the rules solves Prolog goals. Let us append the lists $[a, b, c]$ and $[d, e]$. As the rules are applied, the answer builds up in `?x`.

```
goal Prolog.thy "app(a:b:c:Nil, d:e:Nil, ?x)";
  Level 0
  app(a : b : c : Nil, d : e : Nil, ?x)
   1. app(a : b : c : Nil, d : e : Nil, ?x)

by (resolve_tac [appNil,appCons] 1);
  Level 1
  app(a : b : c : Nil, d : e : Nil, a : ?zs1)
   1. app(b : c : Nil, d : e : Nil, ?zs1)

by (resolve_tac [appNil,appCons] 1);
  Level 2
  app(a : b : c : Nil, d : e : Nil, a : b : ?zs2)
   1. app(c : Nil, d : e : Nil, ?zs2)
```

At this point, the first two elements of the result are a and b.

```
by (resolve_tac [appNil,appCons] 1);
  Level 3
  app(a : b : c : Nil, d : e : Nil, a : b : c : ?zs3)
   1. app(Nil, d : e : Nil, ?zs3)

by (resolve_tac [appNil,appCons] 1);
  Level 4
  app(a : b : c : Nil, d : e : Nil, a : b : c : d : e : Nil)
  No subgoals!
```

Prolog can run functions backwards. Which list can be appended with $[c, d]$ to produce $[a, b, c, d]$? Using `REPEAT`, we find the answer at once, $[a, b]$:

```
goal Prolog.thy "app(?x, c:d:Nil, a:b:c:d:Nil)";
  Level 0
  app(?x, c : d : Nil, a : b : c : d : Nil)
   1. app(?x, c : d : Nil, a : b : c : d : Nil)
by (REPEAT (resolve_tac [appNil,appCons] 1));
  Level 1
  app(a : b : Nil, c : d : Nil, a : b : c : d : Nil)
  No subgoals!
```

3.5.2 Backtracking

Prolog backtracking can answer questions that have multiple solutions. Which lists x and y can be appended to form the list $[a, b, c, d]$? This question has five solutions. Using `REPEAT` to apply the rules, we quickly find the first solution, namely $x = []$ and $y = [a, b, c, d]$:

```
goal Prolog.thy "app(?x, ?y, a:b:c:d:Nil)";
  Level 0
  app(?x, ?y, a : b : c : d : Nil)
   1. app(?x, ?y, a : b : c : d : Nil)
by (REPEAT (resolve_tac [appNil,appCons] 1));
  Level 1
  app(Nil, a : b : c : d : Nil, a : b : c : d : Nil)
  No subgoals!
```

Isabelle can lazily generate all the possibilities. The `back` command returns the tactic's next outcome, namely $x = [a]$ and $y = [b, c, d]$:

```
back();
  Level 1
  app(a : Nil, b : c : d : Nil, a : b : c : d : Nil)
  No subgoals!
```

The other solutions are generated similarly.

```
back();
  Level 1
  app(a : b : Nil, c : d : Nil, a : b : c : d : Nil)
  No subgoals!
back();
  Level 1
  app(a : b : c : Nil, d : Nil, a : b : c : d : Nil)
  No subgoals!
back();
  Level 1
  app(a : b : c : d : Nil, Nil, a : b : c : d : Nil)
  No subgoals!
```

3.5.3 Depth-first search

Now let us try *rev*, reversing a list. Bundle the rules together as the ML identifier `rules`. Naive reverse requires 120 inferences for this 14-element list, but the tactic terminates in a few seconds.

```
goal Prolog.thy "rev(a:b:c:d:e:f:g:h:i:j:k:l:m:n:Nil, ?w)";
  Level 0
  rev(a : b : c : d : e : f : g : h : i : j : k : l : m : n : Nil, ?w)
   1. rev(a : b : c : d : e : f : g : h : i : j : k : l : m : n : Nil,
          ?w)
val rules = [appNil,appCons,revNil,revCons];
by (REPEAT (resolve_tac rules 1));
  Level 1
  rev(a : b : c : d : e : f : g : h : i : j : k : l : m : n : Nil,
      n : m : l : k : j : i : h : g : f : e : d : c : b : a : Nil)
  No subgoals!
```

We may execute *rev* backwards. This, too, should reverse a list. What is the reverse of $[a, b, c]$?

```
goal Prolog.thy "rev(?x, a:b:c:Nil)";
  Level 0
  rev(?x, a : b : c : Nil)
   1. rev(?x, a : b : c : Nil)
by (REPEAT (resolve_tac rules 1));
  Level 1
  rev(?x1 : Nil, a : b : c : Nil)
   1. app(Nil, ?x1 : Nil, a : b : c : Nil)
```

The tactic has failed to find a solution! It reached a dead end at subgoal 1: there is no $?x_1$ such that $[]$ appended with $[?x_1]$ equals $[a, b, c]$. Backtracking explores other outcomes.

```
back();
  Level 1
  rev(?x1 : a : Nil, a : b : c : Nil)
   1. app(Nil, ?x1 : Nil, b : c : Nil)
```

This too is a dead end, but the next outcome is successful.

```
back();
  Level 1
  rev(c : b : a : Nil, a : b : c : Nil)
  No subgoals!
```

`REPEAT` goes wrong because it is only a repetition tactical, not a search tactical. `REPEAT` stops when it cannot continue, regardless of which state is reached. The tactical `DEPTH_FIRST` searches for a satisfactory state, as specified by an ML predicate. Below, `has_fewer_prems` specifies that the proof state should have no subgoals.

```
val prolog_tac = DEPTH_FIRST (has_fewer_prems 1)
                             (resolve_tac rules 1);
```

Since Prolog uses depth-first search, this tactic is a (slow!) Prolog interpreter. We return to the start of the proof using choplev, and apply prolog_tac:

```
choplev 0;
  Level 0
  rev(?x, a : b : c : Nil)
   1. rev(?x, a : b : c : Nil)
by (DEPTH_FIRST (has_fewer_prems 1) (resolve_tac rules 1));
  Level 1
  rev(c : b : a : Nil, a : b : c : Nil)
  No subgoals!
```

Let us try prolog_tac on one more example, containing four unknowns:

```
goal Prolog.thy "rev(a:?x:c:?y:Nil, d:?z:b:?u)";
  Level 0
  rev(a : ?x : c : ?y : Nil, d : ?z : b : ?u)
   1. rev(a : ?x : c : ?y : Nil, d : ?z : b : ?u)
by prolog_tac;
  Level 1
  rev(a : b : c : d : Nil, d : c : b : a : Nil)
  No subgoals!
```

Although Isabelle is much slower than a Prolog system, Isabelle tactics can exploit logic programming techniques.

Part II

The Isabelle Reference Manual

4. Basic Use of Isabelle

The Reference Manual is a comprehensive description of Isabelle, including all commands, functions and packages. It really is intended for reference, perhaps for browsing, but not for reading through. It is not a tutorial, but assumes familiarity with the basic concepts of Isabelle.

When you are looking for a way of performing some task, scan the Table of Contents for a relevant heading. Functions are organized by their purpose, by their operands (subgoals, tactics, theorems), and by their usefulness. In each section, basic functions appear first, then advanced functions, and finally esoteric functions. Use the Index when you are looking for the definition of a particular Isabelle function.

A few examples are presented. Many examples files are distributed with Isabelle, however; please experiment interactively.

4.1 Basic interaction with Isabelle

Isabelle provides no means of storing theorems or proofs on files. Theorems are simply part of the ML state and are named by ML identifiers. To save your work between sessions, you must save a copy of the ML image. The procedure for doing so is compiler-dependent:

- At the end of a session, Poly/ML saves the state, provided you have created a database for your own use. You can create a database by copying an existing one, or by calling the Poly/ML function `make_database`; the latter course uses much less disc space. A Poly/ML database *does not* save the contents of references, such as the current state of a backward proof.
- With New Jersey ML you must save the state explicitly before ending the session. While a Poly/ML database can be small, a New Jersey image occupies several megabytes.

See your ML compiler's documentation for full instructions on saving the state.

Saving the state is not enough. Record, on a file, the top-level commands that generate your theories and proofs. Such a record allows you to replay the proofs whenever required, for instance after making minor changes to the axioms. Ideally, your record will be intelligible to others as a formal description of your work.

Since Isabelle's user interface is the ML top level, some kind of window support is essential. One window displays the Isabelle session, while the other displays a file — your proof record — being edited. The Emacs editor supports windows and can manage interactive sessions.

4.2 Ending a session

```
quit     : unit -> unit
commit   : unit -> unit                              Poly/ML only
exportML : string -> bool                     New Jersey ML only
```

`quit();` aborts the Isabelle session, without saving the state.

`commit();` saves the current state in your Poly/ML database without ending the session. The contents of references are lost, so never do this during an interactive proof!

`exportML "`*file*`";` saves an image of your session to the given *file*.

! Typing control-D also finishes the session, but its effect is compiler-dependent. Poly/ML will then save the state, if you have a private database. New Jersey ML will discard the state!

4.3 Reading ML files

```
cd             : string -> unit
use            : string -> unit
time_use       : string -> unit
```

Section 9.2 describes commands for loading theory files.

`cd "`*dir*`";` changes the current directory to *dir*. This is the default directory for reading files and for writing temporary files.

`use "`*file*`";` reads the given *file* as input to the ML session. Reading a file of Isabelle commands is the usual way of replaying a proof.

`time_use "`*file*`";` performs `use "`*file*`"` and prints the total execution time.

4.4 Printing of terms and theorems

Isabelle's pretty printer is controlled by a number of parameters.

4.4.1 Printing limits

```
Pretty.setdepth  : int -> unit
Pretty.setmargin : int -> unit
print_depth      : int -> unit
```

These set limits for terminal output.

`Pretty.setdepth` d; tells Isabelle's pretty printer to limit the printing depth to d. This affects Isabelle's display of theorems and terms. The default value is 0, which permits printing to an arbitrary depth. Useful values for d are 10 and 20.

`Pretty.setmargin` m; tells Isabelle's pretty printer to assume a right margin (page width) of m. The initial margin is 80.

`print_depth` n; limits the printing depth of complex ML values, such as theorems and terms. This command affects the ML top level and its effect is compiler-dependent. Typically n should be less than 10.

4.4.2 Printing of hypotheses, types and sorts

```
show_hyps  : bool ref                                    initially true
show_types : bool ref                                    initially false
show_sorts : bool ref                                    initially false
```

These flags allow you to control how much information is displayed for terms and theorems. The hypotheses are normally shown; types and sorts are not. Displaying types and sorts may explain why a polymorphic inference rule fails to resolve with some goal.

`show_hyps := false;` makes Isabelle show each meta-level hypothesis as a dot.

`show_types := true;` makes Isabelle show types when printing a term or theorem.

`show_sorts := true;` makes Isabelle show the sorts of type variables. It has no effect unless `show_types` is `true`.

4.4.3 η-contraction before printing

```
eta_contract: bool ref                                   initially false
```

The η-**contraction law** asserts $(\lambda x . f(x)) \equiv f$, provided x is not free in f. It asserts **extensionality** of functions: $f \equiv g$ if $f(x) \equiv g(x)$ for all x. Higher-order unification frequently puts terms into a fully η-expanded form. For example, if F has type $(\tau \Rightarrow \tau) \Rightarrow \tau$ then its expanded form is $\lambda h . F(\lambda x . h(x))$. By default, the user sees this expanded form.

`eta_contract := true;` makes Isabelle perform η-contractions before printing, so that $\lambda h.F(\lambda x.h(x))$ appears simply as F. The distinction between a term and its η-expanded form occasionally matters.

4.5 Displaying exceptions as error messages

```
print_exn: exn -> 'a
```

Certain Isabelle primitives, such as the forward proof functions `RS` and `RSN`, are called both interactively and from programs. They indicate errors not by printing messages, but by raising exceptions. For interactive use, ML's reporting of an uncaught exception is uninformative. The Poly/ML function `exception_trace` can generate a backtrace.

`print_exn` *e* displays the exception *e* in a readable manner, and then re-raises *e*. Typical usage is *EXP* `handle e => print_exn e;`, where *EXP* is an expression that may raise an exception.

`print_exn` can display the following common exceptions, which concern types, terms, theorems and theories, respectively. Each carries a message and related information.

```
exception TYPE   of string * typ list * term list
exception TERM   of string * term list
exception THM    of string * int * thm list
exception THEORY of string * theory list
```

! `print_exn` prints terms by calling `prin`, which obtains pretty printing information from the proof state last stored in the subgoal module. The appearance of the output thus depends upon the theory used in the last interactive proof.

4.6 Shell scripts

The following files are distributed with Isabelle, and work under Unix™. They can be executed as commands to the Unix shell. Some of them depend upon shell environment variables.

`make-all` *switches* compiles the Isabelle system, when executed on the source directory. Environment variables specify which ML compiler to use. These variables, and the *switches*, are documented on the file itself.

`teeinput` *program* executes the *program*, while piping the standard input to a log file designated by the `$LISTEN` environment variable. Normally the program is Isabelle, and the log file receives a copy of all the Isabelle commands.

`xlisten` *program* is a trivial 'user interface' for the X Window System. It creates two windows using `xterm`. One executes an interactive session via `teeinput` *program*, while the other displays the log file. To make a proof record, simply paste lines from the log file into an editor window.

`expandshort` *files* reads the *files* and replaces all occurrences of the shorthand commands `br`, `be`, `brs`, `bes`, etc., with the corresponding full commands. Shorthand commands should appear one per line. The old versions of the files are renamed to have the suffix `~~`.

5. Proof Management: The Subgoal Module

The subgoal module stores the current proof state and many previous states; commands can produce new states or return to previous ones. The *state list* at level n is a list of pairs

$$[(\psi_n, \Psi_n),\ (\psi_{n-1}, \Psi_{n-1}),\ \ldots,\ (\psi_0, [])]$$

where ψ_n is the current proof state, ψ_{n-1} is the previous one, ..., and ψ_0 is the initial proof state. The Ψ_i are sequences (lazy lists) of proof states, storing branch points where a tactic returned a list longer than one. The state lists permit various forms of backtracking.

Chopping elements from the state list reverts to previous proof states. Besides this, the undo command keeps a list of state lists. The module actually maintains a stack of state lists, to support several proofs at the same time.

The subgoal module always contains some proof state. At the start of the Isabelle session, this state consists of a dummy formula.

5.1 Basic commands

Most proofs begin with goal or goalw and require no other commands than by, chop and undo. They typically end with a call to result.

5.1.1 Starting a backward proof

```
goal        : theory -> string -> thm list
goalw       : theory -> thm list -> string -> thm list
goalw_cterm : thm list -> Sign.cterm -> thm list
premises    : unit -> thm list
```

The goal commands start a new proof by setting the goal. They replace the current state list by a new one consisting of the initial proof state. They also empty the undo list; this command cannot be undone!

They all return a list of meta-hypotheses taken from the main goal. If this list is non-empty, bind its value to an ML identifier by typing something like

val prems = goal *theory formula*;

These assumptions serve as the premises when you are deriving a rule. They are also stored internally and can be retrieved later by the function `premises`. When the proof is finished, `result` compares the stored assumptions with the actual assumptions in the proof state.

Some of the commands unfold definitions using meta-rewrite rules. This expansion affects both the initial subgoal and the premises, which would otherwise require use of `rewrite_goals_tac` and `rewrite_rule`.

If the main goal has the form "!!*vars*. ...", with an outermost quantifier, then the list of premises will be empty. Subgoal 1 will contain the meta-quantified *vars* as parameters and the goal's premises as assumptions. This avoids having to call `cut_facts_tac` with the list of premises (Sect. 6.2.3).

`goal` *theory formula*; begins a new proof, where *theory* is usually an ML identifier and the *formula* is written as an ML string.

`goalw` *theory defs formula*; is like `goal` but also applies the list of *defs* as meta-rewrite rules to the first subgoal and the premises.

`goalw_cterm` *theory defs ct*; is a version of `goalw` for programming applications. The main goal is supplied as a cterm, not as a string. Typically, the cterm is created from a term *t* by `Sign.cterm_of (sign_of thy)` *t*.

`premises()` returns the list of meta-hypotheses associated with the current proof (in case you forgot to bind them to an ML identifier).

5.1.2 Applying a tactic

```
by   : tactic -> unit
byev : tactic list -> unit
```

These commands extend the state list. They apply a tactic to the current proof state. If the tactic succeeds, it returns a non-empty sequence of next states. The head of the sequence becomes the next state, while the tail is retained for backtracking (see `back`).

`by` *tactic*; applies the *tactic* to the proof state.

`byev` *tactics*; applies the list of *tactics*, one at a time. It is useful for testing calls to `prove_goal`, and abbreviates `by (EVERY` *tactics*`)`.

Error indications:

- `"by: tactic failed"` means that the tactic returned an empty sequence when applied to the current proof state.
- `"Warning: same as previous level"` means that the new proof state is identical to the previous state.

- "Warning: signature of proof state has changed" means that some rule was applied whose theory is outside the theory of the initial proof state. This could signify a mistake such as expressing the goal in intuitionistic logic and proving it using classical logic.

5.1.3 Extracting the proved theorem

```
result  : unit -> thm
uresult : unit -> thm
```

`result()` returns the final theorem, after converting the free variables to schematics. It discharges the assumptions supplied to the matching `goal` command.

It raises an exception unless the proof state passes certain checks. There must be no assumptions other than those supplied to `goal`. There must be no subgoals. The theorem proved must be a (first-order) instance of the original goal, as stated in the `goal` command. This allows **answer extraction** — instantiation of variables — but no other changes to the main goal. The theorem proved must have the same signature as the initial proof state.

These checks are needed because an Isabelle tactic can return any proof state at all.

`uresult()` is like `result()` but omits the checks. It is needed when the initial goal contains function unknowns, when definitions are unfolded in the main goal (by calling `rewrite_tac`), or when `assume_ax` has been used.

5.1.4 Undoing and backtracking

```
chop    : unit -> unit
choplev : int -> unit
back    : unit -> unit
undo    : unit -> unit
```

`chop();` deletes the top level of the state list, cancelling the last `by` command. It provides a limited undo facility, and the `undo` command can cancel it.

`choplev` n; truncates the state list to level n.

`back();` searches the state list for a non-empty branch point, starting from the top level. The first one found becomes the current proof state — the most recent alternative branch is taken. This is a form of interactive backtracking.

`undo();` cancels the most recent change to the proof state by the commands `by`, `chop`, `choplev`, and `back`. It **cannot** cancel `goal` or `undo` itself. It can be repeated to cancel a series of commands.

Error indications for `back`*:*

- `"Warning: same as previous choice at this level"` means `back` found a non-empty branch point, but that it contained the same proof state as the current one.
- `"Warning: signature of proof state has changed"` means the signature of the alternative proof state differs from that of the current state.
- `"back: no alternatives"` means `back` could find no alternative proof state.

5.1.5 Printing the proof state

```
pr    : unit -> unit
prlev : int -> unit
goals_limit: int ref
```
initially 10

`pr();` prints the current proof state.

`prlev` n; prints the proof state at level n. This allows you to review the previous steps of the proof.

`goals_limit :=` k; specifies k as the maximum number of subgoals to print.

5.1.6 Timing

```
proof_timing: bool ref
```
initially false

`proof_timing := true;` makes the `by` and `prove_goal` commands display how much processor time was spent. This information is compiler-dependent.

5.2 Shortcuts for applying tactics

These commands call `by` with common tactics. Their chief purpose is to minimise typing, although the scanning shortcuts are useful in their own right. Chapter 6 explains the tactics themselves.

5.2.1 Refining a given subgoal

```
ba  :              int -> unit
br  : thm      -> int -> unit
be  : thm      -> int -> unit
bd  : thm      -> int -> unit
brs : thm list -> int -> unit
bes : thm list -> int -> unit
bds : thm list -> int -> unit
```

`ba` *i*; performs `by (assume_tac` *i*`);`

`br` *thm i*; performs `by (resolve_tac [`*thm*`]` *i*`);`

`be` *thm i*; performs `by (eresolve_tac [`*thm*`]` *i*`);`

`bd` *thm i*; performs `by (dresolve_tac [`*thm*`]` *i*`);`

`brs` *thms i*; performs `by (resolve_tac` *thms i*`);`

`bes` *thms i*; performs `by (eresolve_tac` *thms i*`);`

`bds` *thms i*; performs `by (dresolve_tac` *thms i*`);`

5.2.2 Scanning shortcuts

These shortcuts scan for a suitable subgoal (starting from subgoal 1). They refine the first subgoal for which the tactic succeeds. Thus, they require less typing than `br`, etc. They display the selected subgoal's number; please watch this, for it may not be what you expect!

```
fa  : unit     -> unit
fr  : thm      -> unit
fe  : thm      -> unit
fd  : thm      -> unit
frs : thm list -> unit
fes : thm list -> unit
fds : thm list -> unit
```

`fa();` solves some subgoal by assumption.

`fr` *thm*; refines some subgoal using `resolve_tac [`*thm*`]`

`fe` *thm*; refines some subgoal using `eresolve_tac [`*thm*`]`

`fd` *thm*; refines some subgoal using `dresolve_tac [`*thm*`]`

`frs` *thms*; refines some subgoal using `resolve_tac` *thms*

`fes` *thms*; refines some subgoal using `eresolve_tac` *thms*

`fds` *thms*; refines some subgoal using `dresolve_tac` *thms*

5.2.3 Other shortcuts

```
bw  : thm -> unit
bws : thm list -> unit
ren : string -> int -> unit
```

`bw` *def*; performs `by (rewrite_goals_tac [`*def*`]);` It unfolds definitions in the subgoals (but not the main goal), by meta-rewriting with the given definition.

`bws` is like `bw` but takes a list of definitions.

`ren` *names i*; performs `by (rename_tac` *names i*`);` it renames parameters in subgoal *i*. (Ignore the message `Warning: same as previous level.`)

5.3 Executing batch proofs

```
prove_goal  :   theory->          string->(thm list->tactic list)->thm
prove_goalw :   theory->thm list->string->(thm list->tactic list)->thm
prove_goalw_cterm: thm list->Sign.cterm->(thm list->tactic list)->thm
```

These batch functions create an initial proof state, then apply a tactic to it, yielding a sequence of final proof states. The head of the sequence is returned, provided it is an instance of the theorem originally proposed. The forms `prove_goal`, `prove_goalw` and `prove_goalw_cterm` are analogous to `goal`, `goalw` and `goalw_cterm`.

The tactic is specified by a function from theorem lists to tactic lists. The function is applied to the list of meta-assumptions taken from the main goal. The resulting tactics are applied in sequence (using `EVERY`). For example, a proof consisting of the commands

```
val prems = goal  theory  formula;
by tac1;  ...  by tacn;
val my_thm = result();
```

can be transformed to an expression as follows:

```
val my_thm = prove_goal  theory  formula
  (fn prems=> [ tac1, ..., tacn ]);
```

The methods perform identical processing of the initial *formula* and the final proof state. But `prove_goal` executes the tactic as a atomic operation, bypassing the subgoal module; the current interactive proof is unaffected.

`prove_goal` *theory formula tacsf*; executes a proof of the *formula* in the given *theory*, using the given tactic function.

`prove_goalw` *theory defs formula tacsf*; is like `prove_goal` but also applies the list of *defs* as meta-rewrite rules to the first subgoal and the premises.

`prove_goalw_cterm` *theory defs ct tacsf*; is a version of `prove_goalw` for programming applications. The main goal is supplied as a cterm, not as a string. Typically, the cterm is created from a term *t* as follows:

```
Sign.cterm_of (sign_of thy) t
```

5.4 Managing multiple proofs

You may save the current state of the subgoal module and resume work on it later. This serves two purposes.

1. At some point, you may be uncertain of the next step, and wish to experiment.

2. During a proof, you may see that a lemma should be proved first.

Each saved proof state consists of a list of levels; `chop` behaves independently for each of the saved proofs. In addition, each saved state carries a separate `undo` list.

5.4.1 The stack of proof states

```
push_proof   : unit -> unit
pop_proof    : unit -> thm list
rotate_proof : unit -> thm list
```

The subgoal module maintains a stack of proof states. Most subgoal commands affect only the top of the stack. The `goal` command *replaces* the top of the stack; the only command that pushes a proof on the stack is `push_proof`.

To save some point of the proof, call `push_proof`. You may now state a lemma using `goal`, or simply continue to apply tactics. Later, you can return to the saved point by calling `pop_proof` or `rotate_proof`.

To view the entire stack, call `rotate_proof` repeatedly; as it rotates the stack, it prints the new top element.

`push_proof();` duplicates the top element of the stack, pushing a copy of the current proof state on to the stack.

`pop_proof();` discards the top element of the stack. It returns the list of assumptions associated with the new proof; you should bind these to an ML identifier. They can also be obtained by calling `premises`.

`rotate_proof();` rotates the stack, moving the top element to the bottom. It returns the list of assumptions associated with the new proof.

5.4.2 Saving and restoring proof states

```
save_proof    : unit -> proof
restore_proof : proof -> thm list
```

States of the subgoal module may be saved as ML values of type `proof`, and later restored.

`save_proof();` returns the current state, which is on top of the stack.

`restore_proof` *prf*; replaces the top of the stack by *prf*. It returns the list of assumptions associated with the new proof.

5.5 Debugging and inspecting

These specialized operations support the debugging of tactics. They refer to the current proof state of the subgoal module.

5.5.1 Reading and printing terms

```
read    : string -> term
prin    : term -> unit
printyp : typ -> unit
```

These read and print terms (or types) using the syntax associated with the proof state.

`read` *string* reads the *string* as a term, without type checking.

`prin` t; prints the term t at the terminal.

`printyp` T; prints the type T at the terminal.

5.5.2 Inspecting the proof state

```
topthm  : unit -> thm
getgoal : int -> term
gethyps : int -> thm list
```

`topthm()` returns the proof state as an Isabelle theorem. This is what by would supply to a tactic at this point. It omits the post-processing of `result` and `uresult`.

`getgoal` i returns subgoal i of the proof state, as a term. You may print this using `prin`, though you may have to examine the internal data structure in order to locate the problem!

`gethyps` i returns the hypotheses of subgoal i as meta-level assumptions. In these theorems, the subgoal's parameters become free variables. This command is supplied for debugging uses of `METAHYPS`.

5.5.3 Filtering lists of rules

```
filter_goal: (term*term->bool) -> thm list -> int -> thm list
```

`filter_goal` *could ths* i applies `filter_thms` *could* to subgoal i of the proof state and returns the list of theorems that survive the filtering.

6. Tactics

Tactics have type `tactic`. They are essentially functions from theorems to theorem sequences, where the theorems represent states of a backward proof. Tactics seldom need to be coded from scratch, as functions; instead they are expressed using basic tactics and tacticals.

6.1 Resolution and assumption tactics

Resolution is Isabelle's basic mechanism for refining a subgoal using a rule. **Elim-resolution** is particularly suited for elimination rules, while **destruct-resolution** is particularly suited for destruction rules. The `r`, `e`, `d` naming convention is maintained for several different kinds of resolution tactics, as well as the shortcuts in the subgoal module.

All the tactics in this section act on a subgoal designated by a positive integer i. They fail (by returning the empty sequence) if i is out of range.

6.1.1 Resolution tactics

```
resolve_tac  : thm list -> int -> tactic
eresolve_tac : thm list -> int -> tactic
dresolve_tac : thm list -> int -> tactic
forward_tac  : thm list -> int -> tactic
```

These perform resolution on a list of theorems, *thms*, representing a list of object-rules. When generating next states, they take each of the rules in the order given. Each rule may yield several next states, or none: higher-order resolution may yield multiple resolvents.

`resolve_tac` *thms* i refines the proof state using the rules, which should normally be introduction rules. It resolves a rule's conclusion with subgoal i of the proof state.

`eresolve_tac` *thms* i performs elim-resolution with the rules, which should normally be elimination rules. It resolves with a rule, solves its first premise by assumption, and finally *deletes* that assumption from any new subgoals.

`dresolve_tac` *thms i* performs destruct-resolution with the rules, which normally should be destruction rules. This replaces an assumption by the result of applying one of the rules.

`forward_tac` is like `dresolve_tac` except that the selected assumption is not deleted. It applies a rule to an assumption, adding the result as a new assumption.

6.1.2 Assumption tactics

```
assume_tac    : int -> tactic
eq_assume_tac : int -> tactic
```

`assume_tac` *i* attempts to solve subgoal *i* by assumption.

`eq_assume_tac` is like `assume_tac` but does not use unification. It succeeds (with a *unique* next state) if one of the assumptions is identical to the subgoal's conclusion. Since it does not instantiate variables, it cannot make other subgoals unprovable. It is intended to be called from proof strategies, not interactively.

6.1.3 Matching tactics

```
match_tac  : thm list -> int -> tactic
ematch_tac : thm list -> int -> tactic
dmatch_tac : thm list -> int -> tactic
```

These are just like the resolution tactics except that they never instantiate unknowns in the proof state. Flexible subgoals are not updated willy-nilly, but are left alone. Matching — strictly speaking — means treating the unknowns in the proof state as constants; these tactics merely discard unifiers that would update the proof state.

`match_tac` *thms i* refines the proof state using the rules, matching a rule's conclusion with subgoal *i* of the proof state.

`ematch_tac` is like `match_tac`, but performs elim-resolution.

`dmatch_tac` is like `match_tac`, but performs destruct-resolution.

6.1.4 Resolution with instantiation

```
res_inst_tac  : (string*string)list -> thm -> int -> tactic
eres_inst_tac : (string*string)list -> thm -> int -> tactic
dres_inst_tac : (string*string)list -> thm -> int -> tactic
forw_inst_tac : (string*string)list -> thm -> int -> tactic
```

These tactics are designed for applying rules such as substitution and induction, which cause difficulties for higher-order unification. The tactics accept explicit instantiations for unknowns in the rule — typically, in the rule's conclusion. Each instantiation is a pair `(v,e)`, where v is an unknown *without* its leading question mark!

- If v is the type unknown `'a`, then the rule must contain a type unknown `?'a` of some sort s, and e should be a type of sort s.

- If v is the unknown `P`, then the rule must contain an unknown `?P` of some type τ, and e should be a term of some type σ such that τ and σ are unifiable. If the unification of τ and σ instantiates any type unknowns in τ, these instantiations are recorded for application to the rule.

Types are instantiated before terms. Because type instantiations are inferred from term instantiations, explicit type instantiations are seldom necessary — if `?t` has type `?'a`, then the instantiation list `[("'a","bool"),("t","True")]` may be simplified to `[("t","True")]`. Type unknowns in the proof state may cause failure because the tactics cannot instantiate them.

The instantiation tactics act on a given subgoal. Terms in the instantiations are type-checked in the context of that subgoal — in particular, they may refer to that subgoal's parameters. Any unknowns in the terms receive subscripts and are lifted over the parameters; thus, you may not refer to unknowns in the subgoal.

`res_inst_tac` *insts thm i* instantiates the rule *thm* with the instantiations *insts*, as described above, and then performs resolution on subgoal i. Resolution typically causes further instantiations; you need not give explicit instantiations for every unknown in the rule.

`eres_inst_tac` is like `res_inst_tac`, but performs elim-resolution.

`dres_inst_tac` is like `res_inst_tac`, but performs destruct-resolution.

`forw_inst_tac` is like `dres_inst_tac` except that the selected assumption is not deleted. It applies the instantiated rule to an assumption, adding the result as a new assumption.

6.2 Other basic tactics

6.2.1 Definitions and meta-level rewriting

Definitions in Isabelle have the form $t \equiv u$, where t is typically a constant or a constant applied to a list of variables, for example $sqr(n) \equiv n \times n$. (Conditional definitions, $\phi \Longrightarrow t \equiv u$, are not supported.) **Unfolding** the definition $t \equiv u$ means using it as a rewrite rule, replacing t by u throughout a theorem.

Folding $t \equiv u$ means replacing u by t. Rewriting continues until no rewrites are applicable to any subterm.

There are rules for unfolding and folding definitions; Isabelle does not do this automatically. The corresponding tactics rewrite the proof state, yielding a single next state. See also the `goalw` command, which is the easiest way of handling definitions.

```
rewrite_goals_tac : thm list -> tactic
rewrite_tac       : thm list -> tactic
fold_goals_tac    : thm list -> tactic
fold_tac          : thm list -> tactic
```

`rewrite_goals_tac` *defs* unfolds the *defs* throughout the subgoals of the proof state, while leaving the main goal unchanged. Use `SELECT_GOAL` to restrict it to a particular subgoal.

`rewrite_tac` *defs* unfolds the *defs* throughout the proof state, including the main goal — not normally desirable!

`fold_goals_tac` *defs* folds the *defs* throughout the subgoals of the proof state, while leaving the main goal unchanged.

`fold_tac` *defs* folds the *defs* throughout the proof state.

6.2.2 Tactic shortcuts

```
rtac     :            thm -> int -> tactic
etac     :            thm -> int -> tactic
dtac     :            thm -> int -> tactic
atac     :                   int -> tactic
ares_tac : thm list -> int -> tactic
rewtac   :            thm ->        tactic
```

These abbreviate common uses of tactics.

`rtac` *thm i* abbreviates `resolve_tac [`*thm*`]` *i*, doing resolution.

`etac` *thm i* abbreviates `eresolve_tac [`*thm*`]` *i*, doing elim-resolution.

`dtac` *thm i* abbreviates `dresolve_tac [`*thm*`]` *i*, doing destruct-resolution.

`atac` *i* abbreviates `assume_tac` *i*, doing proof by assumption.

`ares_tac` *thms i* tries proof by assumption and resolution; it abbreviates

`assume_tac` *i* `ORELSE resolve_tac` *thms i*

`rewtac` *def* abbreviates `rewrite_goals_tac [`*def*`]`, unfolding a definition.

6.2.3 Inserting premises and facts

```
cut_facts_tac : thm list -> int -> tactic
cut_inst_tac  : (string*string)list -> thm -> int -> tactic
subgoal_tac   : string -> int -> tactic
```

These tactics add assumptions to a given subgoal.

`cut_facts_tac` *thms* i adds the *thms* as new assumptions to subgoal i. Once they have been inserted as assumptions, they become subject to tactics such as `eresolve_tac` and `rewrite_goals_tac`. Only rules with no premises are inserted: Isabelle cannot use assumptions that contain $\Longrightarrow$ or $\bigwedge$. Sometimes the theorems are premises of a rule being derived, returned by `goal`; instead of calling this tactic, you could state the goal with an outermost meta-quantifier.

`cut_inst_tac` *insts thm* i instantiates the *thm* with the instantiations *insts*, as described in Sect. 6.1.4. It adds the resulting theorem as a new assumption to subgoal i.

`subgoal_tac` *formula* i adds the *formula* as a assumption to subgoal i, and inserts the same *formula* as a new subgoal, $i+1$.

6.2.4 Theorems useful with tactics

```
asm_rl: thm
cut_rl: thm
```

`asm_rl` is $\psi \Longrightarrow \psi$. Under elim-resolution it does proof by assumption, and `eresolve_tac (asm_rl::`*thms*`)` i is equivalent to

`assume_tac` i `ORELSE eresolve_tac` *thms* i

`cut_rl` is $[\![\psi \Longrightarrow \theta, \psi]\!] \Longrightarrow \theta$. It is useful for inserting assumptions; it underlies `forward_tac`, `cut_facts_tac` and `subgoal_tac`.

6.3 Obscure tactics

6.3.1 Tidying the proof state

```
prune_params_tac : tactic
flexflex_tac     : tactic
```

`prune_params_tac` removes unused parameters from all subgoals of the proof state. It works by rewriting with the theorem $(\bigwedge x \,.\, V) \equiv V$. This tactic can make the proof state more readable. It is used with `rule_by_tactic` to simplify the resulting theorem.

`flexflex_tac` removes all flex-flex pairs from the proof state by applying the trivial unifier. This drastic step loses information, and should only be done as the last step of a proof.

Flex-flex constraints arise from difficult cases of higher-order unification. To prevent this, use `res_inst_tac` to instantiate some variables in a rule (Sect. 6.1.4). Normally flex-flex constraints can be ignored; they often disappear as unknowns get instantiated.

6.3.2 Renaming parameters in a goal

```
rename_tac            : string -> int -> tactic
rename_last_tac       : string -> string list -> int -> tactic
Logic.set_rename_prefix : string -> unit
Logic.auto_rename       : bool ref
```
initially false

When creating a parameter, Isabelle chooses its name by matching variable names via the object-rule. Given the rule $(\forall I)$ formalized as $(\bigwedge x\,.\,P(x)) \Longrightarrow \forall x\,.\,P(x)$, Isabelle will note that the $\bigwedge$-bound variable in the premise has the same name as the $\forall$-bound variable in the conclusion.

Sometimes there is insufficient information and Isabelle chooses an arbitrary name. The renaming tactics let you override Isabelle's choice. Because renaming parameters has no logical effect on the proof state, the `by` command prints the message `Warning: same as previous level`.

Alternatively, you can suppress the naming mechanism described above and have Isabelle generate uniform names for parameters. These names have the form pa, pb, pc, ..., where p is any desired prefix. They are ugly but predictable.

`rename_tac` *str* i interprets the string *str* as a series of blank-separated variable names, and uses them to rename the parameters of subgoal i. The names must be distinct. If there are fewer names than parameters, then the tactic renames the innermost parameters and may modify the remaining ones to ensure that all the parameters are distinct.

`rename_last_tac` *prefix suffixes* i generates a list of names by attaching each of the *suffixes* to the *prefix*. It is intended for coding structural induction tactics, where several of the new parameters should have related names.

`Logic.set_rename_prefix` *prefix*; sets the prefix for uniform renaming to *prefix*. The default prefix is `"k"`.

`Logic.auto_rename := true;` makes Isabelle generate uniform names for parameters.

6.3.3 Composition: resolution without lifting

```
compose_tac: (bool * thm * int) -> int -> tactic
```

Composing two rules means resolving them without prior lifting or renaming of unknowns. This low-level operation, which underlies the resolution tactics, may occasionally be useful for special effects. A typical application is `res_inst_tac`, which lifts and instantiates a rule, then passes the result to `compose_tac`.

`compose_tac` (*flag*, *rule*, *m*) *i* refines subgoal *i* using *rule*, without lifting. The *rule* is taken to have the form $[\![\psi_1; \ldots; \psi_m]\!] \Longrightarrow \psi$, where ψ need not be atomic; thus *m* determines the number of new subgoals. If *flag* is `true` then it performs elim-resolution — it solves the first premise of *rule* by assumption and deletes that assumption.

6.4 Managing lots of rules

These operations are not intended for interactive use. They are concerned with the processing of large numbers of rules in automatic proof strategies. Higher-order resolution involving a long list of rules is slow. Filtering techniques can shorten the list of rules given to resolution, and can also detect whether a given subgoal is too flexible, with too many rules applicable.

6.4.1 Combined resolution and elim-resolution

```
biresolve_tac   : (bool*thm)list -> int -> tactic
bimatch_tac     : (bool*thm)list -> int -> tactic
subgoals_of_brl : bool*thm -> int
lessb           : (bool*thm) * (bool*thm) -> bool
```

Bi-resolution takes a list of (*flag*, *rule*) pairs. For each pair, it applies resolution if the flag is `false` and elim-resolution if the flag is `true`. A single tactic call handles a mixture of introduction and elimination rules.

`biresolve_tac` *brls* *i* refines the proof state by resolution or elim-resolution on each rule, as indicated by its flag. It affects subgoal *i* of the proof state.

`bimatch_tac` is like `biresolve_tac`, but performs matching: unknowns in the proof state are never updated (see Sect. 6.1.3).

`subgoals_of_brl`(*flag*, *rule*) returns the number of new subgoals that bi-resolution would yield for the pair (if applied to a suitable subgoal). This is *n* if the flag is `false` and $n - 1$ if the flag is `true`, where *n* is the number of premises of the rule. Elim-resolution yields one fewer subgoal than ordinary resolution because it solves the major premise by assumption.

`lessb` (*brl1*, *brl2*) returns the result of

`subgoals_of_brl` *brl1* `<` `subgoals_of_brl` *brl2*

Note that `sort lessb` *brls* sorts a list of (*flag*, *rule*) pairs by the number of new subgoals they will yield. Thus, those that yield the fewest subgoals should be tried first.

6.4.2 Discrimination nets for fast resolution

```
net_resolve_tac  : thm list -> int -> tactic
net_match_tac    : thm list -> int -> tactic
net_biresolve_tac: (bool*thm) list -> int -> tactic
net_bimatch_tac  : (bool*thm) list -> int -> tactic
filt_resolve_tac : thm list -> int -> int -> tactic
could_unify      : term*term->bool
filter_thms      : (term*term->bool) -> int*term*thm list -> thm list
```

The module `Net` implements a discrimination net data structure for fast selection of rules [6, Chapter 14]. A term is classified by the symbol list obtained by flattening it in preorder. The flattening takes account of function applications, constants, and free and bound variables; it identifies all unknowns and also regards λ-abstractions as unknowns, since they could η-contract to anything.

A discrimination net serves as a polymorphic dictionary indexed by terms. The module provides various functions for inserting and removing items from nets. It provides functions for returning all items whose term could match or unify with a target term. The matching and unification tests are overly lax (due to the identifications mentioned above) but they serve as useful filters.

A net can store introduction rules indexed by their conclusion, and elimination rules indexed by their major premise. Isabelle provides several functions for 'compiling' long lists of rules into fast resolution tactics. When supplied with a list of theorems, these functions build a discrimination net; the net is used when the tactic is applied to a goal. To avoid repreatedly constructing the nets, use currying: bind the resulting tactics to ML identifiers.

`net_resolve_tac` *thms* builds a discrimination net to obtain the effect of a similar call to `resolve_tac`.

`net_match_tac` *thms* builds a discrimination net to obtain the effect of a similar call to `match_tac`.

`net_biresolve_tac` *brls* builds a discrimination net to obtain the effect of a similar call to `biresolve_tac`.

`net_bimatch_tac` *brls* builds a discrimination net to obtain the effect of a similar call to `bimatch_tac`.

`filt_resolve_tac` *thms maxr i* uses discrimination nets to extract the *thms* that are applicable to subgoal *i*. If more than *maxr* theorems are applicable then the tactic fails. Otherwise it calls `resolve_tac`.

This tactic helps avoid runaway instantiation of unknowns, for example in type inference.

`could_unify` (t, u) returns `false` if t and u are 'obviously' non-unifiable, and otherwise returns `true`. It assumes all variables are distinct, reporting that `?a=?a` may unify with `0=1`.

`filter_thms` *could* (*limit*, *prem*, *thms*) returns the list of potentially resolvable rules (in *thms*) for the subgoal *prem*, using the predicate *could* to compare the conclusion of the subgoal with the conclusion of each rule. The resulting list is no longer than *limit*.

6.5 Programming tools for proof strategies

Do not consider using the primitives discussed in this section unless you really need to code tactics from scratch.

6.5.1 Operations on type tactic

A tactic maps theorems to theorem sequences (lazy lists). The type constructor for sequences is called `Sequence.seq`. To simplify the types of tactics and tacticals, Isabelle defines a type of tactics:

```
datatype tactic = Tactic of thm -> thm Sequence.seq
```

`Tactic` and `tapply` convert between tactics and functions. The other operations provide means for coding tactics in a clean style.

```
tapply    : tactic * thm -> thm Sequence.seq
Tactic    :       (thm -> thm Sequence.seq) -> tactic
PRIMITIVE :                    (thm -> thm) -> tactic
STATE     :                 (thm -> tactic) -> tactic
SUBGOAL   : ((term*int) -> tactic) -> int -> tactic
```

`tapply(`*tac*`,` *thm*`)` returns the result of applying the tactic, as a function, to *thm*.

`Tactic` f packages f as a tactic.

`PRIMITIVE` f applies f to the proof state and returns the result as a one-element sequence. This packages the meta-rule f as a tactic.

`STATE` f applies f to the proof state and then applies the resulting tactic to the same state. It supports the following style, where the tactic body is expressed using tactics and tacticals, but may peek at the proof state:

```
STATE (fn state => tactic-valued expression)
```

`SUBGOAL` f i extracts subgoal i from the proof state as a term t, and computes a tactic by calling $f(t, i)$. It applies the resulting tactic to the same state.

The tactic body is expressed using tactics and tacticals, but may peek at a particular subgoal:

```
SUBGOAL (fn (t,i) => tactic-valued expression)
```

6.5.2 Tracing

```
pause_tac: tactic
print_tac: tactic
```

These tactics print tracing information when they are applied to a proof state. Their output may be difficult to interpret. Note that certain of the searching tacticals, such as `REPEAT`, have built-in tracing options.

`pause_tac` prints `** Press RETURN to continue:` and then reads a line from the terminal. If this line is blank then it returns the proof state unchanged; otherwise it fails (which may terminate a repetition).

`print_tac` returns the proof state unchanged, with the side effect of printing it at the terminal.

6.6 Sequences

The module `Sequence` declares a type of lazy lists. It uses Isabelle's type `option` to represent the possible presence (`Some`) or absence (`None`) of a value:

```
datatype 'a option = None  |  Some of 'a;
```

For clarity, the module name `Sequence` is omitted from the signature specifications below; for instance, `null` appears instead of `Sequence.null`.

6.6.1 Basic operations on sequences

```
null   : 'a seq
seqof  : (unit -> ('a * 'a seq) option) -> 'a seq
single : 'a -> 'a seq
pull   : 'a seq -> ('a * 'a seq) option
```

`Sequence.null` is the empty sequence.

`Sequence.seqof (fn()=> Some(`x`,`s`))` constructs the sequence with head x and tail s, neither of which is evaluated.

`Sequence.single` x constructs the sequence containing the single element x.

`Sequence.pull` s returns `None` if the sequence is empty and `Some(`x`,`s'`)` if the sequence has head x and tail s'. Warning: calling Sequence.pull s again will *recompute* the value of x; it is not stored!

6.6.2 Converting between sequences and lists

```
chop      : int * 'a seq -> 'a list * 'a seq
list_of_s : 'a seq -> 'a list
s_of_list : 'a list -> 'a seq
```

`Sequence.chop(`n`,`s`)` returns the first n elements of s as a list, paired with the remaining elements of s. If s has fewer than n elements, then so will the list.

`Sequence.list_of_s` s returns the elements of s, which must be finite, as a list.

`Sequence.s_of_list` l creates a sequence containing the elements of l.

6.6.3 Combining sequences

```
append     : 'a seq * 'a seq -> 'a seq
interleave : 'a seq * 'a seq -> 'a seq
flats      : 'a seq seq -> 'a seq
maps       : ('a -> 'b) -> 'a seq -> 'b seq
filters    : ('a -> bool) -> 'a seq -> 'a seq
```

`Sequence.append(`s_1`,`s_2`)` concatenates s_1 to s_2.

`Sequence.interleave(`s_1`,`s_2`)` joins s_1 with s_2 by interleaving their elements. The result contains all the elements of the sequences, even if both are infinite.

`Sequence.flats` ss concatenates a sequence of sequences.

`Sequence.maps` f s applies f to every element of $s = x_1, x_2, \ldots$, yielding the sequence $f(x_1), f(x_2), \ldots$.

`Sequence.filters` p s returns the sequence consisting of all elements x of s such that $p(x)$ is `true`.

7. Tacticals

Tacticals are operations on tactics. Their implementation makes use of functional programming techniques, especially for sequences. Most of the time, you may forget about this and regard tacticals as high-level control structures.

7.1 The basic tacticals

7.1.1 Joining two tactics

The tacticals THEN and ORELSE, which provide sequencing and alternation, underlie most of the other control structures in Isabelle. APPEND and INTLEAVE provide more sophisticated forms of alternation.

```
THEN     : tactic * tactic -> tactic                     infix 1
ORELSE   : tactic * tactic -> tactic                     infix
APPEND   : tactic * tactic -> tactic                     infix
INTLEAVE : tactic * tactic -> tactic                     infix
```

tac_1 THEN tac_2 is the sequential composition of the two tactics. Applied to a proof state, it returns all states reachable in two steps by applying tac_1 followed by tac_2. First, it applies tac_1 to the proof state, getting a sequence of next states; then, it applies tac_2 to each of these and concatenates the results.

tac_1 ORELSE tac_2 makes a choice between the two tactics. Applied to a state, it tries tac_1 and returns the result if non-empty; if tac_1 fails then it uses tac_2. This is a deterministic choice: if tac_1 succeeds then tac_2 is excluded.

tac_1 APPEND tac_2 concatenates the results of tac_1 and tac_2. By not making a commitment to either tactic, APPEND helps avoid incompleteness during search.

tac_1 INTLEAVE tac_2 interleaves the results of tac_1 and tac_2. Thus, it includes all possible next states, even if one of the tactics returns an infinite sequence.

7.1.2 Joining a list of tactics

```
EVERY : tactic list -> tactic
FIRST : tactic list -> tactic
```

EVERY and FIRST are block structured versions of THEN and ORELSE.

EVERY $[tac_1, \ldots, tac_n]$ abbreviates tac_1 THEN ... THEN tac_n. It is useful for writing a series of tactics to be executed in sequence.

FIRST $[tac_1, \ldots, tac_n]$ abbreviates tac_1 ORELSE ... ORELSE tac_n. It is useful for writing a series of tactics to be attempted one after another.

7.1.3 Repetition tacticals

```
TRY           : tactic -> tactic
REPEAT_DETERM : tactic -> tactic
REPEAT        : tactic -> tactic
REPEAT1       : tactic -> tactic
trace_REPEAT  : bool ref                                    initially false
```

TRY *tac* applies *tac* to the proof state and returns the resulting sequence, if non-empty; otherwise it returns the original state. Thus, it applies *tac* at most once.

REPEAT_DETERM *tac* applies *tac* to the proof state and, recursively, to the head of the resulting sequence. It returns the first state to make *tac* fail. It is deterministic, discarding alternative outcomes.

REPEAT *tac* applies *tac* to the proof state and, recursively, to each element of the resulting sequence. The resulting sequence consists of those states that make *tac* fail. Thus, it applies *tac* as many times as possible (including zero times), and allows backtracking over each invocation of *tac*. It is more general than REPEAT_DETERM, but requires more space.

REPEAT1 *tac* is like REPEAT *tac* but it always applies *tac* at least once, failing if this is impossible.

trace_REPEAT := true; enables an interactive tracing mode for the tacticals REPEAT_DETERM and REPEAT. To view the tracing options, type h at the prompt.

7.1.4 Identities for tacticals

```
all_tac : tactic
no_tac  : tactic
```

`all_tac` maps any proof state to the one-element sequence containing that state. Thus, it succeeds for all states. It is the identity element of the tactical `THEN`.

`no_tac` maps any proof state to the empty sequence. Thus it succeeds for no state. It is the identity element of `ORELSE`, `APPEND`, and `INTLEAVE`. Also, it is a zero element for `THEN`, which means that *tac* `THEN no_tac` is equivalent to `no_tac`.

These primitive tactics are useful when writing tacticals. For example, `TRY` and `REPEAT` (ignoring tracing) can be coded as follows:

```
fun TRY tac = tac ORELSE all_tac;

fun REPEAT tac = Tactic
     (fn state => tapply((tac THEN REPEAT tac) ORELSE all_tac,
                         state));
```

If *tac* can return multiple outcomes then so can `REPEAT` *tac*. Since `REPEAT` uses `ORELSE` and not `APPEND` or `INTLEAVE`, it applies *tac* as many times as possible in each outcome.

! Note `REPEAT`'s explicit abstraction over the proof state. Recursive tacticals must be coded in this awkward fashion to avoid infinite recursion. With the following definition, `REPEAT` *tac* would loop due to ML's eager evaluation strategy:

```
fun REPEAT tac = (tac THEN REPEAT tac) ORELSE all_tac;
```

The built-in `REPEAT` avoids `THEN`, handling sequences explicitly and using tail recursion. This sacrifices clarity, but saves much space by discarding intermediate proof states.

7.2 Control and search tacticals

A predicate on theorems, namely a function of type `thm->bool`, can test whether a proof state enjoys some desirable property — such as having no subgoals. Tactics that search for satisfactory states are easy to express. The main search procedures, depth-first, breadth-first and best-first, are provided as tacticals. They generate the search tree by repeatedly applying a given tactic.

7.2.1 Filtering a tactic's results

```
FILTER  : (thm -> bool) -> tactic -> tactic
CHANGED : tactic -> tactic
```

`FILTER` *p tac* applies *tac* to the proof state and returns a sequence consisting of those result states that satisfy *p*.

CHANGED *tac* applies *tac* to the proof state and returns precisely those states that differ from the original state. Thus, CHANGED *tac* always has some effect on the state.

7.2.2 Depth-first search

```
DEPTH_FIRST   : (thm->bool) -> tactic -> tactic
DEPTH_SOLVE   :                tactic -> tactic
DEPTH_SOLVE_1 :                tactic -> tactic
trace_DEPTH_FIRST: bool ref                        initially false
```

DEPTH_FIRST *satp tac* returns the proof state if *satp* returns true. Otherwise it applies *tac*, then recursively searches from each element of the resulting sequence. The code uses a stack for efficiency, in effect applying *tac* THEN DEPTH_FIRST *satp tac* to the state.

DEPTH_SOLVE *tac* uses DEPTH_FIRST to search for states having no subgoals.

DEPTH_SOLVE_1 *tac* uses DEPTH_FIRST to search for states having fewer subgoals than the given state. Thus, it insists upon solving at least one subgoal.

trace_DEPTH_FIRST := true; enables interactive tracing for DEPTH_FIRST. To view the tracing options, type h at the prompt.

7.2.3 Other search strategies

```
BREADTH_FIRST   :              (thm->bool) -> tactic -> tactic
BEST_FIRST      : (thm->bool)*(thm->int) -> tactic -> tactic
THEN_BEST_FIRST : tactic * ((thm->bool) * (thm->int) * tactic)
                  -> tactic                                  infix 1
trace_BEST_FIRST: bool ref                           initially false
```

These search strategies will find a solution if one exists. However, they do not enumerate all solutions; they terminate after the first satisfactory result from *tac*.

BREADTH_FIRST *satp tac* uses breadth-first search to find states for which *satp* is true. For most applications, it is too slow.

BEST_FIRST (*satp*, *distf*) *tac* does a heuristic search, using *distf* to estimate the distance from a satisfactory state. It maintains a list of states ordered by distance. It applies *tac* to the head of this list; if the result contains any satisfactory states, then it returns them. Otherwise, BEST_FIRST adds the new states to the list, and continues.

The distance function is typically size_of_thm, which computes the size of the state. The smaller the state, the fewer and simpler subgoals it has.

tac_0 `THEN_BEST_FIRST` (*satp*, *distf*, *tac*) is like `BEST_FIRST`, except that the priority queue initially contains the result of applying tac_0 to the proof state. This tactical permits separate tactics for starting the search and continuing the search.

`trace_BEST_FIRST := true;` enables an interactive tracing mode for the tactical `BEST_FIRST`. To view the tracing options, type `h` at the prompt.

7.2.4 Auxiliary tacticals for searching

```
COND        : (thm->bool) -> tactic -> tactic -> tactic
IF_UNSOLVED : tactic -> tactic
DETERM      : tactic -> tactic
```

`COND` p tac_1 tac_2 applies tac_1 to the proof state if it satisfies p, and applies tac_2 otherwise. It is a conditional tactical in that only one of tac_1 and tac_2 is applied to a proof state. However, both tac_1 and tac_2 are evaluated because ML uses eager evaluation.

`IF_UNSOLVED` *tac* applies *tac* to the proof state if it has any subgoals, and simply returns the proof state otherwise. Many common tactics, such as `resolve_tac`, fail if applied to a proof state that has no subgoals.

`DETERM` *tac* applies *tac* to the proof state and returns the head of the resulting sequence. `DETERM` limits the search space by making its argument deterministic.

7.2.5 Predicates and functions useful for searching

```
has_fewer_prems : int -> thm -> bool
eq_thm          : thm * thm -> bool
size_of_thm     : thm -> int
```

`has_fewer_prems` n *thm* reports whether *thm* has fewer than n premises. By currying, `has_fewer_prems` n is a predicate on theorems; it may be given to the searching tacticals.

`eq_thm`(*thm*1, *thm*2) reports whether *thm*1 and *thm*2 are equal. Both theorems must have identical signatures. Both theorems must have the same conclusions, and the same hypotheses, in the same order. Names of bound variables are ignored.

`size_of_thm` *thm* computes the size of *thm*, namely the number of variables, constants and abstractions in its conclusion. It may serve as a distance function for `BEST_FIRST`.

7.3 Tacticals for subgoal numbering

When conducting a backward proof, we normally consider one goal at a time. A tactic can affect the entire proof state, but many tactics — such as `resolve_tac` and `assume_tac` — work on a single subgoal. Subgoals are designated by a positive integer, so Isabelle provides tacticals for combining values of type `int->tactic`.

7.3.1 Restricting a tactic to one subgoal

```
SELECT_GOAL : tactic -> int -> tactic
METAHYPS    : (thm list -> tactic) -> int -> tactic
```

`SELECT_GOAL` *tac i* restricts the effect of *tac* to subgoal *i* of the proof state. It fails if there is no subgoal *i*, or if *tac* changes the main goal (do not use `rewrite_tac`). It applies *tac* to a dummy proof state and uses the result to refine the original proof state at subgoal *i*. If *tac* returns multiple results then so does `SELECT_GOAL` *tac i*.

`SELECT_GOAL` works by creating a state of the form $\phi \Longrightarrow \phi$, with the one subgoal ϕ. If subgoal i has the form $\psi \Longrightarrow \theta$ then $(\psi \Longrightarrow \theta) \Longrightarrow (\psi \Longrightarrow \theta)$ is in fact $[\![\psi \Longrightarrow \theta;\ \psi]\!] \Longrightarrow \theta$, a proof state with two subgoals. Such a proof state might cause tactics to go astray. Therefore `SELECT_GOAL` inserts a quantifier to create the state

$$(\bigwedge x\,.\,\psi \Longrightarrow \theta) \Longrightarrow (\bigwedge x\,.\,\psi \Longrightarrow \theta).$$

`METAHYPS` *tacf i* takes subgoal *i*, of the form

$$\bigwedge x_1 \ldots x_l\,.\,[\![\theta_1; \ldots; \theta_k]\!] \Longrightarrow \theta,$$

and creates the list $\theta'_1, \ldots, \theta'_k$ of meta-level assumptions. In these theorems, the subgoal's parameters ($x_1, \ldots, x_l$) become free variables. It supplies the assumptions to *tacf* and applies the resulting tactic to the proof state $\theta \Longrightarrow \theta$.

If the resulting proof state is $[\![\phi_1; \ldots; \phi_n]\!] \Longrightarrow \phi$, possibly containing $\theta'_1, \ldots, \theta'_k$ as assumptions, then it is lifted back into the original context, yielding n subgoals.

Meta-level assumptions may not contain unknowns. Unknowns in the hypotheses $\theta_1, \ldots, \theta_k$ become free variables in $\theta'_1, \ldots, \theta'_k$, and are restored afterwards; the `METAHYPS` call cannot instantiate them. Unknowns in θ may be instantiated. New unknowns in $\phi_1, \ldots, \phi_n$ are lifted over the parameters.

Here is a typical application. Calling `hyp_res_tac` i resolves subgoal i with one of its own assumptions, which may itself have the form of an inference rule (these are called **higher-level assumptions**).

```
val hyp_res_tac = METAHYPS (fn prems => resolve_tac prems 1);
```

The function `gethyps` is useful for debugging applications of `METAHYPS`.

! `METAHYPS` fails if the context or new subgoals contain type unknowns. In principle, the tactical could treat these like ordinary unknowns.

7.3.2 Scanning for a subgoal by number

```
ALLGOALS          : (int -> tactic) -> tactic
TRYALL            : (int -> tactic) -> tactic
SOMEGOAL          : (int -> tactic) -> tactic
FIRSTGOAL         : (int -> tactic) -> tactic
REPEAT_SOME       : (int -> tactic) -> tactic
REPEAT_FIRST      : (int -> tactic) -> tactic
trace_goalno_tac : (int -> tactic) -> int -> tactic
```

These apply a tactic function of type `int -> tactic` to all the subgoal numbers of a proof state, and join the resulting tactics using `THEN` or `ORELSE`. Thus, they apply the tactic to all the subgoals, or to one subgoal.

Suppose that the original proof state has n subgoals.

`ALLGOALS` *tacf* is equivalent to *tacf*(n) `THEN` ... `THEN` *tacf*(1).

It applies *tacf* to all the subgoals, counting downwards (to avoid problems when subgoals are added or deleted).

`TRYALL` *tacf* is equivalent to `TRY`(*tacf*(n)) `THEN` ... `THEN` `TRY`(*tacf*(1)).

It attempts to apply *tacf* to all the subgoals. For instance, the tactic `TRYALL assume_tac` attempts to solve all the subgoals by assumption.

`SOMEGOAL` *tacf* is equivalent to *tacf*(n) `ORELSE` ... `ORELSE` *tacf*(1).

It applies *tacf* to one subgoal, counting downwards. For instance, the tactic `SOMEGOAL assume_tac` solves one subgoal by assumption, failing if this is impossible.

`FIRSTGOAL` *tacf* is equivalent to *tacf*(1) `ORELSE` ... `ORELSE` *tacf*(n).

It applies *tacf* to one subgoal, counting upwards.

`REPEAT_SOME` *tacf* applies *tacf* once or more to a subgoal, counting downwards.

`REPEAT_FIRST` *tacf* applies *tacf* once or more to a subgoal, counting upwards.

`trace_goalno_tac` *tac i* applies *tac i* to the proof state. If the resulting sequence is non-empty, then it is returned, with the side-effect of printing `Subgoal` *i* `selected`. Otherwise, `trace_goalno_tac` returns the empty sequence and prints nothing.

It indicates that 'the tactic worked for subgoal *i*' and is mainly used with `SOMEGOAL` and `FIRSTGOAL`.

7.3.3 Joining tactic functions

```
THEN'     : ('a -> tactic) * ('a -> tactic) -> 'a -> tactic     infix 1
ORELSE'   : ('a -> tactic) * ('a -> tactic) -> 'a -> tactic     infix
APPEND'   : ('a -> tactic) * ('a -> tactic) -> 'a -> tactic     infix
INTLEAVE' : ('a -> tactic) * ('a -> tactic) -> 'a -> tactic     infix
EVERY'    : ('a -> tactic) list -> 'a -> tactic
FIRST'    : ('a -> tactic) list -> 'a -> tactic
```

These help to express tactics that specify subgoal numbers. The tactic

```
SOMEGOAL (fn i => resolve_tac rls i  ORELSE  eresolve_tac erls i)
```

can be simplified to

```
SOMEGOAL (resolve_tac rls  ORELSE'  eresolve_tac erls)
```

Note that `TRY'`, `REPEAT'`, `DEPTH_FIRST'`, etc. are not provided, because function composition accomplishes the same purpose. The tactic

```
ALLGOALS (fn i => REPEAT (etac exE i  ORELSE  atac i))
```

can be simplified to

```
ALLGOALS (REPEAT o (etac exE  ORELSE'  atac))
```

These tacticals are polymorphic; x need not be an integer.

($tacf_1$ `THEN'` $tacf_2$)(x)	yields	$tacf_1(x)$ `THEN` $tacf_2(x)$
($tacf_1$ `ORELSE'` $tacf_2$)(x)	yields	$tacf_1(x)$ `ORELSE` $tacf_2(x)$
($tacf_1$ `APPEND'` $tacf_2$)(x)	yields	$tacf_1(x)$ `APPEND` $tacf_2(x)$
($tacf_1$ `INTLEAVE'` $tacf_2$)(x)	yields	$tacf_1(x)$ `INTLEAVE` $tacf_2(x)$
`EVERY'` $[tacf_1, \ldots, tacf_n]$ (x)	yields	`EVERY` $[tacf_1(x), \ldots, tacf_n(x)]$
`FIRST'` $[tacf_1, \ldots, tacf_n]$ (x)	yields	`FIRST` $[tacf_1(x), \ldots, tacf_n(x)]$

7.3.4 Applying a list of tactics to 1

```
EVERY1: (int -> tactic) list -> tactic
FIRST1: (int -> tactic) list -> tactic
```

A common proof style is to treat the subgoals as a stack, always restricting attention to the first subgoal. Such proofs contain long lists of tactics, each applied to 1. These can be simplified using `EVERY1` and `FIRST1`:

`EVERY1` $[tacf_1, \ldots, tacf_n]$	abbreviates	`EVERY` $[tacf_1(1), \ldots, tacf_n(1)]$
`FIRST1` $[tacf_1, \ldots, tacf_n]$	abbreviates	`FIRST` $[tacf_1(1), \ldots, tacf_n(1)]$

8. Theorems and Forward Proof

Theorems, which represent the axioms, theorems and rules of object-logics, have type `thm`. This chapter begins by describing operations that print theorems and that join them in forward proof. Most theorem operations are intended for advanced applications, such as programming new proof procedures. Many of these operations refer to signatures, certified terms and certified types, which have the ML types `Sign.sg`, `Sign.cterm` and `Sign.ctyp` and are discussed in Chapter 9. Beginning users should ignore such complexities — and skip all but the first section of this chapter.

The theorem operations do not print error messages. Instead, they raise exception `THM`. Use `print_exn` to display exceptions nicely:

```
allI RS mp  handle e => print_exn e;
  Exception THM raised:
  RSN: no unifiers -- premise 1
  (!!x. ?P(x)) ==> ALL x. ?P(x)
  [| ?P --> ?Q; ?P |] ==> ?Q

  uncaught exception THM
```

8.1 Basic operations on theorems

8.1.1 Pretty-printing a theorem

```
prth          : thm -> thm
prths         : thm list -> thm list
prthq         : thm Sequence.seq -> thm Sequence.seq
print_thm     : thm -> unit
print_goals   : int -> thm -> unit
string_of_thm : thm -> string
```

The first three commands are for interactive use. They are identity functions that display, then return, their argument. The ML identifier `it` will refer to the value just displayed.

The others are for use in programs. Functions with result type `unit` are convenient for imperative programming.

`prth` *thm* prints *thm* at the terminal.

`prths` *thms* prints *thms*, a list of theorems.

`prthq` *thmq* prints *thmq*, a sequence of theorems. It is useful for inspecting the output of a tactic.

`print_thm` *thm* prints *thm* at the terminal.

`print_goals` *limit* *thm* prints *thm* in goal style, with the premises as subgoals. It prints at most *limit* subgoals. The subgoal module calls `print_goals` to display proof states.

`string_of_thm` *thm* converts *thm* to a string.

8.1.2 Forward proof: joining rules by resolution

```
RSN : thm * (int * thm) -> thm                          infix
RS  : thm * thm -> thm                                  infix
MRS : thm list * thm -> thm                             infix
RLN : thm list * (int * thm list) -> thm list           infix
RL  : thm list * thm list -> thm list                   infix
MRL : thm list list * thm list -> thm list              infix
```

Joining rules together is a simple way of deriving new rules. These functions are especially useful with destruction rules.

thm_1 `RSN` (i, thm_2) resolves the conclusion of thm_1 with the ith premise of thm_2. Unless there is precisely one resolvent it raises exception `THM`; in that case, use `RLN`.

thm_1 `RS` thm_2 abbreviates thm_1 `RSN` $(1, thm_2)$. Thus, it resolves the conclusion of thm_1 with the first premise of thm_2.

$[thm_1, \ldots, thm_n]$ `MRS` *thm* uses `RSN` to resolve thm_i against premise i of *thm*, for $i = n, \ldots, 1$. This applies thm_n, …, thm_1 to the first n premises of *thm*. Because the theorems are used from right to left, it does not matter if the thm_i create new premises. `MRS` is useful for expressing proof trees.

$thms_1$ `RLN` $(i, thms_2)$ joins lists of theorems. For every thm_1 in $thms_1$ and thm_2 in $thms_2$, it resolves the conclusion of thm_1 with the ith premise of thm_2, accumulating the results.

$thms_1$ `RL` $thms_2$ abbreviates $thms_1$ `RLN` $(1, thms_2)$.

$[thms_1, \ldots, thms_n]$ `MRL` *thms* is analogous to `MRS`, but combines theorem lists rather than theorems. It too is useful for expressing proof trees.

8.1.3 Expanding definitions in theorems

```
rewrite_rule       : thm list -> thm -> thm
rewrite_goals_rule : thm list -> thm -> thm
```

`rewrite_rule` *defs thm* unfolds the *defs* throughout the theorem *thm*.

`rewrite_goals_rule` *defs thm* unfolds the *defs* in the premises of *thm*, but leaves the conclusion unchanged. This rule underlies `rewrite_goals_tac`, but serves little purpose in forward proof.

8.1.4 Instantiating a theorem

```
read_instantiate    :           (string*string)list -> thm -> thm
read_instantiate_sg : Sign.sg -> (string*string)list -> thm -> thm
cterm_instantiate   :      (Sign.cterm*Sign.cterm)list -> thm -> thm
```

These meta-rules instantiate type and term unknowns in a theorem. They are occasionally useful. They can prevent difficulties with higher-order unification, and define specialized versions of rules.

`read_instantiate` *insts thm* processes the instantiations *insts* and instantiates the rule *thm*. The processing of instantiations is described in Sect. 6.1.4, under `res_inst_tac`.

Use `res_inst_tac`, not `read_instantiate`, to instantiate a rule and refine a particular subgoal. The tactic allows instantiation by the subgoal's parameters, and reads the instantiations using the signature associated with the proof state.

Use `read_instantiate_sg` below if *insts* appears to be treated incorrectly.

`read_instantiate_sg` *sg insts thm* resembles `read_instantiate` *insts thm*, but reads the instantiations under signature *sg*. This is necessary to instantiate a rule from a general theory, such as first-order logic, using the notation of some specialized theory. Use the function `sign_of` to get a theory's signature.

`cterm_instantiate` *ctpairs thm* is similar to `read_instantiate`, but the instantiations are provided as pairs of certified terms, not as strings to be read.

8.1.5 Miscellaneous forward rules

```
standard         :           thm -> thm
zero_var_indexes :           thm -> thm
make_elim        :           thm -> thm
rule_by_tactic   : tactic -> thm -> thm
```

`standard` *thm* puts *thm* into the standard form of object-rules. It discharges all meta-assumptions, replaces free variables by schematic variables, and renames schematic variables to have subscript zero.

`zero_var_indexes` *thm* makes all schematic variables have subscript zero, renaming them to avoid clashes.

`make_elim` *thm* converts *thm*, a destruction rule of the form $[\![P_1; \ldots; P_m]\!] \Longrightarrow Q$, to the elimination rule $[\![P_1; \ldots; P_m; Q \Longrightarrow R]\!] \Longrightarrow R$. This is the basis for destruct-resolution: `dresolve_tac`, etc.

`rule_by_tactic` *tac thm* applies *tac* to the *thm*, freezing its variables first, then yields the proof state returned by the tactic. In typical usage, the *thm* represents an instance of a rule with several premises, some with contradictory assumptions (because of the instantiation). The tactic proves those subgoals and does whatever else it can, and returns whatever is left.

8.1.6 Taking a theorem apart

```
concl_of       : thm -> term
prems_of       : thm -> term list
nprems_of      : thm -> int
tpairs_of      : thm -> (term*term)list
stamps_of_thy  : thm -> string ref list
dest_state     : thm*int -> (term*term)list*term list*term*term
rep_thm        : thm -> {prop:term, hyps:term list,
                         maxidx:int, sign:Sign.sg}
```

`concl_of` *thm* returns the conclusion of *thm* as a term.

`prems_of` *thm* returns the premises of *thm* as a list of terms.

`nprems_of` *thm* returns the number of premises in *thm*, and is equivalent to `length(prems_of` *thm*`)`.

`tpairs_of` *thm* returns the flex-flex constraints of *thm*.

`stamps_of_thm` *thm* returns the stamps of the signature associated with *thm*.

`dest_state` (*thm*, i) decomposes *thm* as a tuple containing a list of flex-flex constraints, a list of the subgoals 1 to $i-1$, subgoal i, and the rest of the theorem (this will be an implication if there are more than i subgoals).

`rep_thm` *thm* decomposes *thm* as a record containing the statement of *thm*, its list of meta-assumptions, the maximum subscript of its unknowns, and its signature.

8.1.7 Tracing flags for unification

```
Unify.trace_simp   : bool ref                    initially false
Unify.trace_types  : bool ref                    initially false
Unify.trace_bound  : int ref                     initially 10
Unify.search_bound : int ref                     initially 20
```

Tracing the search may be useful when higher-order unification behaves unexpectedly. Letting `res_inst_tac` circumvent the problem is easier, though.

`Unify.trace_simp := true;` causes tracing of the simplification phase.

`Unify.trace_types := true;` generates warnings of incompleteness, when unification is not considering all possible instantiations of type unknowns.

`Unify.trace_bound :=` n`;` causes unification to print tracing information once it reaches depth n. Use $n = 0$ for full tracing. At the default value of 10, tracing information is almost never printed.

`Unify.search_bound :=` n`;` causes unification to limit its search to depth n. Because of this bound, higher-order unification cannot return an infinite sequence, though it can return a very long one. The search rarely approaches the default value of 20. If the search is cut off, unification prints `***Unification bound exceeded`.

8.2 Primitive meta-level inference rules

These implement the meta-logic in LCF style, as functions from theorems to theorems. They are, rarely, useful for deriving results in the pure theory. Mainly, they are included for completeness, and most users should not bother with them. The meta-rules raise exception `THM` to signal malformed premises, incompatible signatures and similar errors.

The meta-logic uses natural deduction. Each theorem may depend on meta-level assumptions. Certain rules, such as $(\Longrightarrow I)$, discharge assumptions; in most other rules, the conclusion depends on all of the assumptions of the premises. Formally, the system works with assertions of the form

$$\phi \quad [\phi_1, \ldots, \phi_n],$$

where $\phi_1, \ldots, \phi_n$ are the assumptions. Do not confuse meta-level assumptions with the object-level assumptions in a subgoal, which are represented in the meta-logic using $\Longrightarrow$.

Each theorem has a signature. Certified terms have a signature. When a rule takes several premises and certified terms, it merges the signatures to make a signature for the conclusion. This fails if the signatures are incompatible.

The **implication** rules are $(\Longrightarrow I)$ and $(\Longrightarrow E)$:

$$\frac{\begin{matrix}[\phi]\\ \vdots \\ \psi\end{matrix}}{\phi \Longrightarrow \psi}\,(\Longrightarrow I) \qquad \frac{\phi \Longrightarrow \psi \quad \phi}{\psi}\,(\Longrightarrow E)$$

Equality of truth values means logical equivalence:

$$\frac{\begin{matrix}[\phi] & [\psi]\\ \vdots & \vdots \\ \psi & \phi\end{matrix}}{\phi \equiv \psi}\,(\equiv I) \qquad \frac{\phi \equiv \psi \quad \phi}{\psi}\,(\equiv E)$$

The **equality** rules are reflexivity, symmetry, and transitivity:

$$a \equiv a\,(\mathit{refl}) \qquad \frac{a \equiv b}{b \equiv a}\,(\mathit{sym}) \qquad \frac{a \equiv b \quad b \equiv c}{a \equiv c}\,(\mathit{trans})$$

The λ-conversions are α-conversion, β-conversion, and extensionality:[1]

$$(\lambda x\,.\,a) \equiv (\lambda y\,.\,a[y/x]) \qquad ((\lambda x\,.\,a)(b)) \equiv a[b/x] \qquad \frac{f(x) \equiv g(x)}{f \equiv g}\,(\mathit{ext})$$

The **abstraction** and **combination** rules let conversions be applied to subterms:[2]

$$\frac{a \equiv b}{(\lambda x\,.\,a) \equiv (\lambda x\,.\,b)}\,(\mathit{abs}) \qquad \frac{f \equiv g \quad a \equiv b}{f(a) \equiv g(b)}\,(\mathit{comb})$$

The **universal quantification** rules are $(\bigwedge I)$ and $(\bigwedge E)$:[3]

$$\frac{\phi}{\bigwedge x\,.\,\phi}\,(\bigwedge I) \qquad \frac{\bigwedge x\,.\,\phi}{\phi[b/x]}\,(\bigwedge E)$$

8.2.1 Assumption rule

```
assume: Sign.cterm -> thm
```

`assume` *ct* makes the theorem $\phi\ [\phi]$, where ϕ is the value of *ct*. The rule checks that *ct* has type *prop* and contains no unknowns, which are not allowed in assumptions.

[1] α-conversion holds if y is not free in a; (ext) holds if x is not free in the assumptions, f, or g.

[2] Abstraction holds if x is not free in the assumptions.

[3] $(\bigwedge I)$ holds if x is not free in the assumptions.

8.2.2 Implication rules

```
implies_intr      : Sign.cterm -> thm -> thm
implies_intr_list : Sign.cterm list -> thm -> thm
implies_intr_hyps : thm -> thm
implies_elim      : thm -> thm -> thm
implies_elim_list : thm -> thm list -> thm
```

`implies_intr` *ct thm* is $(\Longrightarrow I)$, where *ct* is the assumption to discharge, say ϕ. It maps the premise ψ to the conclusion $\phi \Longrightarrow \psi$, removing all occurrences of ϕ from the assumptions. The rule checks that *ct* has type *prop*.

`implies_intr_list` *cts thm* applies $(\Longrightarrow I)$ repeatedly, on every element of the list *cts*.

`implies_intr_hyps` *thm* applies $(\Longrightarrow I)$ to discharge all the hypotheses (assumptions) of *thm*. It maps the premise $\phi\ [\phi_1, \ldots, \phi_n]$ to the conclusion $[\![\phi_1, \ldots, \phi_n]\!] \Longrightarrow \phi$.

`implies_elim` thm_1 thm_2 applies $(\Longrightarrow E)$ to thm_1 and thm_2. It maps the premises $\phi \Longrightarrow \psi$ and ϕ to the conclusion ψ.

`implies_elim_list` *thm thms* applies $(\Longrightarrow E)$ repeatedly to *thm*, using each element of *thms* in turn. It maps the premises $[\![\phi_1, \ldots, \phi_n]\!] \Longrightarrow \psi$ and $\phi_1, \ldots, \phi_n$ to the conclusion ψ.

8.2.3 Logical equivalence rules

```
equal_intr : thm -> thm -> thm
equal_elim : thm -> thm -> thm
```

`equal_intr` thm_1 thm_2 applies $(\equiv I)$ to thm_1 and thm_2. It maps the premises ψ and ϕ to the conclusion $\phi \equiv \psi$; the assumptions are those of the first premise with ϕ removed, plus those of the second premise with ψ removed.

`equal_elim` thm_1 thm_2 applies $(\equiv E)$ to thm_1 and thm_2. It maps the premises $\phi \equiv \psi$ and ϕ to the conclusion ψ.

8.2.4 Equality rules

```
reflexive  : Sign.cterm -> thm
symmetric  : thm -> thm
transitive : thm -> thm -> thm
```

`reflexive` *ct* makes the theorem $ct \equiv ct$.

`symmetric` *thm* maps the premise $a \equiv b$ to the conclusion $b \equiv a$.

`transitive` thm_1 thm_2 maps the premises $a \equiv b$ and $b \equiv c$ to the conclusion $a \equiv c$.

8.2.5 The λ-conversion rules

```
beta_conversion : Sign.cterm -> thm
extensional     : thm -> thm
abstract_rule   : string -> Sign.cterm -> thm -> thm
combination     : thm -> thm -> thm
```

There is no rule for α-conversion because Isabelle regards α-convertible theorems as equal.

`beta_conversion` *ct* makes the theorem $((\lambda x . a)(b)) \equiv a[b/x]$, where *ct* is the term $(\lambda x . a)(b)$.

`extensional` *thm* maps the premise $f(x) \equiv g(x)$ to the conclusion $f \equiv g$. Parameter x is taken from the premise. It may be an unknown or a free variable (provided it does not occur in the assumptions); it must not occur in f or g.

`abstract_rule` *v x thm* maps the premise $a \equiv b$ to the conclusion $(\lambda x . a) \equiv (\lambda x . b)$, abstracting over all occurrences (if any!) of x. Parameter x is supplied as a cterm. It may be an unknown or a free variable (provided it does not occur in the assumptions). In the conclusion, the bound variable is named v.

`combination` thm_1 thm_2 maps the premises $f \equiv g$ and $a \equiv b$ to the conclusion $f(a) \equiv g(b)$.

8.2.6 Forall introduction rules

```
forall_intr       : Sign.cterm      -> thm -> thm
forall_intr_list  : Sign.cterm list -> thm -> thm
forall_intr_frees :                    thm -> thm
```

`forall_intr` *x thm* applies $(\bigwedge I)$, abstracting over all occurrences (if any!) of x. The rule maps the premise ϕ to the conclusion $\bigwedge x . \phi$. Parameter x is supplied as a cterm. It may be an unknown or a free variable (provided it does not occur in the assumptions).

`forall_intr_list` *xs thm* applies $(\bigwedge I)$ repeatedly, on every element of the list *xs*.

`forall_intr_frees` *thm* applies $(\bigwedge I)$ repeatedly, generalizing over all the free variables of the premise.

8.2.7 Forall elimination rules

```
forall_elim       : Sign.cterm      -> thm -> thm
forall_elim_list  : Sign.cterm list -> thm -> thm
forall_elim_var   :             int -> thm -> thm
forall_elim_vars  :             int -> thm -> thm
```

`forall_elim` *ct thm* applies $(\bigwedge E)$, mapping the premise $\bigwedge x \,.\, \phi$ to the conclusion $\phi[ct/x]$. The rule checks that *ct* and x have the same type.

`forall_elim_list` *cts thm* applies $(\bigwedge E)$ repeatedly, on every element of the list *cts*.

`forall_elim_var` *k thm* applies $(\bigwedge E)$, mapping the premise $\bigwedge x \,.\, \phi$ to the conclusion $\phi[?x_k/x]$. Thus, it replaces the outermost $\bigwedge$-bound variable by an unknown having subscript k.

`forall_elim_vars` *ks thm* applies `forall_elim_var` repeatedly, for every element of the list *ks*.

8.2.8 Instantiation of unknowns

```
instantiate: (indexname*Sign.ctyp)list *
             (Sign.cterm*Sign.cterm)list  -> thm -> thm
```

`instantiate` (*tyinsts*, *insts*) *thm* simultaneously substitutes types for type unknowns (the *tyinsts*) and terms for term unknowns (the *insts*). Instantiations are given as (v, t) pairs, where v is an unknown and t is a term (of the same type as v) or a type (of the same sort as v). All the unknowns must be distinct. The rule normalizes its conclusion.

8.2.9 Freezing/thawing type unknowns

```
freezeT: thm -> thm
varifyT: thm -> thm
```

`freezeT` *thm* converts all the type unknowns in *thm* to free type variables.

`varifyT` *thm* converts all the free type variables in *thm* to type unknowns.

8.3 Derived rules for goal-directed proof

Most of these rules have the sole purpose of implementing particular tactics. There are few occasions for applying them directly to a theorem.

8.3.1 Proof by assumption

```
assumption    : int -> thm -> thm Sequence.seq
eq_assumption : int -> thm -> thm
```

`assumption` *i thm* attempts to solve premise *i* of *thm* by assumption.

`eq_assumption` is like `assumption` but does not use unification.

8.3.2 Resolution

```
biresolution : bool -> (bool*thm)list -> int -> thm
                 -> thm Sequence.seq
```

`biresolution` *match rules i state* performs bi-resolution on subgoal *i* of *state*, using the list of (*flag*, *rule*) pairs. For each pair, it applies resolution if the flag is `false` and elim-resolution if the flag is `true`. If *match* is `true`, the *state* is not instantiated.

8.3.3 Composition: resolution without lifting

```
compose   : thm * int * thm -> thm list
COMP      : thm * thm -> thm
bicompose : bool -> bool * thm * int -> int -> thm
              -> thm Sequence.seq
```

In forward proof, a typical use of composition is to regard an assertion of the form $\phi \Longrightarrow \psi$ as atomic. Schematic variables are not renamed, so beware of clashes!

`compose (`thm_1`, `i`, `thm_2`)` uses thm_1, regarded as an atomic formula, to solve premise i of thm_2. Let thm_1 and thm_2 be ψ and $[\![\phi_1; \ldots; \phi_n]\!] \Longrightarrow \phi$. For each s that unifies ψ and ϕ_i, the result list contains the theorem

$$([\![\phi_1; \ldots; \phi_{i-1}; \phi_{i+1}; \ldots; \phi_n]\!] \Longrightarrow \phi)s.$$

thm_1 `COMP` thm_2 calls `compose (`thm_1`, 1, `thm_2`)` and returns the result, if unique; otherwise, it raises exception `THM`. It is analogous to `RS`.

For example, suppose that thm_1 is $a = b \Longrightarrow b = a$, a symmetry rule, and that thm_2 is $[\![P \Longrightarrow Q; \neg Q]\!] \Longrightarrow \neg P$, which is the principle of contrapositives. Then the result would be the derived rule $\neg(b = a) \Longrightarrow \neg(a = b)$.

`bicompose` *match* `(`*flag*`, `*rule*`, `*m*`)` *i state* refines subgoal *i* of *state* using *rule*, without lifting. The *rule* is taken to have the form $[\![\psi_1; \ldots; \psi_m]\!] \Longrightarrow \psi$, where ψ need not be atomic; thus m determines the number of new subgoals. If *flag* is `true` then it performs elim-resolution — it solves the first premise of *rule* by assumption and deletes that assumption. If *match* is `true`, the *state* is not instantiated.

8.3.4 Other meta-rules

```
trivial             : Sign.cterm -> thm
lift_rule           : (thm * int) -> thm -> thm
rename_params_rule  : string list * int -> thm -> thm
rewrite_cterm       : thm list -> Sign.cterm -> thm
flexflex_rule       : thm -> thm Sequence.seq
```

`trivial` *ct* makes the theorem $\phi \Longrightarrow \phi$, where ϕ is the value of *ct*. This is the initial state for a goal-directed proof of ϕ. The rule checks that *ct* has type *prop*.

`lift_rule (`*state*`,` *i*`)` *rule* prepares *rule* for resolution by lifting it over the parameters and assumptions of subgoal *i* of *state*.

`rename_params_rule (`*names*`,` *i*`)` *thm* uses the *names* to rename the parameters of premise *i* of *thm*. The names must be distinct. If there are fewer names than parameters, then the rule renames the innermost parameters and may modify the remaining ones to ensure that all the parameters are distinct.

`rewrite_cterm` *defs ct* transforms *ct* to *ct'* by repeatedly applying *defs* as rewrite rules; it returns the conclusion $ct \equiv ct'$. This underlies the meta-rewriting tactics and rules.

`flexflex_rule` *thm* removes all flex-flex pairs from *thm* using the trivial unifier.

9. Theories, Terms and Types

Theories organize the syntax, declarations and axioms of a mathematical development. They are built, starting from the Pure theory, by extending and merging existing theories. They have the ML type `theory`. Theory operations signal errors by raising exception `THEORY`, returning a message and a list of theories.

Signatures, which contain information about sorts, types, constants and syntax, have the ML type `Sign.sg`. For identification, each signature carries a unique list of **stamps**, which are ML references to strings. The strings serve as human-readable names; the references serve as unique identifiers. Each primitive signature has a single stamp. When two signatures are merged, their lists of stamps are also merged. Every theory carries a unique signature.

Terms and types are the underlying representation of logical syntax. Their ML definitions are irrelevant to naive Isabelle users. Programmers who wish to extend Isabelle may need to know such details, say to code a tactic that looks for subgoals of a particular form. Terms and types may be 'certified' to be well-formed with respect to a given signature.

9.1 Defining theories

Theories can be defined either using concrete syntax or by calling certain ML functions (see Sect. 9.4.2). Appendix A presents the concrete syntax for theories; here is an explanation of the constituent parts:

theoryDef is the full definition. The new theory is called *id*. It is the union of the named **parent theories**, possibly extended with new classes, etc. The basic theory, which contains only the meta-logic, is called `Pure`.

Normally each *name* is an identifier, the name of the parent theory. Strings can be used to document additional dependencies; see Sect. 9.2 for details.

classes is a series of class declarations. Declaring *id* `<` id_1 ... id_n makes *id* a subclass of the existing classes $id_1 \ldots id_n$. This rules out cyclic class structures. Isabelle automatically computes the transitive closure of subclass hierarchies; it is not necessary to declare `c < e` in addition to `c < d` and `d < e`.

default introduces *sort* as the new default sort for type variables. This applies to unconstrained type variables in an input string but not to type variables created internally. If omitted, the default sort is the union of the default sorts of the parent theories.

sort is a finite set of classes. A single class *id* abbreviates the singleton set $\{id\}$.

types is a series of type declarations. Each declares a new type constructor or type synonym. An n-place type constructor is specified by $(\alpha_1, \ldots, \alpha_n)name$, where the type variables serve only to indicate the number n.

Only 2-place type constructors can have infix status and symbolic names; an example is `('a,'b)"*"`, which expresses binary product types.

A **type synonym** is an abbreviation $(\alpha_1, \ldots, \alpha_n)name$ = "τ", where *name* can be a string and τ must be enclosed in quotation marks.

infix declares a type or constant to be an infix operator of priority *nat* associating to the left (`infixl`) or right (`infixr`).

arities is a series of arity declarations. Each assigns arities to type constructors. The *name* must be an existing type constructor, which is given the additional arity *arity*.

constDecl is a series of constant declarations. Each new constant *name* is given the type specified by the *string*. The optional *mixfix* annotations may attach concrete syntax to the constant.

mixfix annotations can take three forms:

- A mixfix template given as a *string* of the form "..._..._..." where the i-th underscore indicates the position where the i-th argument should go. The list of numbers gives the priority of each argument. The final number gives the priority of the whole construct.
- A constant f of type $\tau_1 \Rightarrow (\tau_2 \Rightarrow \tau)$ can be given **infix** status.
- A constant f of type $(\tau_1 \Rightarrow \tau_2) \Rightarrow \tau$ can be given **binder** status. The declaration `binder` $\mathcal{Q}$ p causes $\mathcal{Q}\, x\, .\, F(x)$ to be treated like $f(F)$, where p is the priority.

trans specifies syntactic translation rules. There are three forms: parse rules (`=>`), print rules (`<=`), and parse/print rules (`==`).

rule is a series of rule declarations. Each has a name *id* and the formula is given by the *string*. Rule names must be distinct.

ml consists of ML code, typically for parse and print translations.

Chapter 11 explains mixfix declarations, translation rules and the ML section in more detail.

9.1.1 *Classes and arities

In order to guarantee principal types [37], arity declarations must obey two conditions:

- There must be no two declarations $ty :: (\mathbf{r})c$ and $ty :: (\mathbf{s})c$ with $\mathbf{r} \neq \mathbf{s}$. For example, the following is forbidden:

```
types 'a ty
arities ty :: ({logic}) logic
        ty :: ({})logic
```

- If there are two declarations $ty :: (s_1, \ldots, s_n)c$ and $ty :: (s'_1, \ldots, s'_n)c'$ such that $c' < c$ then $s'_i \preceq s_i$ must hold for $i = 1, \ldots, n$. The relationship $\preceq$, defined as
$$s' \preceq s \iff \forall c \in s \,.\, \exists c' \in s' \,.\, c' \leq c,$$
expresses that the set of types represented by s' is a subset of the set of types represented by s. For example, the following is forbidden:

```
classes term < logic
types 'a ty
arities ty :: ({logic})logic
        ty :: ({})term
```

9.2 Loading a new theory

```
use_thy         : string -> unit
time_use_thy    : string -> unit
loadpath        : string list ref                    initially ["."]
delete_tmpfiles : bool ref                           initially true
```

`use_thy` *thyname* reads the theory *thyname* and creates an ML structure as described below.

`time_use_thy` *thyname* calls `use_thy` *thyname* and reports the time taken.

`loadpath` contains a list of directories to search when locating the files that define a theory. This list is only used if the theory name in `use_thy` does not specify the path explicitly.

`delete_tmpfiles := false;` suppresses the deletion of temporary files.

Each theory definition must reside in a separate file. Let the file T`.thy` contain the definition of a theory called T, whose parent theories are $TB_1 \ldots TB_n$. Calling `use_thy "`T`"` reads the file T`.thy`, writes a temporary ML file `.`T`.thy.ML`, and reads the latter file. Recursive `use_thy` calls load those parent theories that have not been loaded previously; the recursive calls may continue to any depth. One `use_thy` call can read an entire logic provided all theories are linked appropriately.

The result is an ML structure T containing a component `thy` for the new theory and components for each of the rules. The structure also contains the definitions of the ML section, if present. The file `.`T`.thy.ML` is then deleted if `delete_tmpfiles` is set to `true` and no errors occurred.

Finally the file T`.ML` is read, if it exists. This file normally begins with the declaration `open` T and contains proofs involving the new theory.

Special applications, such as ZF's inductive definition package, construct theories directly by calling the ML function `extend_theory`. In this situation there is no `.thy` file, only an `.ML` file. The `.ML` file must declare an ML structure having the theory's name. See the file `ZF/ex/LList.ML` for an example. Section 9.3.2 below describes a way of linking such theories to their parents.

! Temporary files are written to the current directory, which therefore must be writable. Isabelle inherits the current directory from the operating system; you can change it within Isabelle by typing `cd "`*dir*`";`

9.3 Reloading modified theories

```
update      : unit -> unit
unlink_thy : string -> unit
```

Changing a theory on disk often makes it necessary to reload all theories descended from it. However, `use_thy` reads only one theory, even if some of the parent theories are out of date. In this case you should call `update()`.

Isabelle keeps track of all loaded theories and their files. If `use_thy` finds that the theory to be loaded has been read before, it determines whether to reload the theory as follows. First it looks for the theory's files in their previous location. If it finds them, it compares their modification times to the internal data and stops if they are equal. If the files have been moved, `use_thy` searches for them as it would for a new theory. After `use_thy` reloads a theory, it marks the children as out-of-date.

`update()` reloads all modified theories and their descendants in the correct order.

`unlink_thy` *thyname* informs Isabelle that theory *thyname* no longer exists. If you delete the theory files for *thyname* then you must execute `unlink_thy`; otherwise `update` will complain about a missing file.

9.3.1 Important note for Poly/ML users

The theory mechanism depends upon reference variables. At the end of a Poly/ML session, the contents of references are lost unless they are declared in the current database. Assignments to references in the Pure database are lost, including all information about loaded theories.

To avoid losing such information you must declare a new Readthy structure in the child database:

```
structure Readthy = ReadthyFUN (structure ThySyn = ThySyn);
Readthy.loaded_thys := !loaded_thys;
open Readthy;
```

The assignment copies into the new reference loaded_thys the contents of the original reference, which is the list of already loaded theories. You should not omit the declarations even if the parent database has no loaded theories, since they allocate new references.

9.3.2 *Pseudo theories

Any automatic reloading facility requires complete knowledge of all dependencies. Sometimes theories depend on objects created in ML files with no associated .thy file. These objects may be theories but they could also be theorems, proof procedures, etc.

Unless such dependencies are documented, update fails to reload these ML files and the system is left in a state where some objects, such as theorems, still refer to old versions of theories. This may lead to the error

```
Attempt to merge different versions of theory: T
```

Therefore there is a way to link theories and **orphaned** ML files — those without associated .thy file.

Let us assume we have an orphaned ML file named orphan.ML and a theory B that depends on orphan.ML — for example, B.ML uses theorems proved in orphan.ML. Then B.thy should mention this dependence by means of a string:

```
B = ... + "orphan" + ...
```

Strings stand for ML files rather than .thy files, and merely document additional dependencies. Thus orphan is not a theory and is not used in building the base of theory B, but orphan.ML is loaded automatically whenever B is (re)built.

The orphaned file may have its own dependencies. If orphan.ML depends on theories A_1, ..., A_n, record this by creating a **pseudo theory** in the file orphan.thy:

```
orphan = A1 + ... + An
```

The resulting theory is a dummy, but it ensures that update reloads theory *orphan* whenever it reloads one of the A_i.

For an extensive example of how this technique can be used to link over 30 theory files and load them by just two use_thy calls, consult the source files of

ZF set theory.

9.4 Basic operations on theories

9.4.1 Extracting an axiom from a theory

```
get_axiom: theory -> string -> thm
assume_ax: theory -> string -> thm
```

`get_axiom` *thy name* returns an axiom with the given *name* from *thy*, raising exception `THEORY` if none exists. Merging theories can cause several axioms to have the same name; `get_axiom` returns an arbitrary one.

`assume_ax` *thy formula* reads the *formula* using the syntax of *thy*, following the same conventions as axioms in a theory definition. You can thus pretend that *formula* is an axiom and use the resulting theorem like an axiom. Actually `assume_ax` returns an assumption; `result` complains about additional assumptions, but `uresult` does not.

For example, if *formula* is `a=b ==> b=a` then the resulting theorem has the form `?a=?b ==> ?b=?a  [!!a b. a=b ==> b=a]`

9.4.2 Building a theory

```
pure_thy        : theory
merge_theories : theory * theory -> theory
extend_theory   : theory -> string -> ··· -> theory
```

`pure_thy` contains just the types, constants, and syntax of the meta-logic. There are no axioms: meta-level inferences are carried out by ML functions.

`merge_theories` (thy_1, thy_2) merges the two theories thy_1 and thy_2. The resulting theory contains all types, constants and axioms of the constituent theories; its default sort is the union of the default sorts of the constituent theories.

`extend_theory` *thy* "*T*" ··· extends the theory *thy* with new types, constants, etc. *T* identifies the theory internally. When a theory is redeclared, say to change an incorrect axiom, bindings to the old axiom may persist. Isabelle ensures that the old and new theories are not involved in the same proof. Attempting to combine different theories having the same name *T* yields the fatal error

```
Attempt to merge different versions of theory: T
```

9.4.3 Inspecting a theory

```
print_theory   : theory -> unit
axioms_of      : theory -> (string*thm)list
parents_of     : theory -> theory list
sign_of        : theory -> Sign.sg
stamps_of_thy  : theory -> string ref list
```

These provide a simple means of viewing a theory's components. Unfortunately, there is no direct connection between a theorem and its theory.

`print_theory` *thy* prints the theory *thy* at the terminal as a set of identifiers.

`axioms_of` *thy* returns the axioms of *thy* and its ancestors.

`parents_of` *thy* returns the parents of *thy*. This list contains zero, one or two elements, depending upon whether *thy* is `pure_thy`, `extend_theory` *thy* or `merge_theories` `(`thy_1`,` thy_2`)`.

`stamps_of_thy` *thy* returns the stamps of the signature associated with *thy*.

`sign_of` *thy* returns the signature associated with *thy*. It is useful with functions like `read_instantiate_sg`, which take a signature as an argument.

9.5 Terms

Terms belong to the ML type `term`, which is a concrete datatype with six constructors: there are six kinds of term.

```
type indexname = string * int;
infix 9 $;
datatype term = Const of string * typ
              | Free  of string * typ
              | Var   of indexname * typ
              | Bound of int
              | Abs   of string * typ * term
              | op $  of term * term;
```

`Const(`a`,` T`)` is the **constant** with name a and type T. Constants include connectives like $\wedge$ and $\forall$ as well as constants like 0 and *Suc*. Other constants may be required to define a logic's concrete syntax.

`Free(`a`,` T`)` is the **free variable** with name a and type T.

`Var(`v`,` T`)` is the **scheme variable** with indexname v and type T. An **indexname** is a string paired with a non-negative index, or subscript; a term's scheme variables can be systematically renamed by incrementing their subscripts. Scheme variables are essentially free variables, but may be instantiated during unification.

`Bound` i is the **bound variable** with de Bruijn index i, which counts the number of lambdas, starting from zero, between a variable's occurrence and its binding. The representation prevents capture of variables. For more information see de Bruijn [12] or Paulson [46, page 336].

`Abs(`a`,` T`,` u`)` is the **λ-abstraction** with body u, and whose bound variable has name a and type T. The name is used only for parsing and printing; it has no logical significance.

t `$` u is the **application** of t to u.

Application is written as an infix operator to aid readability. Here is an ML pattern to recognize `FOL` formulae of the form $A \to B$, binding the subformulae to A and B:

```
Const("Trueprop",_) $ (Const("op -->",_) $ A $ B)
```

9.6 Variable binding

```
loose_bnos      : term -> int list
incr_boundvars  : int -> term -> term
abstract_over   : term*term -> term
variant_abs     : string * typ * term -> string * term
aconv           : term*term -> bool                    infix
```

These functions are all concerned with the de Bruijn representation of bound variables.

`loose_bnos` t returns the list of all dangling bound variable references. In particular, `Bound 0` is loose unless it is enclosed in an abstraction. Similarly `Bound 1` is loose unless it is enclosed in at least two abstractions; if enclosed in just one, the list will contain the number 0. A well-formed term does not contain any loose variables.

`incr_boundvars` j increases a term's dangling bound variables by the offset j. This is required when moving a subterm into a context where it is enclosed by a different number of abstractions. Bound variables with a matching abstraction are unaffected.

`abstract_over` (v, t) forms the abstraction of t over v, which may be any well-formed term. It replaces every occurrence of v by a `Bound` variable with the correct index.

`variant_abs` (a, T, u) substitutes into u, which should be the body of an abstraction. It replaces each occurrence of the outermost bound variable by a free variable. The free variable has type T and its name is a variant of a chosen to be distinct from all constants and from all variables free in u.

t `aconv` u tests whether terms t and u are α-convertible: identical up to renaming of bound variables.

- Two constants, `Free`s, or `Var`s are α-convertible if their names and types are equal. (Variables having the same name but different types are thus distinct. This confusing situation should be avoided!)
- Two bound variables are α-convertible if they have the same number.
- Two abstractions are α-convertible if their bodies are, and their bound variables have the same type.
- Two applications are α-convertible if the corresponding subterms are.

9.7 Certified terms

A term t can be **certified** under a signature to ensure that every type in t is declared in the signature and every constant in t is declared as a constant of the same type in the signature. It must be well-typed and its use of bound variables must be well-formed. Meta-rules such as `forall_elim` take certified terms as arguments.

Certified terms belong to the abstract type `cterm`. Elements of the type can only be created through the certification process. In case of error, Isabelle raises exception `TERM`.

9.7.1 Printing terms

```
     string_of_cterm :               cterm -> string
Sign.string_of_term  : Sign.sg -> term -> string
```

`string_of_cterm` *ct* displays *ct* as a string.

`Sign.string_of_term` *sign* t displays t as a string, using the syntax of *sign*.

9.7.2 Making and inspecting certified terms

```
cterm_of   : Sign.sg -> term -> cterm
read_cterm : Sign.sg -> string * typ -> cterm
rep_cterm  : cterm -> {T:typ, t:term, sign:Sign.sg, maxidx:int}
```

`cterm_of` *sign* t certifies t with respect to signature *sign*.

`read_cterm` *sign* `(`*s*`,` T`)` reads the string *s* using the syntax of *sign*, creating a certified term. The term is checked to have type T; this type also tells the parser what kind of phrase to parse.

`rep_cterm` *ct* decomposes *ct* as a record containing its type, the term itself, its signature, and the maximum subscript of its unknowns. The type and maximum subscript are computed during certification.

9.8 Types

Types belong to the ML type `typ`, which is a concrete datatype with three constructor functions. These correspond to type constructors, free type variables and schematic type variables. Types are classified by sorts, which are lists of classes. A class is represented by a string.

```
type class = string;
type sort  = class list;

datatype typ = Type  of string * typ list
             | TFree of string * sort
             | TVar  of indexname * sort;

infixr 5 -->;
fun S --> T = Type("fun",[S,T]);
```

`Type(`*a*`, `*Ts*`)` applies the **type constructor** named *a* to the type operands *Ts*. Type constructors include `fun`, the binary function space constructor, as well as nullary type constructors such as `prop`. Other type constructors may be introduced. In expressions, but not in patterns, *S*`-->`*T* is a convenient shorthand for function types.

`TFree(`*a*`, `*s*`)` is the **type variable** with name *a* and sort *s*.

`TVar(`*v*`, `*s*`)` is the **type unknown** with indexname *v* and sort *s*. Type unknowns are essentially free type variables, but may be instantiated during unification.

9.9 Certified types

Certified types, which are analogous to certified terms, have type `ctyp`.

9.9.1 Printing types

```
     string_of_ctyp :              ctyp -> string
Sign.string_of_typ  : Sign.sg -> typ -> string
```

`string_of_ctyp` *cT* displays *cT* as a string.

`Sign.string_of_typ` *sign T* displays *T* as a string, using the syntax of *sign*.

9.9.2 Making and inspecting certified types

```
ctyp_of  : Sign.sg -> typ -> ctyp
rep_ctyp : ctyp -> {T: typ, sign: Sign.sg}
```

`ctyp_of` *sign* T certifies T with respect to signature *sign*.

`rep_ctyp` cT decomposes cT as a record containing the type itself and its signature.

10. Defining Logics

This chapter explains how to define new formal systems — in particular, their concrete syntax. While Isabelle can be regarded as a theorem prover for set theory, higher-order logic or the sequent calculus, its distinguishing feature is support for the definition of new logics.

Isabelle logics are hierarchies of theories, which are described and illustrated in Sect. 3.2. That material, together with the theory files provided in the examples directories, should suffice for all simple applications. The easiest way to define a new theory is by modifying a copy of an existing theory.

This chapter documents the meta-logic syntax, mixfix declarations and pretty printing. The extended examples in Sect. 10.4 demonstrate the logical aspects of the definition of theories.

10.1 Priority grammars

A context-free grammar contains a set of **nonterminal symbols**, a set of **terminal symbols** and a set of **productions**. Productions have the form $A = \gamma$, where A is a nonterminal and γ is a string of terminals and nonterminals. One designated nonterminal is called the **start symbol**. The language defined by the grammar consists of all strings of terminals that can be derived from the start symbol by applying productions as rewrite rules.

The syntax of an Isabelle logic is specified by a **priority grammar**. Each nonterminal is decorated by an integer priority, as in $A^{(p)}$. A nonterminal $A^{(p)}$ in a derivation may be rewritten using a production $A^{(q)} = \gamma$ only if $p \leq q$. Any priority grammar can be translated into a normal context free grammar by introducing new nonterminals and productions.

Formally, a set of context free productions G induces a derivation relation $\longrightarrow_G$. Let α and β denote strings of terminal or nonterminal symbols. Then

$$\alpha\, A^{(p)}\, \beta \;\longrightarrow_G\; \alpha\, \gamma\, \beta$$

if and only if G contains some production $A^{(q)} = \gamma$ for $p \leq q$.

The following simple grammar for arithmetic expressions demonstrates how binding power and associativity of operators can be enforced by priorities.

$$\begin{array}{rcl} A^{(9)} & = & 0 \\ A^{(9)} & = & (\ A^{(0)}\ ) \\ A^{(0)} & = & A^{(0)} + A^{(1)} \\ A^{(2)} & = & A^{(3)} * A^{(2)} \\ A^{(3)} & = & -\ A^{(3)} \end{array}$$

The choice of priorities determines that - binds tighter than *, which binds tighter than +. Furthermore + associates to the left and * to the right.

For clarity, grammars obey these conventions:

- All priorities must lie between 0 and `max_pri`, which is a some fixed integer. Sometimes `max_pri` is written as ∞.
- Priority 0 on the right-hand side and priority `max_pri` on the left-hand side may be omitted.
- The production $A^{(p)} = \alpha$ is written as $A = \alpha\ (p)$; the priority of the left-hand side actually appears in a column on the far right.
- Alternatives are separated by |.
- Repetition is indicated by dots (...) in an informal but obvious way.

Using these conventions and assuming $\infty = 9$, the grammar takes the form

$$\begin{array}{rcll} A & = & 0 & \\ & | & (\ A\ ) & \\ & | & A + A^{(1)} & (0) \\ & | & A^{(3)} * A^{(2)} & (2) \\ & | & -\ A^{(3)} & (3) \end{array}$$

10.2 The Pure syntax

At the root of all object-logics lies the theory `Pure`. It contains, among many other things, the Pure syntax. An informal account of this basic syntax (types, terms and formulae) appears in Sect. 2.1. A more precise description using a priority grammar appears in Fig. 10.1. It defines the following nonterminals:

`prop` denotes terms of type `prop`. These are formulae of the meta-logic.

`aprop` denotes atomic propositions. These typically include the judgement forms of the object-logic; its definition introduces a meta-level predicate for each judgement form.

`logic` denotes terms whose type belongs to class `logic`. As the syntax is extended by new object-logics, more productions for `logic` are added automatically (see below).

prop	=	`PROP` *aprop* | `(` *prop* `)`	
	|	$logic^{(3)}$ `==` $logic^{(2)}$	(2)
	|	$logic^{(3)}$ `=?=` $logic^{(2)}$	(2)
	|	$prop^{(2)}$ `==>` $prop^{(1)}$	(1)
	|	`[|` *prop* `;` ... `;` *prop* `|]` `==>` $prop^{(1)}$	(1)
	|	`!!` *idts* `.` *prop*	(0)
logic	=	*prop* | *fun*	
aprop	=	*id* | *var* | $fun^{(\infty)}$ `(` *logic* `,` ... `,` *logic* `)`	
fun	=	*id* | *var* | `(` *fun* `)`	
	|	$fun^{(\infty)}$ `(` *logic* `,` ... `,` *logic* `)`	
	|	$fun^{(\infty)}$ `::` *type*	
	|	`%` *idts* `.` *logic*	(0)
idts	=	*idt* | $idt^{(1)}$ *idts*	
idt	=	*id* | `(` *idt* `)`	
	|	*id* `::` *type*	(0)
type	=	*tid* | *tvar* | *tid* `::` *sort* | *tvar* `::` *sort*	
	|	*id* | $type^{(\infty)}$ *id* | `(` *type* `,` ... `,` *type* `)` *id*	
	|	$type^{(1)}$ `=>` *type*	(0)
	|	`[` *type* `,` ... `,` *type* `]` `=>` *type*	(0)
	|	`(` *type* `)`	
sort	=	*id* | `{}` | `{` *id* `,` ... `,` *id* `}`	

Fig. 10.1. Meta-logic syntax

`fun` denotes terms potentially of function type.

`type` denotes types of the meta-logic.

`idts` denotes a list of identifiers, possibly constrained by types.

! In `idts`, note that `x::nat y` is parsed as `x::(nat y)`, treating `y` like a type constructor applied to `nat`. The likely result is an error message. To avoid this interpretation, use parentheses and write `(x::nat) y`.

Similarly, `x::nat y::nat` is parsed as `x::(nat y::nat)` and yields an error. The correct form is `(x::nat) (y::nat)`.

10.2.1 Logical types and default syntax

Isabelle's representation of mathematical languages is based on the simply typed λ-calculus. All logical types, namely those of class `logic`, are automatically equipped with a basic syntax of types, identifiers, variables, parentheses, λ-abstractions and applications.

More precisely, for each type constructor *ty* with arity $(\mathtt{s})c$, where c is a subclass of `logic`, several productions are added:

$$\begin{array}{rcl} ty & = & id \;\mid\; var \;\mid\; \texttt{(}\; ty\; \texttt{)} \\ & \mid & fun^{(\infty)}\; \texttt{(}\; logic\; \texttt{,}\; \ldots\; \texttt{,}\; logic\; \texttt{)} \\ & \mid & ty^{(\infty)}\; \texttt{::}\; type \\ \\ logic & = & ty \end{array}$$

10.2.2 Lexical matters

The parser does not process input strings directly. It operates on token lists provided by Isabelle's **lexer**. There are two kinds of tokens: **delimiters** and **name tokens**.

Delimiters can be regarded as reserved words of the syntax. You can add new ones when extending theories. In Fig. 10.1 they appear in typewriter font, for example `==`, `=?=` and `PROP`.

Name tokens have a predefined syntax. The lexer distinguishes four disjoint classes of names: identifiers, unknowns, type identifiers, type unknowns. They are denoted by `id`, `var`, `tid`, `tvar`, respectively. Typical examples are `x`, `?x7`, `'a`, `?'a3`. Here is the precise syntax:

$$\begin{array}{rcl} id & = & letter\; quasiletter^* \\ var & = & \texttt{?}id \;\mid\; \texttt{?}id\texttt{.}nat \\ tid & = & \texttt{'}id \\ tvar & = & \texttt{?}tid \;\mid\; \texttt{?}tid\texttt{.}nat \end{array}$$

$$
\begin{aligned}
letter &= \text{one of } \texttt{a}\ldots\texttt{z}\ \texttt{A}\ldots\texttt{Z} \\
digit &= \text{one of } \texttt{0}\ldots\texttt{9} \\
quasiletter &= letter \mid digit \mid \_ \mid {}' \\
nat &= digit^{+}
\end{aligned}
$$

A `var` or `tvar` describes an unknown, which is internally a pair of base name and index (ML type `indexname`). These components are either separated by a dot as in `?x.1` or `?x7.3` or run together as in `?x1`. The latter form is possible if the base name does not end with digits. If the index is 0, it may be dropped altogether: `?x` abbreviates both `?x0` and `?x.0`.

The lexer repeatedly takes the maximal prefix of the input string that forms a valid token. A maximal prefix that is both a delimiter and a name is treated as a delimiter. Spaces, tabs and newlines are separators; they never occur within tokens.

Delimiters need not be separated by white space. For example, if `-` is a delimiter but `--` is not, then the string `--` is treated as two consecutive occurrences of the token `-`. In contrast, ML treats `--` as a single symbolic name. The consequence of Isabelle's more liberal scheme is that the same string may be parsed in different ways after extending the syntax: after adding `--` as a delimiter, the input `--` is treated as a single token.

Although name tokens are returned from the lexer rather than the parser, it is more logical to regard them as nonterminals. Delimiters, however, are terminals; they are just syntactic sugar and contribute nothing to the abstract syntax tree.

10.2.3 *Inspecting the syntax

```
syn_of              : theory -> Syntax.syntax
Syntax.print_syntax : Syntax.syntax -> unit
Syntax.print_gram   : Syntax.syntax -> unit
Syntax.print_trans  : Syntax.syntax -> unit
```

The abstract type `Syntax.syntax` allows manipulation of syntaxes in ML. You can display values of this type by calling the following functions:

`syn_of` *thy* returns the syntax of the Isabelle theory *thy* as an ML value.

`Syntax.print_syntax` *syn* shows virtually all information contained in the syntax *syn*. The displayed output can be large. The following two functions are more selective.

`Syntax.print_gram` *syn* shows the grammar part of *syn*, namely the lexicon, roots and productions. These are discussed below.

`Syntax.print_trans` *syn* shows the translation part of *syn*, namely the constants, parse/print macros and parse/print translations.

Let us demonstrate these functions by inspecting Pure's syntax. Even that is too verbose to display in full.

```
Syntax.print_syntax (syn_of Pure.thy);
  lexicon: "!!" "%" "(" ")" "," "." "::" ";" "==" "==>" ...
  roots: logic type fun prop
  prods:
    type = tid  (1000)
    type = tvar  (1000)
    type = id  (1000)
    type = tid "::" sort[0]  => "_ofsort" (1000)
    type = tvar "::" sort[0]  => "_ofsort" (1000)
    :
  consts: "_K" "_appl" "_aprop" "_args" "_asms" "_bigimpl" ...
  parse_ast_translation: "_appl" "_bigimpl" "_bracket"
    "_idtyp" "_lambda" "_tapp" "_tappl"
  parse_rules:
  parse_translation: "!!" "_K" "_abs" "_aprop"
  print_translation: "all"
  print_rules:
  print_ast_translation: "==>" "_abs" "_idts" "fun"
```

As you can see, the output is divided into labelled sections. The grammar is represented by `lexicon`, `roots` and `prods`. The rest refers to syntactic translations and macro expansion. Here is an explanation of the various sections.

`lexicon` lists the delimiters used for lexical analysis.

`roots` lists the grammar's nonterminal symbols. You must name the desired root when calling lower level functions or specifying macros. Higher level functions usually expect a type and derive the actual root as described in Sect. 10.3.1.

`prods` lists the productions of the priority grammar. The nonterminal $A^{(n)}$ is rendered in ASCII as `A[n]`. Each delimiter is quoted. Some productions are shown with `=>` and an attached string. These strings later become the heads of parse trees; they also play a vital role when terms are printed (see Sect. 11.1).

Productions with no strings attached are called **copy productions**. Their right-hand side must have exactly one nonterminal symbol (or name token). The parser does not create a new parse tree node for copy productions, but simply returns the parse tree of the right-hand symbol.

If the right-hand side consists of a single nonterminal with no delimiters, then the copy production is called a **chain production**. Chain productions act as abbreviations: conceptually, they are removed from the grammar by adding new productions. Priority information attached to chain productions is ignored; only the dummy value -1 is displayed.

`consts`, `parse_rules`, `print_rules` relate to macros (see Sect. 11.5).

`parse_ast_translation, print_ast_translation` list sets of constants that invoke translation functions for abstract syntax trees. Section Sect. 11.1 below discusses this obscure matter.

`parse_translation, print_translation` list sets of constants that invoke translation functions for terms (see Sect. 11.6).

10.3 Mixfix declarations

When defining a theory, you declare new constants by giving their names, their type, and an optional **mixfix annotation**. Mixfix annotations allow you to extend Isabelle's basic λ-calculus syntax with readable notation. They can express any context-free priority grammar. Isabelle syntax definitions are inspired by OBJ [20]; they are more general than the priority declarations of ML and Prolog.

A mixfix annotation defines a production of the priority grammar. It describes the concrete syntax, the translation to abstract syntax, and the pretty printing. Special case annotations provide a simple means of specifying infix operators, binders and so forth.

10.3.1 Grammar productions

Let us examine the treatment of the production

$$A^{(p)} = w_0\, A_1^{(p_1)}\, w_1\, A_2^{(p_2)}\, \ldots\, A_n^{(p_n)}\, w_n.$$

Here $A_i^{(p_i)}$ is a nonterminal with priority p_i for $i = 1, \ldots, n$, while $w_0, \ldots, w_n$ are strings of terminals. In the corresponding mixfix annotation, the priorities are given separately as $[p_1, \ldots, p_n]$ and p. The nonterminal symbols are identified with types $\tau, \tau_1, \ldots, \tau_n$ respectively, and the production's effect on nonterminals is expressed as the function type

$$[\tau_1, \ldots, \tau_n] \Rightarrow \tau.$$

Finally, the template

$$w_0\ \_\ w_1\ \_\ \ldots\ \_\ w_n$$

describes the strings of terminals.

A simple type is typically declared for each nonterminal symbol. In first-order logic, type i stands for terms and o for formulae. Only the outermost type constructor is taken into account. For example, any type of the form $\sigma\mathit{list}$ stands for a list; productions may refer to the symbol `list` and will apply to lists of any type.

The symbol associated with a type is called its **root** since it may serve as the root of a parse tree. Precisely, the root of $(\tau_1, \ldots, \tau_n)\mathit{ty}$ is ty, where $\tau_1, \ldots, \tau_n$ are types and ty is a type constructor. Type infixes are a special case of this;

in particular, the root of $\tau_1 \Rightarrow \tau_2$ is `fun`. Finally, the root of a type variable is `logic`; general productions might refer to this nonterminal.

Identifying nonterminals with types allows a constant's type to specify syntax as well. We can declare the function f to have type $[\tau_1, \ldots, \tau_n] \Rightarrow \tau$ and, through a mixfix annotation, specify the layout of the function's n arguments. The constant's name, in this case f, will also serve as the label in the abstract syntax tree. There are two exceptions to this treatment of constants:

1. A production need not map directly to a logical function. In this case, you must declare a constant whose purpose is purely syntactic. By convention such constants begin with the symbol `@`, ensuring that they can never be written in formulae.
2. A copy production has no associated constant.

There is something artificial about this representation of productions, but it is convenient, particularly for simple theory extensions.

10.3.2 The general mixfix form

Here is a detailed account of mixfix declarations. Suppose the following line occurs within the `consts` section of a `.thy` file:

$$c \;::\; "\sigma" \;\; ("\mathit{template}" \;\; ps \;\; p)$$

This constant declaration and mixfix annotation are interpreted as follows:

- The string c is the name of the constant associated with the production; unless it is a valid identifier, it must be enclosed in quotes. If c is empty (given as `""`) then this is a copy production. Otherwise, parsing an instance of the phrase *template* generates the AST `("`c`"` a_1 ... a_n`)`, where a_i is the AST generated by parsing the i-th argument.
- The constant c, if non-empty, is declared to have type σ.
- The string *template* specifies the right-hand side of the production. It has the form

  $$w_0 \,\_\, w_1 \,\_\, \ldots \,\_\, w_n,$$

  where each occurrence of `_` denotes an argument position and the w_i do not contain `_`. (If you want a literal `_` in the concrete syntax, you must escape it as described below.) The w_i may consist of delimiters, spaces or pretty printing annotations (see below).
- The type σ specifies the production's nonterminal symbols (or name tokens). If *template* is of the form above then σ must be a function type with at least n argument positions, say $\sigma = [\tau_1, \ldots, \tau_n] \Rightarrow \tau$. Nonterminal symbols are derived from the types $\tau_1, \ldots, \tau_n, \tau$ as described above. Any of these may be function types; the corresponding root is then `fun`.

- The optional list *ps* may contain at most *n* integers, say `[`p_1`,` ...`,` p_m`]`, where p_i is the minimal priority required of any phrase that may appear as the i-th argument. Missing priorities default to 0.
- The integer *p* is the priority of this production. If omitted, it defaults to the maximal priority. Priorities range between 0 and `max_pri` (= 1000).

The declaration *c* `::` `"`σ`"` `("`*template*`")` specifies no priorities. The resulting production puts no priority constraints on any of its arguments and has maximal priority itself. Omitting priorities in this manner will introduce syntactic ambiguities unless the production's right-hand side is fully bracketed, as in `"if _ then _ else _ fi"`.

Omitting the mixfix annotation completely, as in *c* `::` `"`σ`"`, is sensible only if *c* is an identifier. Otherwise you will be unable to write terms involving *c*.

! Theories must sometimes declare types for purely syntactic purposes. One example is `type`, the built-in type of types. This is a 'type of all types' in the syntactic sense only. Do not declare such types under `arities` as belonging to class `logic`, for that would allow their use in arbitrary Isabelle expressions (Sect. 10.2.1).

10.3.3 Example: arithmetic expressions

This theory specification contains a `consts` section with mixfix declarations encoding the priority grammar from Sect. 10.1:

```
EXP = Pure +
types
  exp
arities
  exp :: logic
consts
  "0" :: "exp"                    ("0"      9)
  "+" :: "[exp, exp] => exp"     ("_ + _"  [0, 1] 0)
  "*" :: "[exp, exp] => exp"     ("_ * _"  [3, 2] 2)
  "-" :: "exp => exp"            ("- _"    [3] 3)
end
```

The `arities` declaration causes `exp` to be added as a new root. If you put this into a file `EXP.thy` and load it via `use_thy "EXP"`, you can run some tests:

```
val read_exp = Syntax.test_read (syn_of EXP.thy) "exp";
  val it = fn : string -> unit
read_exp "0 * 0 * 0 * 0 + 0 + 0 + 0";
  tokens: "0" "*" "0" "*" "0" "*" "0" "+" "0" "+" "0" "+" "0"
  raw: ("+" ("+" ("+" ("*" "0" ("*" "0" ("*" "0" "0"))) "0") "0") "0")
  ⋮
read_exp "0 + - 0 + 0";
  tokens: "0" "+" "-" "0" "+" "0"
  raw: ("+" ("+" "0" ("-" "0")) "0")
  ⋮
```

The output of `Syntax.test_read` includes the token list (`tokens`) and the raw AST directly derived from the parse tree, ignoring parse AST translations. The rest is tracing information provided by the macro expander (see Sect. 11.5).

Executing `Syntax.print_gram` reveals the productions derived from our mixfix declarations (lots of additional information deleted):

```
Syntax.print_gram (syn_of EXP.thy);
  exp = "0"  => "0" (9)
  exp = exp[0] "+" exp[1]  => "+" (0)
  exp = exp[3] "*" exp[2]  => "*" (2)
  exp = "-" exp[3]  => "-" (3)
```

10.3.4 The mixfix template

Let us take a closer look at the string *template* appearing in mixfix annotations. This string specifies a list of parsing and printing directives: delimiters, arguments, spaces, blocks of indentation and line breaks. These are encoded by the following character sequences:

d is a delimiter, namely a non-empty sequence of characters other than the special characters `_`, `(`, `)` and `/`. Even these characters may appear if escaped; this means preceding it with a `'` (single quote). Thus you have to write `''` if you really want a single quote. Delimiters may never contain spaces.

`_` is an argument position, which stands for a nonterminal symbol or name token.

s is a non-empty sequence of spaces for printing. This and the following specifications do not affect parsing at all.

`(`*n* opens a pretty printing block. The optional number *n* specifies how much indentation to add when a line break occurs within the block. If `(` is not followed by digits, the indentation defaults to 0.

`)` closes a pretty printing block.

`//` forces a line break.

`/`*s* allows a line break. Here *s* stands for the string of spaces (zero or more) right after the `/` character. These spaces are printed if the break is not taken.

For example, the template `"(_ +/ _)"` specifies an infix operator. There are two argument positions; the delimiter `+` is preceded by a space and followed by a space or line break; the entire phrase is a pretty printing block. Other examples appear in Fig. 11.4 below. Isabelle's pretty printer resembles the one described in Paulson [46].

10.3.5 Infixes

Infix operators associating to the left or right can be declared using `infixl` or `infixr`. Roughly speaking, the form c `::` `"`σ`"` `(infixl` p`)` abbreviates the constant declarations

```
"op c" :: "σ"   ("op c")
"op c" :: "σ"   ("(_ c/ _)" [p, p+1] p)
```

and c `::` `"`σ`"` `(infixr` p`)` abbreviates the constant declarations

```
"op c" :: "σ"   ("op c")
"op c" :: "σ"   ("(_ c/ _)" [p+1, p] p)
```

The infix operator is declared as a constant with the prefix `op`. Thus, prefixing infixes with `op` makes them behave like ordinary function symbols, as in ML. Special characters occurring in c must be escaped, as in delimiters, using a single quote.

The expanded forms above would be actually illegal in a `.thy` file because they declare the constant `"op` c`"` twice.

10.3.6 Binders

A **binder** is a variable-binding construct such as a quantifier. The constant declaration

```
c :: "σ"   (binder "Q" p)
```

introduces a constant c of type σ, which must have the form $(\tau_1 \Rightarrow \tau_2) \Rightarrow \tau_3$. Its concrete syntax is $\mathcal{Q}\ x\ .\ P$, where x is a bound variable of type τ_1, the body P has type τ_2 and the whole term has type τ_3. Special characters in $\mathcal{Q}$ must be escaped using a single quote.

The declaration is expanded internally to

```
c    :: "(τ1 => τ2) => τ3"
"Q"  :: "[idts, τ2] => τ3"   ("(3Q_./ _)" p)
```

Here `idts` is the nonterminal symbol for a list of identifiers with optional type constraints (see Fig. 10.1). The declaration also installs a parse translation for $\mathcal{Q}$ and a print translation for c to translate between the internal and external forms.

A binder of type $(\sigma \Rightarrow \tau) \Rightarrow \tau$ can be nested by giving a list of variables. The external form $\mathcal{Q}\ x_1\ x_2 \ldots x_n\ .\ P$ corresponds to the internal form

$$c(\lambda x_1\ .\ c(\lambda x_2\ .\ \ldots\ c(\lambda x_n\ .\ P) \ldots)).$$

For example, let us declare the quantifier $\forall$:

```
All :: "('a => o) => o"   (binder "ALL " 10)
```

This lets us write $\forall x\ .\ P$ as either `All(%`x`.`P`)` or `ALL` x`.`P. When printing, Isabelle prefers the latter form, but must fall back on `All(`P`)` if P is not an abstraction. Both P and `ALL` x`.`P have type o, the type of formulae, while the bound variable can be polymorphic.

10.4 Example: some minimal logics

This section presents some examples that have a simple syntax. They demonstrate how to define new object-logics from scratch.

First we must define how an object-logic syntax embedded into the meta-logic. Since all theorems must conform to the syntax for `prop` (see Fig. 10.1), that syntax has to be extended with the object-level syntax. Assume that the syntax of your object-logic defines a nonterminal symbol `o` of formulae. These formulae can now appear in axioms and theorems wherever `prop` does if you add the production

$$prop \;=\; o.$$

This is not a copy production but a coercion from formulae to propositions:

```
Base = Pure +
types
  o
arities
  o :: logic
consts
  Trueprop :: "o => prop"    ("_" 5)
end
```

The constant `Trueprop` (the name is arbitrary) acts as an invisible coercion function. Assuming this definition resides in a file `Base.thy`, you have to load it with the command `use_thy "Base"`.

One of the simplest nontrivial logics is **minimal logic** of implication. Its definition in Isabelle needs no advanced features but illustrates the overall mechanism nicely:

```
Hilbert = Base +
consts
  "-->" :: "[o, o] => o"   (infixr 10)
rules
  K     "P --> Q --> P"
  S     "(P --> Q --> R) --> (P --> Q) --> P --> R"
  MP    "[| P --> Q; P |] ==> Q"
end
```

After loading this definition from the file `Hilbert.thy`, you can start to prove theorems in the logic:

```
goal Hilbert.thy "P --> P";
  Level 0
  P --> P
   1.  P --> P
by (resolve_tac [Hilbert.MP] 1);
  Level 1
  P --> P
   1.  ?P --> P --> P
   2.  ?P
```

```
by (resolve_tac [Hilbert.MP] 1);
  Level 2
  P --> P
   1. ?P1 --> ?P --> P --> P
   2. ?P1
   3. ?P
by (resolve_tac [Hilbert.S] 1);
  Level 3
  P --> P
   1. P --> ?Q2 --> P
   2. P --> ?Q2
by (resolve_tac [Hilbert.K] 1);
  Level 4
  P --> P
   1. P --> ?Q2
by (resolve_tac [Hilbert.K] 1);
  Level 5
  P --> P
  No subgoals!
```

As we can see, this Hilbert-style formulation of minimal logic is easy to define but difficult to use. The following natural deduction formulation is better:

```
MinI = Base +
consts
  "-->" :: "[o, o] => o"   (infixr 10)
rules
  impI  "(P ==> Q) ==> P --> Q"
  impE  "[| P --> Q; P |] ==> Q"
end
```

Note, however, that although the two systems are equivalent, this fact cannot be proved within Isabelle. Axioms `S` and `K` can be derived in `MinI` (exercise!), but `impI` cannot be derived in `Hilbert`. The reason is that `impI` is only an **admissible** rule in `Hilbert`, something that can only be shown by induction over all possible proofs in `Hilbert`.

We may easily extend minimal logic with falsity:

```
MinIF = MinI +
consts
  False :: "o"
rules
  FalseE "False ==> P"
end
```

On the other hand, we may wish to introduce conjunction only:

```
MinC = Base +
consts
  "&" :: "[o, o] => o"   (infixr 30)
rules
  conjI  "[| P; Q |] ==> P & Q"
  conjE1 "P & Q ==> P"
  conjE2 "P & Q ==> Q"
end
```

And if we want to have all three connectives together, we create and load a theory file consisting of a single line:[1]

```
MinIFC = MinIF + MinC
```

Now we can prove mixed theorems like

```
goal MinIFC.thy "P & False --> Q";
by (resolve_tac [MinI.impI] 1);
by (dresolve_tac [MinC.conjE2] 1);
by (eresolve_tac [MinIF.FalseE] 1);
```

Try this as an exercise!

[1]We can combine the theories without creating a theory file using the ML declaration

```
val MinIFC_thy = merge_theories(MinIF,MinC)
```

11. Syntax Transformations

This chapter is intended for experienced Isabelle users who need to define macros or code their own translation functions. It describes the transformations between parse trees, abstract syntax trees and terms.

11.1 Abstract syntax trees

The parser, given a token list from the lexer, applies productions to yield a parse tree. By applying some internal transformations the parse tree becomes an abstract syntax tree, or AST. Macro expansion, further translations and finally type inference yields a well-typed term. The printing process is the reverse, except for some subtleties to be discussed later.

Figure 11.1 outlines the parsing and printing process. Much of the complexity is due to the macro mechanism. Using macros, you can specify most forms of concrete syntax without writing any ML code.

Abstract syntax trees are an intermediate form between the raw parse trees and the typed λ-terms. An AST is either an atom (constant or variable) or a list of *at least two* subtrees. Internally, they have type `Syntax.ast`:

```
datatype ast = Constant of string
             | Variable of string
             | Appl of ast list
```

Isabelle uses an S-expression syntax for abstract syntax trees. Constant atoms are shown as quoted strings, variable atoms as non-quoted strings and applications as a parenthesised list of subtrees. For example, the AST

```
Appl [Constant "_constrain",
      Appl [Constant "_abs", Variable "x", Variable "t"],
      Appl [Constant "fun", Variable "'a", Variable "'b"]]
```

is shown as `("_constrain" ("_abs" x t) ("fun" 'a 'b))`. Both `()` and `(f)` are illegal because they have too few subtrees.

The resemblance to Lisp's S-expressions is intentional, but there are two kinds of atomic symbols: `Constant` x and `Variable` x. Do not take the names `Constant` and `Variable` too literally; in the later translation to terms, `Variable` x may become a constant, free or bound variable, even a type constructor or class name; the actual outcome depends on the context.

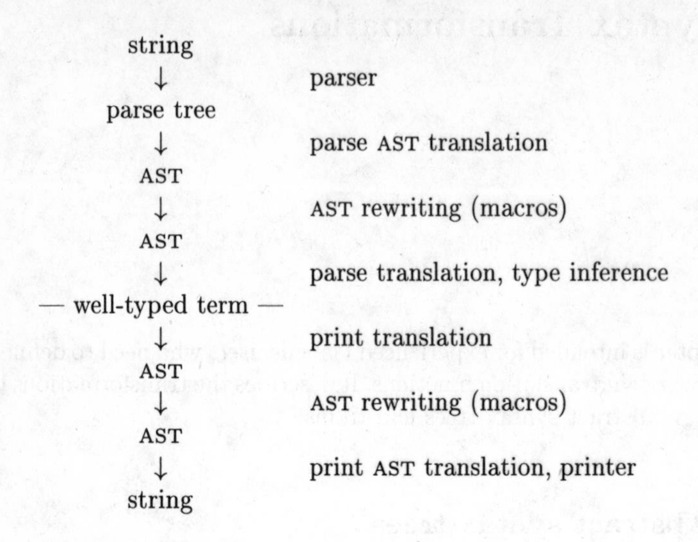

Fig. 11.1. Parsing and printing

Similarly, you can think of (f x_1 ... x_n) as the application of f to the arguments $x_1, \ldots, x_n$. But the kind of application is determined later by context; it could be a type constructor applied to types.

Forms like `(("_abs" x t) u)` are legal, but ASTs are first-order: the `"_abs"` does not bind the `x` in any way. Later at the term level, `("_abs" x t)` will become an `Abs` node and occurrences of `x` in t will be replaced by bound variables (the term constructor `Bound`).

11.2 Transforming parse trees to ASTs

The parse tree is the raw output of the parser. Translation functions, called **parse AST translations**, transform the parse tree into an abstract syntax tree.

The parse tree is constructed by nesting the right-hand sides of the productions used to recognize the input. Such parse trees are simply lists of tokens and constituent parse trees, the latter representing the nonterminals of the productions. Let us refer to the actual productions in the form displayed by `Syntax.print_syntax`.

Ignoring parse AST translations, parse trees are transformed to ASTs by stripping out delimiters and copy productions. More precisely, the mapping $[\![-]\!]$ is derived from the productions as follows:

- Name tokens: $[\![t]\!]$ = `Variable` s, where t is an `id`, `var`, `tid` or `tvar` token, and s its associated string.

input string	AST
"f"	f
"'a"	'a
"t == u"	("==" t u)
"f(x)"	("_appl" f x)
"f(x, y)"	("_appl" f ("_args" x y))
"f(x, y, z)"	("_appl" f ("_args" x ("_args" y z)))
"%x y. t"	("_lambda" ("_idts" x y) t)

Fig. 11.2. Parsing examples using the Pure syntax

input string	AST
"f(x, y, z)"	(f x y z)
"'a ty"	(ty 'a)
"('a, 'b) ty"	(ty 'a 'b)
"%x y z. t"	("_abs" x ("_abs" y ("_abs" z t)))
"%x :: 'a. t"	("_abs" ("_constrain" x 'a) t)
"[| P; Q; R |] => S"	("==>" P ("==>" Q ("==>" R S)))
"['a, 'b, 'c] => 'd"	("fun" 'a ("fun" 'b ("fun" 'c 'd)))

Fig. 11.3. Built-in parse AST translations

- Copy productions: $[\![\ldots P \ldots]\!] = [\![P]\!]$. Here ... stands for strings of delimiters, which are discarded. P stands for the single constituent that is not a delimiter; it is either a nonterminal symbol or a name token.
- 0-ary productions: $[\![\ldots \texttt{=>} c]\!] = \texttt{Constant}\ c$. Here there are no constituents other than delimiters, which are discarded.
- n-ary productions, where $n \geq 1$: delimiters are discarded and the remaining constituents $P_1, \ldots, P_n$ are built into an application whose head constant is c:

$$[\![\ldots P_1 \ldots P_n \ldots \texttt{=>} c]\!] = \texttt{Appl}\ [\texttt{Constant}\ c, [\![P_1]\!], \ldots, [\![P_n]\!]]$$

Figure 11.2 presents some simple examples, where ==, _appl, _args, and so forth name productions of the Pure syntax. These examples illustrate the need for further translations to make ASTs closer to the typed λ-calculus. The Pure syntax provides predefined parse AST translations for ordinary applications, type applications, nested abstractions, meta implications and function types. Figure 11.3 shows their effect on some representative input strings.

The names of constant heads in the AST control the translation process. The list of constants invoking parse AST translations appears in the output of `Syntax.print_syntax` under `parse_ast_translation`.

11.3 Transforming ASTs to terms

The AST, after application of macros (see Sect. 11.5), is transformed into a term. This term is probably ill-typed since type inference has not occurred yet. The term may contain type constraints consisting of applications with head `"_constrain"`; the second argument is a type encoded as a term. Type inference later introduces correct types or rejects the input.

Another set of translation functions, namely parse translations, may affect this process. If we ignore parse translations for the time being, then ASTs are transformed to terms by mapping AST constants to constants, AST variables to schematic or free variables and AST applications to applications.

More precisely, the mapping $[\![-]\!]$ is defined by

- Constants: $[\![\texttt{Constant}\ x]\!] = \texttt{Const}(x, \texttt{dummyT})$.
- Schematic variables: $[\![\texttt{Variable}\ \texttt{"?}xi\texttt{"}]\!] = \texttt{Var}((x, i), \texttt{dummyT})$, where x is the base name and i the index extracted from xi.
- Free variables: $[\![\texttt{Variable}\ x]\!] = \texttt{Free}(x, \texttt{dummyT})$.
- Function applications with n arguments:

$$[\![\texttt{Appl}\ [f, x_1, \ldots, x_n]]\!] = [\![f]\!]\ \texttt{\$}\ [\![x_1]\!]\ \texttt{\$}\ \ldots\ \texttt{\$}\ [\![x_n]\!]$$

Here `Const`, `Var`, `Free` and `$` are constructors of the datatype `term`, while `dummyT` stands for some dummy type that is ignored during type inference.

So far the outcome is still a first-order term. Abstractions and bound variables (constructors `Abs` and `Bound`) are introduced by parse translations. Such translations are attached to `"_abs"`, `"!!"` and user-defined binders.

11.4 Printing of terms

The output phase is essentially the inverse of the input phase. Terms are translated via abstract syntax trees into strings. Finally the strings are pretty printed.

Print translations (Sect. 11.6) may affect the transformation of terms into ASTs. Ignoring those, the transformation maps term constants, variables and applications to the corresponding constructs on ASTs. Abstractions are mapped to applications of the special constant `_abs`.

More precisely, the mapping $[\![-]\!]$ is defined as follows:

- $[\![\texttt{Const}(x, \tau)]\!] = \texttt{Constant}\ x$.

- $[\![\texttt{Free}(x,\tau)]\!] = \mathit{constrain}(\texttt{Variable}\ x, \tau)$.
- $[\![\texttt{Var}((x,i),\tau)]\!] = \mathit{constrain}(\texttt{Variable "?}xi\texttt{"}, \tau)$, where $?xi$ is the string representation of the `indexname` (x, i).
- For the abstraction $\lambda x :: \tau \,.\, t$, let x' be a variant of x renamed to differ from all names occurring in t, and let t' be obtained from t by replacing all bound occurrences of x by the free variable x'. This replaces corresponding occurrences of the constructor `Bound` by the term $\texttt{Free}(x', \texttt{dummyT})$:

$$[\![\texttt{Abs}(x,\tau,t)]\!] = \texttt{Appl [Constant "\_abs"}, \mathit{constrain}(\texttt{Variable}\ x',\tau), [\![t']\!]].$$

- $[\![\texttt{Bound}\ i]\!] = \texttt{Variable "B.}i\texttt{"}$. The occurrence of constructor `Bound` should never happen when printing well-typed terms; it indicates a de Bruijn index with no matching abstraction.
- Where f is not an application,

$$[\![f\ \$\ x_1\ \$\ \ldots\ \$\ x_n]\!] = \texttt{Appl}\,[[\![f]\!], [\![x_1]\!], \ldots, [\![x_n]\!]]$$

Type constraints are inserted to allow the printing of types. This is governed by the boolean variable `show_types`:

- $\mathit{constrain}(x,\tau) = x$ if $\tau = \texttt{dummyT}$ or `show_types` is set to `false`.
- $\mathit{constrain}(x,\tau) = \texttt{Appl [Constant "\_constrain"}, x, [\![\tau]\!]]$ otherwise.
  Here, $[\![\tau]\!]$ is the AST encoding of τ: type constructors go to `Constants`; type identifiers go to `Variables`; type applications go to `Appls` with the type constructor as the first element. If `show_sorts` is set to `true`, some type variables are decorated with an AST encoding of their sort.

The AST, after application of macros (see Sect. 11.5), is transformed into the final output string. The built-in **print AST translations** reverse the parse AST translations of Fig. 11.3.

For the actual printing process, the names attached to productions of the form $\ldots A_1^{(p_1)} \ldots A_n^{(p_n)} \ldots$ `=>` c play a vital role. Each AST with constant head c, namely "c" or ("c" $x_1 \ldots x_n$), is printed according to the production for c. Each argument x_i is converted to a string, and put in parentheses if its priority (p_i) requires this. The resulting strings and their syntactic sugar (denoted by ... above) are joined to make a single string.

If an application ("c" $x_1 \ldots x_m$) has more arguments than the corresponding production, it is first split into (("c" $x_1 \ldots x_n$) $x_{n+1} \ldots x_m$). Applications with too few arguments or with non-constant head or without a corresponding production are printed as $f(x_1, \ldots, x_l)$ or $(\alpha_1, \ldots, \alpha_l)ty$. An occurrence of `Variable` x is simply printed as x.

Blanks are *not* inserted automatically. If blanks are required to separate tokens, specify them in the mixfix declaration, possibly preceded by a slash (/) to allow a line break.

```
SET = Pure +
types
  i, o
arities
  i, o :: logic
consts
  Trueprop     :: "o => prop"                  ("_" 5)
  Collect      :: "[i, i => o] => i"
  "@Collect"   :: "[idt, i, o] => i"           ("(1{_:_./ _})")
  Replace      :: "[i, [i, i] => o] => i"
  "@Replace"   :: "[idt, idt, i, o] => i"      ("(1{_./ _:_, _})")
  Ball         :: "[i, i => o] => o"
  "@Ball"      :: "[idt, i, o] => o"           ("(3ALL _:_./ _)" 10)
translations
  "{x:A. P}"      == "Collect(A, %x. P)"
  "{y. x:A, Q}"   == "Replace(A, %x y. Q)"
  "ALL x:A. P"    == "Ball(A, %x. P)"
end
```

Fig. 11.4. Macro example: set theory

11.5 Macros: Syntactic rewriting

Mixfix declarations alone can handle situations where there is a direct connection between the concrete syntax and the underlying term. Sometimes we require a more elaborate concrete syntax, such as quantifiers and list notation. Isabelle's **macros** and **translation functions** can perform translations such as

```
ALL x:A.P  ⇌  Ball(A, %x.P)
[x, y, z]  ⇌  Cons(x, Cons(y, Cons(z, Nil)))
```

Translation functions (see Sect. 11.6) must be coded in ML; they are the most powerful translation mechanism but are difficult to read or write. Macros are specified by first-order rewriting systems that operate on abstract syntax trees. They are usually easy to read and write, and can express all but the most obscure translations.

Figure 11.4 defines a fragment of first-order logic and set theory.[1] Theory `SET` defines constants for set comprehension (`Collect`), replacement (`Replace`) and bounded universal quantification (`Ball`). Each of these binds some variables. Without additional syntax we should have to write $\forall x \in A . P$ as `Ball(A,%x.P)`, and similarly for the others.

The theory specifies a variable-binding syntax through additional productions that have mixfix declarations. Each non-copy production must specify some constant, which is used for building ASTs. The additional constants are

[1] This and the following theories are complete working examples, though they specify only syntax, no axioms. The file `ZF/ZF.thy` presents the full set theory definition, including many macro rules.

decorated with `@` to stress their purely syntactic purpose; they should never occur within the final well-typed terms. Furthermore, they cannot be written in formulae because they are not legal identifiers.

The translations cause the replacement of external forms by internal forms after parsing, and vice versa before printing of terms. As a specification of the set theory notation, they should be largely self-explanatory. The syntactic constants, `@Collect`, `@Replace` and `@Ball`, appear implicitly in the macro rules via their mixfix forms.

Macros can define variable-binding syntax because they operate on ASTs, which have no inbuilt notion of bound variable. The macro variables `x` and `y` have type `idt` and therefore range over identifiers, in this case bound variables. The macro variables `P` and `Q` range over formulae containing bound variable occurrences.

Other applications of the macro system can be less straightforward, and there are peculiarities. The rest of this section will describe in detail how Isabelle macros are preprocessed and applied.

11.5.1 Specifying macros

Macros are basically rewrite rules on ASTs. But unlike other macro systems found in programming languages, Isabelle's macros work in both directions. Therefore a syntax contains two lists of rewrites: one for parsing and one for printing.

The **`translations`** section specifies macros. The syntax for a macro is

$$(root)\ string \quad \left\{ \begin{array}{c} \texttt{=>} \\ \texttt{<=} \\ \texttt{==} \end{array} \right\} \quad (root)\ string$$

This specifies a parse rule (`=>`), a print rule (`<=`), or both (`==`). The two strings specify the left and right-hand sides of the macro rule. The *(root)* specification is optional; it specifies the nonterminal for parsing the *string* and if omitted defaults to `logic`. AST rewrite rules (l, r) must obey certain conditions:

- Rules must be left linear: l must not contain repeated variables.
- Rules must have constant heads, namely $l = \texttt{"}c\texttt{"}$ or $l = (\texttt{"}c\texttt{"}\ x_1 \ldots x_n)$.
- Every variable in r must also occur in l.

Macro rules may refer to any syntax from the parent theories. They may also refer to anything defined before the the `.thy` file's **`translations`** section — including any mixfix declarations.

Upon declaration, both sides of the macro rule undergo parsing and parse AST translations (see Sect. 11.1), but do not themselves undergo macro expansion. The lexer runs in a different mode that additionally accepts identifiers of the form _ *letter quasiletter** (like `_idt`, `_K`). Thus, a constant whose name starts with an underscore can appear in macro rules but not in ordinary terms.

Some atoms of the macro rule's AST are designated as constants for matching. These are all names that have been declared as classes, types or constants.

The result of this preprocessing is two lists of macro rules, each stored as a pair of ASTs. They can be viewed using `Syntax.print_syntax` (sections `parse_rules` and `print_rules`). For theory SET of Fig. 11.4 these are

```
parse_rules:
  ("@Collect" x A P)  ->  ("Collect" A ("_abs" x P))
  ("@Replace" y x A Q)  -> ("Replace" A ("_abs" x ("_abs" y Q)))
  ("@Ball" x A P)  ->  ("Ball" A ("_abs" x P))
print_rules:
  ("Collect" A ("_abs" x P))  ->  ("@Collect" x A P)
  ("Replace" A ("_abs" x ("_abs" y Q)))  ->  ("@Replace" y x A Q)
  ("Ball" A ("_abs" x P))  ->  ("@Ball" x A P)
```

! Avoid choosing variable names that have previously been used as constants, types or type classes; the `consts` section in the output of `Syntax.print_syntax` lists all such names. If a macro rule works incorrectly, inspect its internal form as shown above, recalling that constants appear as quoted strings and variables without quotes.

! If `eta_contract` is set to `true`, terms will be η-contracted *before* the AST rewriter sees them. Thus some abstraction nodes needed for print rules to match may vanish. For example, `Ball(A, %x. P(x))` contracts to `Ball(A, P)`; the print rule does not apply and the output will be `Ball(A, P)`. This problem would not occur if ML translation functions were used instead of macros (as is done for binder declarations).

! Another trap concerns type constraints. If `show_types` is set to `true`, bound variables will be decorated by their meta types at the binding place (but not at occurrences in the body). Matching with `Collect(A, %x. P)` binds `x` to something like `("_constrain" y "i")` rather than only `y`. AST rewriting will cause the constraint to appear in the external form, say `{y::i:A::i. P::o}`.

To allow such constraints to be re-read, your syntax should specify bound variables using the nonterminal `idt`. This is the case in our example. Choosing `id` instead of `idt` is a common error, especially since it appears in former versions of most of Isabelle's object-logics.

11.5.2 Applying rules

As a term is being parsed or printed, an AST is generated as an intermediate form (recall Fig. 11.1). The AST is normalised by applying macro rules in the manner of a traditional term rewriting system. We first examine how a single rule is applied.

Let t be the abstract syntax tree to be normalised and (l, r) some translation rule. A subtree u of t is a **redex** if it is an instance of l; in this case l is said to **match** u. A redex matched by l may be replaced by the corresponding instance of r, thus **rewriting** the AST t. Matching requires some notion of **place-holders** that may occur in rule patterns but not in ordinary ASTs; `Variable` atoms serve this purpose.

The matching of the object u by the pattern l is performed as follows:

- Every constant matches itself.
- `Variable` x in the object matches `Constant` x in the pattern. This point is discussed further below.
- Every AST in the object matches `Variable` x in the pattern, binding x to u.
- One application matches another if they have the same number of subtrees and corresponding subtrees match.
- In every other case, matching fails. In particular, `Constant` x can only match itself.

A successful match yields a substitution that is applied to r, generating the instance that replaces u.

The second case above may look odd. This is where `Variable`s of non-rule ASTs behave like `Constant`s. Recall that ASTs are not far removed from parse trees; at this level it is not yet known which identifiers will become constants, bounds, frees, types or classes. As Sect. 11.1 describes, former parse tree heads appear in ASTs as `Constant`s, while the name tokens `id`, `var`, `tid` and `tvar` become `Variable`s. On the other hand, when ASTs generated from terms for printing, all constants and type constructors become `Constant`s; see Sect. 11.1. Thus ASTs may contain a messy mixture of `Variable`s and `Constant`s. This is insignificant at macro level because matching treats them alike.

Because of this behaviour, different kinds of atoms with the same name are indistinguishable, which may make some rules prone to misbehaviour. Example:

```
types
  Nil
consts
  Nil     :: "'a list"
  "[]"    :: "'a list"     ("[]")
translations
  "[]"    == "Nil"
```

The term `Nil` will be printed as `[]`, just as expected. The term `%Nil.t` will be printed as `%[].t`, which might not be expected! How is the type `Nil` printed?

Normalizing an AST involves repeatedly applying macro rules until none are applicable. Macro rules are chosen in the order that they appear in the `translations` section. You can watch the normalization of ASTs during parsing and printing by setting `Syntax.trace_norm_ast` to `true`. Alternatively, use `Syntax.test_read`. The information displayed when tracing includes the AST before normalization (`pre`), redexes with results (`rewrote`), the normal form finally reached (`post`) and some statistics (`normalize`). If tracing is off,

Syntax.stat_norm_ast can be set to true in order to enable printing of the normal form and statistics only.

11.5.3 Example: the syntax of finite sets

This example demonstrates the use of recursive macros to implement a convenient notation for finite sets.

```
FINSET = SET +
types
  is
consts
  ""          :: "i => is"                ("_")
  "@Enum"     :: "[i, is] => is"          ("_,/ _")
  empty       :: "i"                      ("{}")
  insert      :: "[i, i] => i"
  "@Finset"   :: "is => i"                ("{(_)}")
translations
  "{x, xs}"    == "insert(x, {xs})"
  "{x}"        == "insert(x, {})"
end
```

Finite sets are internally built up by empty and insert. The declarations above specify {x, y, z} as the external representation of

```
insert(x, insert(y, insert(z, empty)))
```

The nonterminal symbol is stands for one or more objects of type i separated by commas. The mixfix declaration "_,/ _" allows a line break after the comma for pretty printing; if no line break is required then a space is printed instead.

The nonterminal is declared as the type is, but with no arities declaration. Hence is is not a logical type and no default productions are added. If we had needed enumerations of the nonterminal logic, which would include all the logical types, we could have used the predefined nonterminal symbol args and skipped this part altogether. The nonterminal is can later be reused for other enumerations of type i like lists or tuples.

Next follows empty, which is already equipped with its syntax {}, and insert without concrete syntax. The syntactic constant @Finset provides concrete syntax for enumerations of i enclosed in curly braces. Remember that a pair of parentheses, as in "{(_)}", specifies a block of indentation for pretty printing.

The translations may look strange at first. Macro rules are best understood in their internal forms:

```
parse_rules:
  ("@Finset" ("@Enum" x xs))  ->  ("insert" x ("@Finset" xs))
  ("@Finset" x)  ->  ("insert" x "empty")
print_rules:
  ("insert" x ("@Finset" xs))  ->  ("@Finset" ("@Enum" x xs))
  ("insert" x "empty")  ->  ("@Finset" x)
```

This shows that {x, xs} indeed matches any set enumeration of at least two elements, binding the first to x and the rest to xs. Likewise, {xs} and {x} represent any set enumeration. The parse rules only work in the order given.

! The AST rewriter cannot distinguish constants from variables and looks only for names of atoms. Thus the names of Constants occurring in the (internal) left-hand side of translation rules should be regarded as reserved words. Choose non-identifiers like @Finset or sufficiently long and strange names. If a bound variable's name gets rewritten, the result will be incorrect; for example, the term

```
%empty insert. insert(x, empty)
```

is printed as `%empty insert. {x}`.

11.5.4 Example: a parse macro for dependent types

As stated earlier, a macro rule may not introduce new Variables on the right-hand side. Something like "K(B)" => "%x. B" is illegal; if allowed, it could cause variable capture. In such cases you usually must fall back on translation functions. But a trick can make things readable in some cases: *calling translation functions by parse macros*:

```
PROD = FINSET +
consts
  Pi            :: "[i, i => i] => i"
  "@PROD"       :: "[idt, i, i] => i"     ("(3PROD _:_./ _)" 10)
  "@->"         :: "[i, i] => i"          ("(_ ->/ _)" [51, 50] 50)

translations
  "PROD x:A. B" => "Pi(A, %x. B)"
  "A -> B"      => "Pi(A, _K(B))"
end
ML
  val print_translation = [("Pi", dependent_tr' ("@PROD", "@->"))];
```

Here Pi is an internal constant for constructing general products. Two external forms exist: the general case PROD x:A.B and the function space A -> B, which abbreviates Pi(A, %x.B) when B does not depend on x.

The second parse macro introduces _K(B), which later becomes %x.B due to a parse translation associated with _K. The order of the parse rules is critical. Unfortunately there is no such trick for printing, so we have to add a ML section for the print translation dependent_tr'.

Recall that identifiers with a leading _ are allowed in translation rules, but not in ordinary terms. Thus we can create ASTs containing names that are not directly expressible.

The parse translation for _K is already installed in Pure, and dependent_tr' is exported by the syntax module for public use. See Sect. 11.6 below for more of the arcane lore of translation functions.

11.6 Translation functions

This section describes the translation function mechanism. By writing ML functions, you can do almost everything with terms or ASTs during parsing and printing. The logic LK is a good example of sophisticated transformations between internal and external representations of sequents; here, macros would be useless.

A full understanding of translations requires some familiarity with Isabelle's internals, especially the datatypes `term`, `typ`, `Syntax.ast` and the encodings of types and terms as such at the various stages of the parsing or printing process. Most users should never need to use translation functions.

11.6.1 Declaring translation functions

There are four kinds of translation functions. Each such function is associated with a name, which triggers calls to it. Such names can be constants (logical or syntactic) or type constructors.

`Syntax.print_syntax` displays the sets of names associated with the translation functions of a `Syntax.syntax` under `parse_ast_translation`, `parse_translation`, `print_translation` and `print_ast_translation`. You can add new ones via the `ML` section of a `.thy` file. There may never be more than one function of the same kind per name. Conceptually, the `ML` section should appear between `consts` and `translations`; newly installed translation functions are already effective when macros and logical rules are parsed.

The `ML` section is copied verbatim into the ML file generated from a `.thy` file. Definitions made here are accessible as components of an ML structure; to make some definitions private, use an ML `local` declaration. The `ML` section may install translation functions by declaring any of the following identifiers:

```
val parse_ast_translation : (string * (ast list -> ast)) list
val print_ast_translation : (string * (ast list -> ast)) list
val parse_translation     : (string * (term list -> term)) list
val print_translation     : (string * (term list -> term)) list
```

11.6.2 The translation strategy

All four kinds of translation functions are treated similarly. They are called during the transformations between parse trees, ASTs and terms (recall Fig. 11.1). Whenever a combination of the form ("c" $x_1 \ldots x_n$) is encountered, and a translation function f of appropriate kind exists for c, the result is computed by the ML function call f `[`$x_1, \ldots, x_n$`]`.

For AST translations, the arguments $x_1, \ldots, x_n$ are ASTs. A combination has the form `Constant` c or `Appl`[`Constant` $c, x_1, \ldots, x_n$]. For term translations, the arguments are terms and a combination has the form `Const`(c, τ)

or Const(c, τ) \$ x_1 \$ ... \$ x_n. Terms allow more sophisticated transformations than ASTs do, typically involving abstractions and bound variables.

Regardless of whether they act on terms or ASTs, parse translations differ from print translations fundamentally:

Parse translations are applied bottom-up. The arguments are already in translated form. The translations must not fail; exceptions trigger an error message.

Print translations are applied top-down. They are supplied with arguments that are partly still in internal form. The result again undergoes translation; therefore a print translation should not introduce as head the very constant that invoked it. The function may raise exception `Match` to indicate failure; in this event it has no effect.

Only constant atoms — constructor `Constant` for ASTs and `Const` for terms — can invoke translation functions. This causes another difference between parsing and printing.

Parsing starts with a string and the constants are not yet identified. Only parse tree heads create `Constants` in the resulting AST, as described in Sect. 11.2. Macros and parse AST translations may introduce further `Constants`. When the final AST is converted to a term, all `Constants` become `Consts`, as described in Sect. 11.3.

Printing starts with a well-typed term and all the constants are known. So all logical constants and type constructors may invoke print translations. These, and macros, may introduce further constants.

11.6.3 Example: a print translation for dependent types

Let us continue the dependent type example (page 149) by examining the parse translation for `_K` and the print translation `dependent_tr'`, which are both built-in. By convention, parse translations have names ending with `_tr` and print translations have names ending with `_tr'`. Search for such names in the Isabelle sources to locate more examples.

Here is the parse translation for `_K`:

```
fun k_tr [t] = Abs ("x", dummyT, incr_boundvars 1 t)
  | k_tr ts = raise TERM("k_tr",ts);
```

If `k_tr` is called with exactly one argument t, it creates a new `Abs` node with a body derived from t. Since terms given to parse translations are not yet typed, the type of the bound variable in the new `Abs` is simply `dummyT`. The function increments all `Bound` nodes referring to outer abstractions by calling `incr_boundvars`, a basic term manipulation function defined in `Pure/term.ML`.

Here is the print translation for dependent types:

```
fun dependent_tr' (q,r) (A :: Abs (x, T, B) :: ts) =
      if 0 mem (loose_bnos B) then
        let val (x', B') = variant_abs (x, dummyT, B);
        in list_comb (Const (q, dummyT) $ Free (x', T) $ A $ B', ts)
        end
      else list_comb (Const (r, dummyT) $ A $ B, ts)
  | dependent_tr' _ _ = raise Match;
```

The argument `(q,r)` is supplied to the curried function `dependent_tr'` by a partial application during its installation. We can set up print translations for both `Pi` and `Sigma` by including

```
val print_translation =
  [("Pi",    dependent_tr' ("@PROD", "@->")),
   ("Sigma", dependent_tr' ("@SUM", "@*"))];
```

within the `ML` section. The first of these transforms `Pi`$(A,$ `Abs`$(x, T, B))$ into `@PROD`(x', A, B') or `@->`(A, B), choosing the latter form if B does not depend on x. It checks this using `loose_bnos`, yet another function from `Pure/term.ML`. Note that x' is a version of x renamed away from all names in B, and B' is the body B with `Bound` nodes referring to the `Abs` node replaced by `Free`$(x',$ `dummyT`$)$.

We must be careful with types here. While types of `Consts` are ignored, type constraints may be printed for some `Frees` and `Vars` if `show_types` is set to `true`. Variables of type `dummyT` are never printed with constraint, though. The line

```
let val (x', B') = variant_abs (x, dummyT, B);
```

replaces bound variable occurrences in B by the free variable x' with type `dummyT`. Only the binding occurrence of x' is given the correct type `T`, so this is the only place where a type constraint might appear.

12. Substitution Tactics

Replacing equals by equals is a basic form of reasoning. Isabelle supports several kinds of equality reasoning. **Substitution** means replacing free occurrences of t by u in a subgoal. This is easily done, given an equality $t = u$, provided the logic possesses the appropriate rule. The tactic `hyp_subst_tac` performs substitution even in the assumptions. But it works via object-level implication, and therefore must be specially set up for each suitable object-logic.

Substitution should not be confused with object-level **rewriting**. Given equalities of the form $t = u$, rewriting replaces instances of t by corresponding instances of u, and continues until it reaches a normal form. Substitution handles 'one-off' replacements by particular equalities while rewriting handles general equations. Chapter 13 discusses Isabelle's rewriting tactics.

12.1 Substitution rules

Many logics include a substitution rule of the form

$$[\![?a = ?b; ?P(?a)]\!] \Longrightarrow ?P(?b) \qquad (subst)$$

In backward proof, this may seem difficult to use: the conclusion $?P(?b)$ admits far too many unifiers. But, if the theorem `eqth` asserts $t = u$, then `eqth RS subst` is the derived rule

$$?P(t) \Longrightarrow ?P(u).$$

Provided u is not an unknown, resolution with this rule is well-behaved.[1] To replace u by t in subgoal i, use

```
resolve_tac [eqth RS subst] i.
```

To replace t by u in subgoal i, use

```
resolve_tac [eqth RS ssubst] i,
```

where `ssubst` is the 'swapped' substitution rule

$$[\![?a = ?b; ?P(?b)]\!] \Longrightarrow ?P(?a). \qquad (ssubst)$$

[1] Unifying $?P(u)$ with a formula Q expresses Q in terms of its dependence upon u. There are still 2^k unifiers, if Q has k occurrences of u, but Isabelle ensures that the first unifier includes all the occurrences.

If sym denotes the symmetry rule $?a = ?b \Longrightarrow ?b = ?a$, then ssubst is just sym RS subst. Many logics with equality include the rules subst and ssubst, as well as refl, sym and trans (for the usual equality laws). Examples include FOL and HOL, but not CTT (Constructive Type Theory).

Elim-resolution is well-behaved with assumptions of the form $t = u$. To replace u by t or t by u in subgoal i, use

eresolve_tac [subst] i or eresolve_tac [ssubst] i.

12.2 Substitution in the hypotheses

Substitution rules, like other rules of natural deduction, do not affect the assumptions. This can be inconvenient. Consider proving the subgoal

$$[\![c = a; c = b]\!] \Longrightarrow a = b.$$

Calling eresolve_tac [ssubst] i simply discards the assumption $c = a$, since c does not occur in $a = b$. Of course, we can work out a solution. First apply eresolve_tac [subst] i, replacing a by c:

$$[\![c = b]\!] \Longrightarrow c = b$$

Equality reasoning can be difficult, but this trivial proof requires nothing more sophisticated than substitution in the assumptions. Object-logics that include the rule (*subst*) provide tactics for this purpose:

```
hyp_subst_tac       : int -> tactic
bound_hyp_subst_tac : int -> tactic
```

hyp_subst_tac i selects an equality assumption of the form $t = u$ or $u = t$, where t is a free variable or parameter. Deleting this assumption, it replaces t by u throughout subgoal i, including the other assumptions.

bound_hyp_subst_tac i is similar but only substitutes for parameters (bound variables). Uses for this are discussed below.

The term being replaced must be a free variable or parameter. Substitution for constants is usually unhelpful, since they may appear in other theorems. For instance, the best way to use the assumption $0 = 1$ is to contradict a theorem that states $0 \neq 1$, rather than to replace 0 by 1 in the subgoal!

Substitution for free variables is also unhelpful if they appear in the premises of a rule being derived — the substitution affects object-level assumptions, not meta-level assumptions. For instance, replacing a by b could make the premise $P(a)$ worthless. To avoid this problem, use bound_hyp_subst_tac; alternatively, call cut_facts_tac to insert the atomic premises as object-level assumptions.

12.3 Setting up hyp_subst_tac

Many Isabelle object-logics, such as FOL, HOL and their descendants, come with hyp_subst_tac already defined. A few others, such as CTT, do not support this tactic because they lack the rule (*subst*). When defining a new logic that includes a substitution rule and implication, you must set up hyp_subst_tac yourself. It is packaged as the ML functor HypsubstFun, which takes the argument signature HYPSUBST_DATA:

```
signature HYPSUBST_DATA =
  sig
  val subst       : thm
  val sym         : thm
  val rev_cut_eq  : thm
  val imp_intr    : thm
  val rev_mp      : thm
  val dest_eq     : term -> term*term
  end;
```

Thus, the functor requires the following items:

subst should be the substitution rule $[\![?a = ?b;\ ?P(?a)]\!] \Longrightarrow ?P(?b)$.

sym should be the symmetry rule $?a = ?b \Longrightarrow ?b = ?a$.

rev_cut_eq should have the form $[\![?a = ?b;\ ?a = ?b \Longrightarrow ?R]\!] \Longrightarrow ?R$.

imp_intr should be the implies introduction rule $(?P \Longrightarrow ?Q) \Longrightarrow ?P \to ?Q$.

rev_mp should be the 'reversed' implies elimination rule $[\![?P;\ ?P \to ?Q]\!] \Longrightarrow ?Q$.

dest_eq should return the pair (t, u) when applied to the ML term that represents $t = u$. For other terms, it should raise exception Match.

The functor resides in file Provers/hypsubst.ML in the Isabelle distribution directory. It is not sensitive to the precise formalization of the object-logic. It is not concerned with the names of the equality and implication symbols, or the types of formula and terms. Coding the function dest_eq requires knowledge of Isabelle's representation of terms. For FOL it is defined by

```
fun dest_eq (Const("Trueprop",_) $ (Const("op =",_)$t$u)) = (t,u)
```

Here Trueprop is the coercion from type o to type *prop*, while op = is the internal name of the infix operator =. Pattern-matching expresses the function concisely, using wildcards (_) for the types.

Here is how hyp_subst_tac works. Given a subgoal of the form

$$[\![P_1; \cdots; t = u; \cdots; P_n]\!] \Longrightarrow Q,$$

it locates a suitable equality assumption and moves it to the last position using elim-resolution on rev_cut_eq (possibly re-orienting it using sym):

$$[\![P_1; \cdots; P_n; t = u]\!] \Longrightarrow Q$$

Using n calls of `eresolve_tac [rev_mp]`, it creates the subgoal

$$[\![t = u]\!] \Longrightarrow P_1 \rightarrow \cdots \rightarrow P_n \rightarrow Q$$

By `eresolve_tac [ssubst]`, it replaces t by u throughout:

$$P'_1 \rightarrow \cdots \rightarrow P'_n \rightarrow Q'$$

Finally, using n calls of `resolve_tac [imp_intr]`, it restores P'_1, ..., P'_n as assumptions:

$$[\![P'_n; \cdots; P'_1]\!] \Longrightarrow Q'$$

13. Simplification

This chapter describes Isabelle's generic simplification package, which provides a suite of simplification tactics. It performs conditional and unconditional rewriting and uses contextual information ('local assumptions'). It provides a few general hooks, which can provide automatic case splits during rewriting, for example. The simplifier is set up for many of Isabelle's logics: FOL, ZF, LCF and HOL.

13.1 Simplification sets

The simplification tactics are controlled by **simpsets**. These consist of five components: rewrite rules, congruence rules, the subgoaler, the solver and the looper. The simplifier should be set up with sensible defaults so that most simplifier calls specify only rewrite rules. Experienced users can exploit the other components to streamline proofs.

13.1.1 Rewrite rules

Rewrite rules are theorems expressing some form of equality:

$$\begin{aligned} Suc(?m) + ?n &= ?m + Suc(?n) \\ ?P \land ?P &\leftrightarrow ?P \\ ?A \bigcup ?B &\equiv \{x \,.\, x \in ?A \lor x \in ?B\} \end{aligned}$$

Conditional rewrites such as $?m < ?n \Longrightarrow ?m/?n = 0$ are permitted; the conditions can be arbitrary terms. The infix operation `addsimps` adds new rewrite rules, while `delsimps` deletes rewrite rules from a simpset.

Internally, all rewrite rules are translated into meta-equalities, theorems with conclusion $lhs \equiv rhs$. Each simpset contains a function for extracting equalities from arbitrary theorems. For example, $\neg(?x \in \{\})$ could be turned into $?x \in \{\} \equiv False$. This function can be installed using `setmksimps` but only the definer of a logic should need to do this; see Sect. 13.5.2. The function processes theorems added by `addsimps` as well as local assumptions.

! The left-hand side of a rewrite rule must look like a first-order term: none of its unknowns should have arguments. Hence $?i + (?j + ?k) = (?i + ?j) + ?k$ is acceptable. Even $\neg(\forall x . ?P(x)) \leftrightarrow (\exists x . \neg ?P(x))$ is acceptable because its left-hand side is $\neg(All(?P))$ after η-contraction. But $?f(?x) \in \mathtt{range}(?f) = True$ is not acceptable. However, you can replace the offending subterms by adding new variables and conditions: $?y = ?f(?x) \Longrightarrow ?y \in \mathtt{range}(?f) = True$ is acceptable as a conditional rewrite rule.

13.1.2 *Congruence rules

Congruence rules are meta-equalities of the form

$$[\![\ldots]\!] \Longrightarrow f(?x_1, \ldots, ?x_n) \equiv f(?y_1, \ldots, ?y_n).$$

This governs the simplification of the arguments of f. For example, some arguments can be simplified under additional assumptions:

$$[\![?P_1 \leftrightarrow ?Q_1;\ ?Q_1 \Longrightarrow ?P_2 \leftrightarrow ?Q_2]\!] \Longrightarrow (?P_1 \rightarrow ?P_2) \equiv (?Q_1 \rightarrow ?Q_2)$$

Given this rule, the simplifier assumes Q_1 and extracts rewrite rules from it when simplifying P_2. Such local assumptions are effective for rewriting formulae such as $x = 0 \rightarrow y + x = y$. The congruence rule for bounded quantifiers can also supply contextual information:

$$\begin{array}{c}[\![?A = ?B;\ \bigwedge x . x \in ?B \Longrightarrow ?P(x) = ?Q(x)]\!] \Longrightarrow \\ (\forall x \in ?A . ?P(x)) = (\forall x \in ?B . ?Q(x))\end{array}$$

The congruence rule for conditional expressions can supply contextual information for simplifying the arms:

$$[\![?p = ?q;\ ?q \Longrightarrow ?a = ?c;\ \neg ?q \Longrightarrow ?b = ?d]\!] \Longrightarrow \mathit{if}(?p, ?a, ?b) \equiv \mathit{if}(?q, ?c, ?d)$$

A congruence rule can also suppress simplification of certain arguments. Here is an alternative congruence rule for conditional expressions:

$$?p = ?q \Longrightarrow \mathit{if}(?p, ?a, ?b) \equiv \mathit{if}(?q, ?a, ?b)$$

Only the first argument is simplified; the others remain unchanged. This can make simplification much faster, but may require an extra case split to prove the goal.

Congruence rules are added using **addeqcongs**. Their conclusion must be a meta-equality, as in the examples above. It is more natural to derive the rules with object-logic equality, for example

$$[\![?P_1 \leftrightarrow ?Q_1;\ ?Q_1 \Longrightarrow ?P_2 \leftrightarrow ?Q_2]\!] \Longrightarrow (?P_1 \rightarrow ?P_2) \leftrightarrow (?Q_1 \rightarrow ?Q_2),$$

Each object-logic should define an operator called **addcongs** that expects object-equalities and translates them into meta-equalities.

13.1.3 *The subgoaler

The subgoaler is the tactic used to solve subgoals arising out of conditional rewrite rules or congruence rules. The default should be simplification itself. Occasionally this strategy needs to be changed. For example, if the premise of a conditional rule is an instance of its conclusion, as in $Suc(?m) < ?n \Longrightarrow ?m < ?n$, the default strategy could loop.

The subgoaler can be set explicitly with `setsubgoaler`. For example, the subgoaler

```
fun subgoal_tac ss = assume_tac ORELSE'
                     resolve_tac (prems_of_ss ss) ORELSE'
                     asm_simp_tac ss;
```

tries to solve the subgoal by assumption or with one of the premises, calling simplification only if that fails; here `prems_of_ss` extracts the current premises from a simpset.

13.1.4 *The solver

The solver is a tactic that attempts to solve a subgoal after simplification. Typically it just proves trivial subgoals such as `True` and $t = t$. It could use sophisticated means such as `fast_tac`, though that could make simplification expensive. The solver is set using `setsolver`.

Rewriting does not instantiate unknowns. For example, rewriting cannot prove $a \in ?A$ since this requires instantiating $?A$. The solver, however, is an arbitrary tactic and may instantiate unknowns as it pleases. This is the only way the simplifier can handle a conditional rewrite rule whose condition contains extra variables.

The tactic is presented with the full goal, including the asssumptions. Hence it can use those assumptions (say by calling `assume_tac`) even inside `simp_tac`, which otherwise does not use assumptions. The solver is also supplied a list of theorems, namely assumptions that hold in the local context.

The subgoaler is also used to solve the premises of congruence rules, which are usually of the form $s = ?x$, where s needs to be simplified and $?x$ needs to be instantiated with the result. Hence the subgoaler should call the simplifier at some point. The simplifier will then call the solver, which must therefore be prepared to solve goals of the form $t = ?x$, usually by reflexivity. In particular, reflexivity should be tried before any of the fancy tactics like `fast_tac`.

It may even happen that, due to simplification, the subgoal is no longer an equality. For example $False \leftrightarrow ?Q$ could be rewritten to $\neg ?Q$. To cover this case, the solver could try resolving with the theorem $\neg False$.

! If the simplifier aborts with the message `Failed congruence proof!`, then the subgoaler or solver has failed to prove a premise of a congruence rule. This should never occur and indicates that the subgoaler or solver is faulty.

```
infix addsimps addeqcongs delsimps
      setsubgoaler setsolver setloop setmksimps;

signature SIMPLIFIER =
sig
  type simpset
  val simp_tac:           simpset -> int -> tactic
  val asm_simp_tac:       simpset -> int -> tactic
  val asm_full_simp_tac: simpset -> int -> tactic
  val addeqcongs:    simpset * thm list -> simpset
  val addsimps:      simpset * thm list -> simpset
  val delsimps:      simpset * thm list -> simpset
  val empty_ss:      simpset
  val merge_ss:      simpset * simpset -> simpset
  val setsubgoaler: simpset * (simpset -> int -> tactic) -> simpset
  val setsolver:     simpset * (thm list -> int -> tactic) -> simpset
  val setloop:       simpset * (int -> tactic) -> simpset
  val setmksimps:    simpset * (thm -> thm list) -> simpset
  val prems_of_ss:   simpset -> thm list
  val rep_ss:        simpset -> {simps: thm list, congs: thm list}
end;
```

Fig. 13.1. The simplifier primitives

13.1.5 *The looper

The looper is a tactic that is applied after simplification, in case the solver failed to solve the simplified goal. If the looper succeeds, the simplification process is started all over again. Each of the subgoals generated by the looper is attacked in turn, in reverse order. A typical looper is case splitting: the expansion of a conditional. Another possibility is to apply an elimination rule on the assumptions. More adventurous loopers could start an induction. The looper is set with `setloop`.

13.2 The simplification tactics

The actual simplification work is performed by the following tactics. The rewriting strategy is strictly bottom up, except for congruence rules, which are applied while descending into a term. Conditions in conditional rewrite rules are solved recursively before the rewrite rule is applied.

There are three basic simplification tactics:

`simp_tac` *ss* *i* simplifies subgoal *i* using the rules in *ss*. It may solve the subgoal completely if it has become trivial, using the solver.

`asm_simp_tac` is like `simp_tac`, but extracts additional rewrite rules from the assumptions.

`asm_full_simp_tac` is like `asm_simp_tac`, but also simplifies the assumptions one by one. Working from left to right, it uses each assumption in the simplification of those following.

Using the simplifier effectively may take a bit of experimentation. The tactics can be traced with the ML command `trace_simp := true`. To remind yourself of what is in a simpset, use the function `rep_ss` to return its simplification and congruence rules.

13.3 Examples using the simplifier

Assume we are working within `FOL` and that

`Nat.thy` is a theory including the constants 0, *Suc* and +,

`add_0` is the rewrite rule $0 + ?n = ?n$,

`add_Suc` is the rewrite rule $Suc(?m) + ?n = Suc(?m + ?n)$,

`induct` is the induction rule $[\![?P(0); \bigwedge x \,.\, ?P(x) \Longrightarrow ?P(Suc(x))]\!] \Longrightarrow ?P(?n)$.

`FOL_ss` is a basic simpset for `FOL`.[1]

We create a simpset for natural numbers by extending `FOL_ss`:

```
val add_ss = FOL_ss addsimps [add_0, add_Suc];
```

13.3.1 A trivial example

Proofs by induction typically involve simplification. Here is a proof that 0 is a right identity:

```
goal Nat.thy "m+0 = m";
  Level 0
  m + 0 = m
   1. m + 0 = m
```

The first step is to perform induction on the variable m. This returns a base case and inductive step as two subgoals:

```
by (res_inst_tac [("n","m")] induct 1);
  Level 1
  m + 0 = m
   1. 0 + 0 = 0
   2. !!x. x + 0 = x ==> Suc(x) + 0 = Suc(x)
```

[1] These examples reside on the file `FOL/ex/Nat.ML`.

Simplification solves the first subgoal trivially:

```
by (simp_tac add_ss 1);
  Level 2
  m + 0 = m
   1. !!x. x + 0 = x ==> Suc(x) + 0 = Suc(x)
```

The remaining subgoal requires `asm_simp_tac` in order to use the induction hypothesis as a rewrite rule:

```
by (asm_simp_tac add_ss 1);
  Level 3
  m + 0 = m
  No subgoals!
```

13.3.2 An example of tracing

Let us prove a similar result involving more complex terms. The two equations together can be used to prove that addition is commutative.

```
goal Nat.thy "m+Suc(n) = Suc(m+n)";
  Level 0
  m + Suc(n) = Suc(m + n)
   1. m + Suc(n) = Suc(m + n)
```

We again perform induction on m and get two subgoals:

```
by (res_inst_tac [("n","m")] induct 1);
  Level 1
  m + Suc(n) = Suc(m + n)
   1. 0 + Suc(n) = Suc(0 + n)
   2. !!x. x + Suc(n) = Suc(x + n) ==>
               Suc(x) + Suc(n) = Suc(Suc(x) + n)
```

Simplification solves the first subgoal, this time rewriting two occurrences of 0:

```
by (simp_tac add_ss 1);
  Level 2
  m + Suc(n) = Suc(m + n)
   1. !!x. x + Suc(n) = Suc(x + n) ==>
               Suc(x) + Suc(n) = Suc(Suc(x) + n)
```

Switching tracing on illustrates how the simplifier solves the remaining subgoal:

```
trace_simp := true;
by (asm_simp_tac add_ss 1);
  Rewriting:
  Suc(x) + Suc(n) == Suc(x + Suc(n))

  Rewriting:
  x + Suc(n) == Suc(x + n)

  Rewriting:
  Suc(x) + n == Suc(x + n)

  Rewriting:
  Suc(Suc(x + n)) = Suc(Suc(x + n)) == True
```

```
Level 3
m + Suc(n) = Suc(m + n)
No subgoals!
```

Many variations are possible. At Level 1 (in either example) we could have solved both subgoals at once using the tactical `ALLGOALS`:

```
by (ALLGOALS (asm_simp_tac add_ss));
  Level 2
  m + Suc(n) = Suc(m + n)
  No subgoals!
```

13.3.3 Free variables and simplification

Here is a conjecture to be proved for an arbitrary function f satisfying the law $f(Suc(?n)) = Suc(f(?n))$:

```
val [prem] = goal Nat.thy
    "(!!n. f(Suc(n)) = Suc(f(n))) ==> f(i+j) = i+f(j)";
  Level 0
  f(i + j) = i + f(j)
   1. f(i + j) = i + f(j)
  val prem = "f(Suc(?n)) = Suc(f(?n))
              [!!n. f(Suc(n)) = Suc(f(n))]" : thm
```

In the theorem `prem`, note that f is a free variable while $?n$ is a schematic variable.

```
by (res_inst_tac [("n","i")] induct 1);
  Level 1
  f(i + j) = i + f(j)
   1. f(0 + j) = 0 + f(j)
   2. !!x. f(x + j) = x + f(j) ==> f(Suc(x) + j) = Suc(x) + f(j)
```

We simplify each subgoal in turn. The first one is trivial:

```
by (simp_tac add_ss 1);
  Level 2
  f(i + j) = i + f(j)
   1. !!x. f(x + j) = x + f(j) ==> f(Suc(x) + j) = Suc(x) + f(j)
```

The remaining subgoal requires rewriting by the premise, so we add it to `add_ss`:[2]

```
by (asm_simp_tac (add_ss addsimps [prem]) 1);
  Level 3
  f(i + j) = i + f(j)
  No subgoals!
```

---

[2]The previous simplifier required congruence rules for function variables like f in order to simplify their arguments. It was more general than the current simplifier, but harder to use and slower. The present simplifier can be given congruence rules to realize non-standard simplification of a function's arguments, but this is seldom necessary.

13.4 Permutative rewrite rules

A rewrite rule is **permutative** if the left-hand side and right-hand side are the same up to renaming of variables. The most common permutative rule is commutativity: $x + y = y + x$. Other examples include $(x - y) - z = (x - z) - y$ in arithmetic and $insert(x, insert(y, A)) = insert(y, insert(x, A))$ for sets. Such rules are common enough to merit special attention.

Because ordinary rewriting loops given such rules, the simplifier employs a special strategy, called **ordered rewriting**. There is a built-in lexicographic ordering on terms. A permutative rewrite rule is applied only if it decreases the given term with respect to this ordering. For example, commutativity rewrites $b + a$ to $a + b$, but then stops because $a + b$ is strictly less than $b + a$. The Boyer-Moore theorem prover [4] also employs ordered rewriting.

Permutative rewrite rules are added to simpsets just like other rewrite rules; the simplifier recognizes their special status automatically. They are most effective in the case of associative-commutative operators. (Associativity by itself is not permutative.) When dealing with an AC-operator f, keep the following points in mind:

- The associative law must always be oriented from left to right, namely $f(f(x, y), z) = f(x, f(y, z))$. The opposite orientation, if used with commutativity, leads to looping! Future versions of Isabelle may remove this restriction.

- To complete your set of rewrite rules, you must add not just associativity (A) and commutativity (C) but also a derived rule, **left-commutativity** (LC): $f(x, f(y, z)) = f(y, f(x, z))$.

Ordered rewriting with the combination of A, C, and LC sorts a term lexicographically:

$$(b + c) + a \overset{A}{\longmapsto} b + (c + a) \overset{C}{\longmapsto} b + (a + c) \overset{LC}{\longmapsto} a + (b + c)$$

Martin and Nipkow [32] discuss the theory and give many examples; other algebraic structures are amenable to ordered rewriting, such as boolean rings.

13.4.1 Example: sums of integers

This example is set in Isabelle's higher-order logic. Theory `Arith` contains the theory of arithmetic. The simpset `arith_ss` contains many arithmetic laws including distributivity of $\times$ over $+$, while `add_ac` is a list consisting of the A, C and LC laws for $+$. Let us prove the theorem

$$\sum_{i=1}^{n} i = n \times (n + 1)/2.$$

A functional sum represents the summation operator under the interpretation $\text{sum}(f, n+1) = \sum_{i=0}^{n} f(i)$. We extend Arith using a theory file:

```
NatSum = Arith +
consts sum     :: "[nat=>nat, nat] => nat"
rules  sum_0      "sum(f,0) = 0"
       sum_Suc    "sum(f,Suc(n)) = f(n) + sum(f,n)"
end
```

After declaring open NatSum, we make the required simpset by adding the AC-rules for + and the axioms for sum:

```
val natsum_ss = arith_ss addsimps ([sum_0,sum_Suc] @ add_ac);
  val natsum_ss = SS {...} : simpset
```

Our desired theorem now reads $\text{sum}(\lambda i\,.\,i, n+1) = n \times (n+1)/2$. The Isabelle goal has both sides multiplied by 2:

```
goal NatSum.thy "Suc(Suc(0))*sum(%i.i,Suc(n)) = n*Suc(n)";
  Level 0
  Suc(Suc(0)) * sum(%i. i, Suc(n)) = n * Suc(n)
   1. Suc(Suc(0)) * sum(%i. i, Suc(n)) = n * Suc(n)
```

Induction should not be applied until the goal is in the simplest form:

```
by (simp_tac natsum_ss 1);
  Level 1
  Suc(Suc(0)) * sum(%i. i, Suc(n)) = n * Suc(n)
   1. n + (n + (sum(%i. i, n) + sum(%i. i, n))) = n + n * n
```

Ordered rewriting has sorted the terms in the left-hand side. The subgoal is now ready for induction:

```
by (nat_ind_tac "n" 1);
  Level 2
  Suc(Suc(0)) * sum(%i. i, Suc(n)) = n * Suc(n)
   1. 0 + (0 + (sum(%i. i, 0) + sum(%i. i, 0))) = 0 + 0 * 0

   2. !!n1. n1 + (n1 + (sum(%i. i, n1) + sum(%i. i, n1))) =
            n1 + n1 * n1 ==>
            Suc(n1) +
            (Suc(n1) + (sum(%i. i, Suc(n1)) + sum(%i. i, Suc(n1)))) =
            Suc(n1) + Suc(n1) * Suc(n1)
```

Simplification proves both subgoals immediately:

```
by (ALLGOALS (asm_simp_tac natsum_ss));
  Level 3
  Suc(Suc(0)) * sum(%i. i, Suc(n)) = n * Suc(n)
  No subgoals!
```

If we had omitted add_ac from the simpset, simplification would have failed to prove the induction step:

```
Suc(Suc(0)) * sum(%i. i, Suc(n)) = n * Suc(n)
 1. !!n1. n1 + (n1 + (sum(%i. i, n1) + sum(%i. i, n1))) =
          n1 + n1 * n1 ==>
          n1 + (n1 + (n1 + sum(%i. i, n1) + (n1 + sum(%i. i, n1)))) =
          n1 + (n1 + (n1 + n1 * n1))
```

Ordered rewriting proves this by sorting the left-hand side. Proving arithmetic theorems without ordered rewriting requires explicit use of commutativity. This is tedious; try it and see!

Ordered rewriting is equally successful in proving $\sum_{i=1}^{n} i^3 = n^2 \times (n+1)^2/4$.

13.4.2 Re-orienting equalities

Ordered rewriting with the derived rule `symmetry` can reverse equality signs:

```
val symmetry = prove_goal HOL.thy "(x=y) = (y=x)"
                 (fn _ => [fast_tac HOL_cs 1]);
```

This is frequently useful. Assumptions of the form $s = t$, where t occurs in the conclusion but not s, can often be brought into the right form. For example, ordered rewriting with `symmetry` can prove the goal

$$f(a) = b \land f(a) = c \to b = c.$$

Here `symmetry` reverses both $f(a) = b$ and $f(a) = c$ because $f(a)$ is lexicographically greater than b and c. These re-oriented equations, as rewrite rules, replace b and c in the conclusion by $f(a)$.

Another example is the goal $\neg(t = u) \to \neg(u = t)$. The differing orientations make this appear difficult to prove. Ordered rewriting with `symmetry` makes the equalities agree. (Without knowing more about t and u we cannot say whether they both go to $t = u$ or $u = t$.) Then the simplifier can prove the goal outright.

13.5 *Setting up the simplifier

Setting up the simplifier for new logics is complicated. This section describes how the simplifier is installed for intuitionistic first-order logic; the code is largely taken from `FOL/simpdata.ML`.

The simplifier and the case splitting tactic, which reside on separate files, are not part of Pure Isabelle. They must be loaded explicitly:

```
use "../Provers/simplifier.ML";
use "../Provers/splitter.ML";
```

Simplification works by reducing various object-equalities to meta-equality. It requires rules stating that equal terms and equivalent formulae are also equal at the meta-level. The rule declaration part of the file `FOL/ifol.thy` contains the two lines

```
eq_reflection   "(x=y)   ==> (x==y)"
iff_reflection  "(P<->Q) ==> (P==Q)"
```

Of course, you should only assert such rules if they are true for your particular logic. In Constructive Type Theory, equality is a ternary relation of the form $a = b \in A$; the type A determines the meaning of the equality essentially as a partial equivalence relation. The present simplifier cannot be used. Rewriting in CTT uses another simplifier, which resides in the file typedsimp.ML and is not documented. Even this does not work for later variants of Constructive Type Theory that use intensional equality [39].

13.5.1 A collection of standard rewrite rules

The file begins by proving lots of standard rewrite rules about the logical connectives. These include cancellation and associative laws. To prove them easily, it defines a function that echoes the desired law and then supplies it the theorem prover for intuitionistic FOL:

```
fun int_prove_fun s =
 (writeln s;
  prove_goal IFOL.thy s
   (fn prems => [ (cut_facts_tac prems 1),
                  (Int.fast_tac 1) ]));
```

The following rewrite rules about conjunction are a selection of those proved on FOL/simpdata.ML. Later, these will be supplied to the standard simpset.

```
val conj_rews = map int_prove_fun
 ["P & True <-> P",      "True & P <-> P",
  "P & False <-> False", "False & P <-> False",
  "P & P <-> P",
  "P & ~P <-> False",    "~P & P <-> False",
  "(P & Q) & R <-> P & (Q & R)"];
```

The file also proves some distributive laws. As they can cause exponential blowup, they will not be included in the standard simpset. Instead they are merely bound to an ML identifier, for user reference.

```
val distrib_rews = map int_prove_fun
 ["P & (Q | R) <-> P&Q | P&R",
  "(Q | R) & P <-> Q&P | R&P",
  "(P | Q --> R) <-> (P --> R) & (Q --> R)"];
```

13.5.2 Functions for preprocessing the rewrite rules

The next step is to define the function for preprocessing rewrite rules. This will be installed by calling setmksimps below. Preprocessing occurs whenever rewrite rules are added, whether by user command or automatically. Preprocessing involves extracting atomic rewrites at the object-level, then reflecting them to the meta-level.

To start, the function gen_all strips any meta-level quantifiers from the front of the given theorem. Usually there are none anyway.

```
fun gen_all th = forall_elim_vars (#maxidx(rep_thm th)+1) th;
```

The function atomize analyses a theorem in order to extract atomic rewrite rules. The head of all the patterns, matched by the wildcard _, is the coercion function Trueprop.

```
fun atomize th = case concl_of th of
      _ $ (Const("op &",_) $ _ $ _)   => atomize(th RS conjunct1) @
                                         atomize(th RS conjunct2)
    | _ $ (Const("op -->",_) $ _ $ _) => atomize(th RS mp)
    | _ $ (Const("All",_) $ _)        => atomize(th RS spec)
    | _ $ (Const("True",_))           => []
    | _ $ (Const("False",_))          => []
    | _                               => [th];
```

There are several cases, depending upon the form of the conclusion:

- Conjunction: extract rewrites from both conjuncts.
- Implication: convert $P \to Q$ to the meta-implication $P \Longrightarrow Q$ and extract rewrites from Q; these will be conditional rewrites with the condition P.
- Universal quantification: remove the quantifier, replacing the bound variable by a schematic variable, and extract rewrites from the body.
- `True` and `False` contain no useful rewrites.
- Anything else: return the theorem in a singleton list.

The resulting theorems are not literally atomic — they could be disjunctive, for example — but are broken down as much as possible. See the file `ZF/simpdata.ML` for a sophisticated translation of set-theoretic formulae into rewrite rules.

The simplified rewrites must now be converted into meta-equalities. The rule `eq_reflection` converts equality rewrites, while `iff_reflection` converts if-and-only-if rewrites. The latter possibility can arise in two other ways: the negative theorem $\neg P$ is converted to $P \equiv$ `False`, and any other theorem P is converted to $P \equiv$ `True`. The rules `iff_reflection_F` and `iff_reflection_T` accomplish this conversion.

```
val P_iff_F = int_prove_fun "~P ==> (P <-> False)";
val iff_reflection_F = P_iff_F RS iff_reflection;

val P_iff_T = int_prove_fun "P ==> (P <-> True)";
val iff_reflection_T = P_iff_T RS iff_reflection;
```

The function `mk_meta_eq` converts a theorem to a meta-equality using the case analysis described above.

```
fun mk_meta_eq th = case concl_of th of
      _ $ (Const("op =",_)$_$_)   => th RS eq_reflection
    | _ $ (Const("op <->",_)$_$_) => th RS iff_reflection
    | _ $ (Const("Not",_)$_)      => th RS iff_reflection_F
    | _                           => th RS iff_reflection_T;
```

The three functions `gen_all`, `atomize` and `mk_meta_eq` will be composed together and supplied below to `setmksimps`.

13.5.3 Making the initial simpset

It is time to assemble these items. We open module `Simplifier` to gain access to its components. We define the infix operator `addcongs` to insert congruence rules; given a list of theorems, it converts their conclusions into meta-equalities and passes them to `addeqcongs`.

```
open Simplifier;
infix addcongs;
fun ss addcongs congs =
    ss addeqcongs (congs RL [eq_reflection,iff_reflection]);
```

The list `IFOL_rews` contains the default rewrite rules for first-order logic. The first of these is the reflexive law expressed as the equivalence $(a = a) \leftrightarrow$ `True`; the rewrite rule $a = a$ is clearly useless.

```
val IFOL_rews =
   [refl RS P_iff_T] @ conj_rews @ disj_rews @ not_rews @
    imp_rews @ iff_rews @ quant_rews;
```

The list `triv_rls` contains trivial theorems for the solver. Any subgoal that is simplified to one of these will be removed.

```
val notFalseI = int_prove_fun "~False";
val triv_rls = [TrueI,refl,iff_refl,notFalseI];
```

The basic simpset for intuitionistic FOL starts with `empty_ss`. It preprocess rewrites using `gen_all`, `atomize` and `mk_meta_eq`. It solves simplified subgoals using `triv_rls` and assumptions. It uses `asm_simp_tac` to tackle subgoals of conditional rewrites. It takes `IFOL_rews` as rewrite rules. Other simpsets built from `IFOL_ss` will inherit these items. In particular, `FOL_ss` extends `IFOL_ss` with classical rewrite rules such as $\neg\neg P \leftrightarrow P$.

```
val IFOL_ss =
  empty_ss
  setmksimps (map mk_meta_eq o atomize o gen_all)
  setsolver  (fn prems => resolve_tac (triv_rls @ prems) ORELSE'
                          assume_tac)
  setsubgoaler asm_simp_tac
  addsimps IFOL_rews
  addcongs [imp_cong];
```

This simpset takes `imp_cong` as a congruence rule in order to use contextual information to simplify the conclusions of implications:

$$[\![?P \leftrightarrow ?P';\ ?P' \Longrightarrow ?Q \leftrightarrow ?Q']\!] \Longrightarrow (?P \rightarrow ?Q) \leftrightarrow (?P' \rightarrow ?Q')$$

By adding the congruence rule `conj_cong`, we could obtain a similar effect for conjunctions.

13.5.4 Case splitting

To set up case splitting, we must prove the theorem below and pass it to `mk_case_split_tac`. The tactic `split_tac` uses `mk_meta_eq`, defined above, to convert the splitting rules to meta-equalities.

```
val meta_iffD =
    prove_goal FOL.thy "[| P==Q; Q |] ==> P"
        (fn [prem1,prem2] => [rewtac prem1, rtac prem2 1])
fun split_tac splits =
    mk_case_split_tac meta_iffD (map mk_meta_eq splits);
```

The splitter replaces applications of a given function; the right-hand side of the replacement can be anything. For example, here is a splitting rule for conditional expressions:

$$?P(if(?Q, ?x, ?y)) \leftrightarrow (?Q \rightarrow ?P(?x)) \wedge (\neg ?Q \rightarrow ?P(?y))$$

Another example is the elimination operator (which happens to be called *split*) for Cartesian products:

$$?P(split(?f, ?p)) \leftrightarrow (\forall a\ b\ .\ ?p = \langle a, b\rangle \rightarrow ?P(?f(a, b)))$$

Case splits should be allowed only when necessary; they are expensive and hard to control. Here is a typical example of use, where `expand_if` is the first rule above:

```
by (simp_tac (prop_rec_ss setloop (split_tac [expand_if])) 1);
```

14. The Classical Reasoner

Although Isabelle is generic, many users will be working in some extension of classical first-order logic. Isabelle's set theory ZF is built upon theory FOL, while higher-order logic contains first-order logic as a fragment. Theorem-proving in predicate logic is undecidable, but many researchers have developed strategies to assist in this task.

Isabelle's classical reasoner is an ML functor that accepts certain information about a logic and delivers a suite of automatic tactics. Each tactic takes a collection of rules and executes a simple, non-clausal proof procedure. They are slow and simplistic compared with resolution theorem provers, but they can save considerable time and effort. They can prove theorems such as Pelletier's [52] problems 40 and 41 in seconds:

$$(\exists y . \forall x . J(y,x) \leftrightarrow \neg J(x,x)) \rightarrow \neg(\forall x . \exists y . \forall z . J(z,y) \leftrightarrow \neg J(z,x))$$

$$(\forall z . \exists y . \forall x . F(x,y) \leftrightarrow F(x,z) \land \neg F(x,x)) \rightarrow \neg(\exists z . \forall x . F(x,z))$$

The tactics are generic. They are not restricted to first-order logic, and have been heavily used in the development of Isabelle's set theory. Few interactive proof assistants provide this much automation. The tactics can be traced, and their components can be called directly; in this manner, any proof can be viewed interactively.

We shall first discuss the underlying principles, then consider how to use the classical reasoner. Finally, we shall see how to instantiate it for new logics. The logics FOL, HOL and ZF have it already installed.

14.1 The sequent calculus

Isabelle supports natural deduction, which is easy to use for interactive proof. But natural deduction does not easily lend itself to automation, and has a bias towards intuitionism. For certain proofs in classical logic, it can not be called natural. The **sequent calculus**, a generalization of natural deduction, is easier to automate.

A **sequent** has the form $\Gamma \vdash \Delta$, where Γ and Δ are sets of formulae.[1] The sequent

$$P_1, \ldots, P_m \vdash Q_1, \ldots, Q_n$$

[1] For first-order logic, sequents can equivalently be made from lists or multisets of formulae.

is **valid** if $P_1 \wedge \ldots \wedge P_m$ implies $Q_1 \vee \ldots \vee Q_n$. Thus $P_1, \ldots, P_m$ represent assumptions, each of which is true, while $Q_1, \ldots, Q_n$ represent alternative goals. A sequent is **basic** if its left and right sides have a common formula, as in $P, Q \vdash Q, R$; basic sequents are trivially valid.

Sequent rules are classified as **right** or **left**, indicating which side of the $\vdash$ symbol they operate on. Rules that operate on the right side are analogous to natural deduction's introduction rules, and left rules are analogous to elimination rules. Recall the natural deduction rules for first-order logic, Fig. 1.1. The sequent calculus analogue of $(\to I)$ is the rule

$$\frac{P, \Gamma \vdash \Delta, Q}{\Gamma \vdash \Delta, P \to Q} \qquad (\to R)$$

This breaks down some implication on the right side of a sequent; Γ and Δ stand for the sets of formulae that are unaffected by the inference. The analogue of the pair $(\vee I1)$ and $(\vee I2)$ is the single rule

$$\frac{\Gamma \vdash \Delta, P, Q}{\Gamma \vdash \Delta, P \vee Q} \qquad (\vee R)$$

This breaks down some disjunction on the right side, replacing it by both disjuncts. Thus, the sequent calculus is a kind of multiple-conclusion logic.

To illustrate the use of multiple formulae on the right, let us prove the classical theorem $(P \to Q) \vee (Q \to P)$. Working backwards, we reduce this formula to a basic sequent:

$$\cfrac{\cfrac{\cfrac{P, Q \vdash Q, P}{P \vdash Q, (Q \to P)}\,(\to)R}{\vdash (P \to Q), (Q \to P)}\,(\to)R}{\vdash (P \to Q) \vee (Q \to P)}\,(\vee)R$$

This example is typical of the sequent calculus: start with the desired theorem and apply rules backwards in a fairly arbitrary manner. This yields a surprisingly effective proof procedure. Quantifiers add few complications, since Isabelle handles parameters and schematic variables. See Chapter 10 of *ML for the Working Programmer* [46] for further discussion.

14.2 Simulating sequents by natural deduction

Isabelle can represent sequents directly, as in the object-logic LK. But natural deduction is easier to work with, and most object-logics employ it. Fortunately, we can simulate the sequent $P_1, \ldots, P_m \vdash Q_1, \ldots, Q_n$ by the Isabelle formula

$$[\![P_1; \ldots; P_m; \neg Q_2; \ldots; \neg Q_n]\!] \Longrightarrow Q_1,$$

where the order of the assumptions and the choice of Q_1 are arbitrary. Elim-resolution plays a key role in simulating sequent proofs.

We can easily handle reasoning on the left. As discussed in Sect. 1.6.2, elim-resolution with the rules $(\vee E)$, $(\bot E)$ and $(\exists E)$ achieves a similar effect as the corresponding sequent rules. For the other connectives, we use sequent-style elimination rules instead of destruction rules such as $(\wedge E1, 2)$ and $(\forall E)$. But note that the rule $(\neg L)$ has no effect under our representation of sequents!

$$\frac{\Gamma \vdash \Delta, P}{\neg P, \Gamma \vdash \Delta} \qquad (\neg L)$$

What about reasoning on the right? Introduction rules can only affect the formula in the conclusion, namely Q_1. The other right-side formulae are represented as negated assumptions, $\neg Q_2$, ..., $\neg Q_n$. In order to operate on one of these, it must first be exchanged with Q_1. Elim-resolution with the **swap** rule has this effect:

$$[\![\neg P; \neg R \Longrightarrow P]\!] \Longrightarrow R \qquad (swap)$$

To ensure that swaps occur only when necessary, each introduction rule is converted into a swapped form: it is resolved with the second premise of $(swap)$. The swapped form of $(\wedge I)$, which might be called $(\neg\wedge E)$, is

$$[\![\neg(P \wedge Q); \neg R \Longrightarrow P; \neg R \Longrightarrow Q]\!] \Longrightarrow R.$$

Similarly, the swapped form of $(\rightarrow I)$ is

$$[\![\neg(P \rightarrow Q); [\![\neg R; P]\!] \Longrightarrow Q]\!] \Longrightarrow R$$

Swapped introduction rules are applied using elim-resolution, which deletes the negated formula. Our representation of sequents also requires the use of ordinary introduction rules. If we had no regard for readability, we could treat the right side more uniformly by representing sequents as

$$[\![P_1; \ldots; P_m; \neg Q_1; \ldots; \neg Q_n]\!] \Longrightarrow \bot.$$

14.3 Extra rules for the sequent calculus

As mentioned, destruction rules such as $(\wedge E1, 2)$ and $(\forall E)$ must be replaced by sequent-style elimination rules. In addition, we need rules to embody the classical equivalence between $P \rightarrow Q$ and $\neg P \vee Q$. The introduction rules $(\vee I1, 2)$ are replaced by a rule that simulates $(\vee R)$:

$$(\neg Q \Longrightarrow P) \Longrightarrow P \vee Q$$

The destruction rule $(\rightarrow E)$ is replaced by

$$[\![P \rightarrow Q; \neg P \Longrightarrow R; Q \Longrightarrow R]\!] \Longrightarrow R.$$

Quantifier replication also requires special rules. In classical logic, $\exists x.P$ is equivalent to $\neg\forall x.\neg P$; the rules $(\exists R)$ and $(\forall L)$ are dual:

$$\frac{\Gamma\vdash\Delta,\exists x.P,P[t/x]}{\Gamma\vdash\Delta,\exists x.P}\,(\exists R)\qquad\frac{P[t/x],\forall x.P,\Gamma\vdash\Delta}{\forall x.P,\Gamma\vdash\Delta}\,(\forall L)$$

Thus both kinds of quantifier may be replicated. Theorems requiring multiple uses of a universal formula are easy to invent; consider

$$(\forall x\,.\,P(x)\rightarrow P(f(x)))\wedge P(a)\rightarrow P(f^n(a)),$$

for any $n>1$. Natural examples of the multiple use of an existential formula are rare; a standard one is $\exists x\,.\,\forall y\,.\,P(x)\rightarrow P(y)$.

Forgoing quantifier replication loses completeness, but gains decidability, since the search space becomes finite. Many useful theorems can be proved without replication, and the search generally delivers its verdict in a reasonable time. To adopt this approach, represent the sequent rules $(\exists R)$, $(\exists L)$ and $(\forall R)$ by $(\exists I)$, $(\exists E)$ and $(\forall I)$, respectively, and put $(\forall E)$ into elimination form:

$$[\![\forall x.P(x);P(t)\Longrightarrow Q]\!]\Longrightarrow Q\qquad(\forall E_2)$$

Elim-resolution with this rule will delete the universal formula after a single use. To replicate universal quantifiers, replace the rule by

$$[\![\forall x.P(x);\;[\![P(t);\forall x.P(x)]\!]\Longrightarrow Q]\!]\Longrightarrow Q.\qquad(\forall E_3)$$

To replicate existential quantifiers, replace $(\exists I)$ by

$$[\![\neg(\exists x.P(x))\Longrightarrow P(t)]\!]\Longrightarrow\exists x.P(x).$$

All introduction rules mentioned above are also useful in swapped form.

Replication makes the search space infinite; we must apply the rules with care. The classical reasoner distinguishes between safe and unsafe rules, applying the latter only when there is no alternative. Depth-first search may well go down a blind alley; best-first search is better behaved in an infinite search space. However, quantifier replication is too expensive to prove any but the simplest theorems.

14.4 Classical rule sets

Each automatic tactic takes a **classical set** — a collection of rules, classified as introduction or elimination and as **safe** or **unsafe**. In general, safe rules can be attempted blindly, while unsafe rules must be used with care. A safe rule must never reduce a provable goal to an unprovable set of subgoals.

The rule $(\vee I1)$ is unsafe because it reduces $P\vee Q$ to P. Any rule is unsafe whose premises contain new unknowns. The elimination rule $(\forall E_2)$ is unsafe, since it is applied via elim-resolution, which discards the assumption $\forall x.P(x)$

and replaces it by the weaker assumption $P(?t)$. The rule $(\exists I)$ is unsafe for similar reasons. The rule $(\forall E_3)$ is unsafe in a different sense: since it keeps the assumption $\forall x.P(x)$, it is prone to looping. In classical first-order logic, all rules are safe except those mentioned above.

The safe/unsafe distinction is vague, and may be regarded merely as a way of giving some rules priority over others. One could argue that $(\vee E)$ is unsafe, because repeated application of it could generate exponentially many subgoals. Induction rules are unsafe because inductive proofs are difficult to set up automatically. Any inference is unsafe that instantiates an unknown in the proof state — thus `match_tac` must be used, rather than `resolve_tac`. Even proof by assumption is unsafe if it instantiates unknowns shared with other subgoals — thus `eq_assume_tac` must be used, rather than `assume_tac`.

Classical rule sets belong to the abstract type `claset`, which supports the following operations (provided the classical reasoner is installed!):

```
empty_cs : claset
addSIs   : claset * thm list -> claset                    infix 4
addSEs   : claset * thm list -> claset                    infix 4
addSDs   : claset * thm list -> claset                    infix 4
addIs    : claset * thm list -> claset                    infix 4
addEs    : claset * thm list -> claset                    infix 4
addDs    : claset * thm list -> claset                    infix 4
print_cs : claset -> unit
```

There are no operations for deletion from a classical set. The add operations do not check for repetitions.

`empty_cs` is the empty classical set.

cs `addSIs` *rules* adds safe introduction *rules* to *cs*.

cs `addSEs` *rules* adds safe elimination *rules* to *cs*.

cs `addSDs` *rules* adds safe destruction *rules* to *cs*.

cs `addIs` *rules* adds unsafe introduction *rules* to *cs*.

cs `addEs` *rules* adds unsafe elimination *rules* to *cs*.

cs `addDs` *rules* adds unsafe destruction *rules* to *cs*.

`print_cs` *cs* prints the rules of *cs*.

Introduction rules are those that can be applied using ordinary resolution. The classical set automatically generates their swapped forms, which will be applied using elim-resolution. Elimination rules are applied using elim-resolution. In a classical set, rules are sorted by the number of new subgoals they will yield; rules that generate the fewest subgoals will be tried first (see Sect. 6.4.1).

For a given classical set, the proof strategy is simple. Perform as many safe inferences as possible; or else, apply certain safe rules, allowing instantiation of unknowns; or else, apply an unsafe rule. The tactics may also apply `hyp_subst_tac`, if they have been set up to do so (see below). They may perform a form of Modus Ponens: if there are assumptions $P \to Q$ and P, then replace $P \to Q$ by Q.

14.5 The classical tactics

If installed, the classical module provides several tactics (and other operations) for simulating the classical sequent calculus.

14.5.1 The automatic tactics

```
fast_tac : claset -> int -> tactic
best_tac : claset -> int -> tactic
```

Both of these tactics work by applying `step_tac` repeatedly. Their effect is restricted (by `SELECT_GOAL`) to one subgoal; they either solve this subgoal or fail.

`fast_tac` *cs i* applies `step_tac` using depth-first search, to solve subgoal i.

`best_tac` *cs i* applies `step_tac` using best-first search, to solve subgoal i. A heuristic function — typically, the total size of the proof state — guides the search. This function is supplied when the classical reasoner is set up.

14.5.2 Single-step tactics

```
safe_step_tac : claset -> int -> tactic
safe_tac      : claset        -> tactic
inst_step_tac : claset -> int -> tactic
step_tac      : claset -> int -> tactic
slow_step_tac : claset -> int -> tactic
```

The automatic proof procedures call these tactics. By calling them yourself, you can execute these procedures one step at a time.

`safe_step_tac` *cs i* performs a safe step on subgoal i. This may include proof by assumption or Modus Ponens (taking care not to instantiate unknowns), or `hyp_subst_tac`.

`safe_tac` *cs* repeatedly performs safe steps on all subgoals. It is deterministic, with at most one outcome. If the automatic tactics fail, try using `safe_tac` to open up your formula; then you can replicate certain quantifiers explicitly by applying appropriate rules.

`inst_step_tac` *cs i* is like `safe_step_tac`, but allows unknowns to be instantiated.

`step_tac` *cs i* tries `safe_tac`. If this fails, it tries `inst_step_tac`, or applies an unsafe rule from *cs*. This is the basic step of the proof procedure.

`slow_step_tac` resembles `step_tac`, but allows backtracking between using safe rules with instantiation (`inst_step_tac`) and using unsafe rules. The resulting search space is too large for use in the standard proof procedures, but `slow_step_tac` is worth considering in special situations.

14.5.3 Other useful tactics

```
contr_tac    :                int -> tactic
mp_tac       :                int -> tactic
eq_mp_tac    :                int -> tactic
swap_res_tac : thm list -> int -> tactic
```

These can be used in the body of a specialized search.

`contr_tac` *i* solves subgoal *i* by detecting a contradiction among two assumptions of the form P and $\neg P$, or fail. It may instantiate unknowns. The tactic can produce multiple outcomes, enumerating all possible contradictions.

`mp_tac` *i* is like `contr_tac`, but also attempts to perform Modus Ponens in subgoal *i*. If there are assumptions $P \to Q$ and P, then it replaces $P \to Q$ by Q. It may instantiate unknowns. It fails if it can do nothing.

`eq_mp_tac` *i* is like `mp_tac` *i*, but may not instantiate unknowns — thus, it is safe.

`swap_res_tac` *thms i* refines subgoal *i* of the proof state using *thms*, which should be a list of introduction rules. First, it attempts to solve the goal using `assume_tac` or `contr_tac`. It then attempts to apply each rule in turn, attempting resolution and also elim-resolution with the swapped form.

14.5.4 Creating swapped rules

```
swapify   : thm list -> thm list
joinrules : thm list * thm list -> (bool * thm) list
```

`swapify` *thms* returns a list consisting of the swapped versions of *thms*, regarded as introduction rules.

`joinrules` (*intrs*, *elims*) joins introduction rules, their swapped versions, and elimination rules for use with `biresolve_tac`. Each rule is paired with `false` (indicating ordinary resolution) or `true` (indicating elim-resolution).

14.6 Setting up the classical reasoner

Isabelle's classical object-logics, including FOL and HOL, have the classical reasoner already set up. When defining a new classical logic, you should set up the reasoner yourself. It consists of the ML functor ClassicalFun, which takes the argument signature CLASSICAL_DATA:

```
signature CLASSICAL_DATA =
  sig
  val mp              : thm
  val not_elim        : thm
  val swap            : thm
  val sizef           : thm -> int
  val hyp_subst_tacs : (int -> tactic) list
  end;
```

Thus, the functor requires the following items:

mp should be the Modus Ponens rule $[\![?P \rightarrow ?Q;\ ?P]\!] \Longrightarrow ?Q$.

not_elim should be the contradiction rule $[\![\neg ?P;\ ?P]\!] \Longrightarrow ?R$.

swap should be the swap rule $[\![\neg ?P;\ \neg ?R \Longrightarrow ?P]\!] \Longrightarrow ?R$.

sizef is the heuristic function used for best-first search. It should estimate the size of the remaining subgoals. A good heuristic function is size_of_thm, which measures the size of the proof state. Another size function might ignore certain subgoals (say, those concerned with type checking). A heuristic function might simply count the subgoals.

hyp_subst_tacs is a list of tactics for substitution in the hypotheses, typically created by HypsubstFun (see Chapter 12). This list can, of course, be empty. The tactics are assumed to be safe!

The functor is not at all sensitive to the formalization of the object-logic. It does not even examine the rules, but merely applies them according to its fixed strategy. The functor resides in Provers/classical.ML in the Isabelle distribution directory.

Part III

Isabelle's Object-Logics

15. Basic Concepts

Several logics come with Isabelle. Many of them are sufficiently developed to serve as comfortable reasoning environments. They are also good starting points for defining new logics. Each logic is distributed with sample proofs, some of which are described in this document.

`FOL` is many-sorted first-order logic with natural deduction. It comes in both constructive and classical versions.

`ZF` is axiomatic set theory, using the Zermelo-Fraenkel axioms [56]. It is built upon classical `FOL`.

`CCL` is Martin Coen's Classical Computational Logic, which is the basis of a preliminary method for deriving programs from proofs [8]. It is built upon classical `FOL`.

`LCF` is a version of Scott's Logic for Computable Functions, which is also implemented by the LCF system [42]. It is built upon classical `FOL`.

`HOL` is the higher-order logic of Church [7], which is also implemented by Gordon's HOL system [22]. This object-logic should not be confused with Isabelle's meta-logic, which is also a form of higher-order logic.

`HOLCF` is an alternative version of LCF, defined as an extension of `HOL`.

`CTT` is a version of Martin-Löf's Constructive Type Theory [39], with extensional equality. Universes are not included.

`LK` is another version of first-order logic, a classical sequent calculus. Sequents have the form $A_1, \ldots, A_m \vdash B_1, \ldots, B_n$; rules are applied using associative matching.

`Modal` implements the modal logics T, $S4$, and $S43$. It is built upon `LK`.

`Cube` is Barendregt's λ-cube.

The logics `CCL`, `LCF`, `HOLCF`, `Modal` and `Cube` are currently undocumented.

You should not read this before reading *Introduction to Isabelle* and performing some Isabelle proofs. Consult the *Reference Manual* for more information on tactics, packages, etc.

15.1 Syntax definitions

The syntax of each logic is presented using a context-free grammar. These grammars obey the following conventions:

- identifiers denote nonterminal symbols
- `typewriter` font denotes terminal symbols
- parentheses (...) express grouping
- constructs followed by a Kleene star, such as id^* and $(\ldots)^*$ can be repeated 0 or more times
- alternatives are separated by a vertical bar, |
- the symbol for alphanumeric identifiers is *id*
- the symbol for scheme variables is *var*

To reduce the number of nonterminals and grammar rules required, Isabelle's syntax module employs **priorities**, or precedences. Each grammar rule is given by a mixfix declaration, which has a priority, and each argument place has a priority. This general approach handles infix operators that associate either to the left or to the right, as well as prefix and binding operators.

In a syntactically valid expression, an operator's arguments never involve an operator of lower priority unless brackets are used. Consider first-order logic, where $\exists$ has lower priority than $\lor$, which has lower priority than $\land$. There, $P \land Q \lor R$ abbreviates $(P \land Q) \lor R$ rather than $P \land (Q \lor R)$. Also, $\exists x . P \lor Q$ abbreviates $\exists x . (P \lor Q)$ rather than $(\exists x . P) \lor Q$. Note especially that $P \lor (\exists x . Q)$ becomes syntactically invalid if the brackets are removed.

A **binder** is a symbol associated with a constant of type $(\sigma \Rightarrow \tau) \Rightarrow \tau'$. For instance, we may declare $\forall$ as a binder for the constant *All*, which has type $(\alpha \Rightarrow o) \Rightarrow o$. This defines the syntax $\forall x . t$ to mean $All(\lambda x . t)$. We can also write $\forall x_1 \ldots x_m . t$ to abbreviate $\forall x_1 . \ldots \forall x_m . t$; this is possible for any constant provided that τ and τ' are the same type. `HOL`'s description operator $\epsilon x . P(x)$ has type $(\alpha \Rightarrow bool) \Rightarrow \alpha$ and can bind only one variable, except when α is *bool*. `ZF`'s bounded quantifier $\forall x \in A . P(x)$ cannot be declared as a binder because it has type $[i, i \Rightarrow o] \Rightarrow o$. The syntax for binders allows type constraints on bound variables, as in

$$\forall (x{::}\alpha)\ (y{::}\beta) . R(x, y)$$

To avoid excess detail, the logic descriptions adopt a semi-formal style. Infix operators and binding operators are listed in separate tables, which include their priorities. Grammar descriptions do not include numeric priorities; instead, the rules appear in order of decreasing priority. This should suffice for most purposes; for full details, please consult the actual syntax definitions in the `.thy` files.

Each nonterminal symbol is associated with some Isabelle type. For example, the formulae of first-order logic have type o. Every Isabelle expression of type o is therefore a formula. These include atomic formulae such as P, where P is a variable of type o, and more generally expressions such as $P(t, u)$, where P, t and u have suitable types. Therefore, 'expression of type o' is listed as a separate possibility in the grammar for formulae.

15.2 Proof procedures

Most object-logics come with simple proof procedures. These are reasonably powerful for interactive use, though often simplistic and incomplete. You can do single-step proofs using `resolve_tac` and `assume_tac`, referring to the inference rules of the logic by ML identifiers.

For theorem proving, rules can be classified as **safe** or **unsafe**. A rule is safe if applying it to a provable goal always yields provable subgoals. If a rule is safe then it can be applied automatically to a goal without destroying our chances of finding a proof. For instance, all the rules of the classical sequent calculus LK are safe. Universal elimination is unsafe if the formula $\forall x \,.\, P(x)$ is deleted after use. Other unsafe rules include the following:

$$\frac{P}{P \lor Q}\ (\lor I1) \qquad \frac{P \to Q \quad P}{Q}\ (\to E) \qquad \frac{P[t/x]}{\exists x \,.\, P}\ (\exists I)$$

Proof procedures use safe rules whenever possible, delaying the application of unsafe rules. Those safe rules are preferred that generate the fewest subgoals. Safe rules are (by definition) deterministic, while the unsafe rules require search. The design of a suitable set of rules can be as important as the strategy for applying them.

Many of the proof procedures use backtracking. Typically they attempt to solve subgoal i by repeatedly applying a certain tactic to it. This tactic, which is known as a **step tactic**, resolves a selection of rules with subgoal i. This may replace one subgoal by many; the search persists until there are fewer subgoals in total than at the start. Backtracking happens when the search reaches a dead end: when the step tactic fails. Alternative outcomes are then searched by a depth-first or best-first strategy.

16. First-Order Logic

Isabelle implements Gentzen's natural deduction systems NJ and NK. Intuitionistic first-order logic is defined first, as theory `IFOL`. Classical logic, theory `FOL`, is obtained by adding the double negation rule. Basic proof procedures are provided. The intuitionistic prover works with derived rules to simplify implications in the assumptions. Classical `FOL` employs Isabelle's classical reasoner, which simulates a sequent calculus.

16.1 Syntax and rules of inference

The logic is many-sorted, using Isabelle's type classes. The class of first-order terms is called `term` and is a subclass of `logic`. No types of individuals are provided, but extensions can define types such as `nat::term` and type constructors such as `list::(term)term` (see the examples directory, `FOL/ex`). Below, the type variable α ranges over class `term`; the equality symbol and quantifiers are polymorphic (many-sorted). The type of formulae is `o`, which belongs to class `logic`. Figure 16.1 gives the syntax. Note that `a~=b` is translated to $\neg(a = b)$.

Figure 16.2 shows the inference rules with their ML names. Negation is defined in the usual way for intuitionistic logic; $\neg P$ abbreviates $P \to \bot$. The biconditional ($\leftrightarrow$) is defined through $\land$ and $\to$; introduction and elimination rules are derived for it.

The unique existence quantifier, $\exists! x \,.\, P(x)$, is defined in terms of $\exists$ and $\forall$. An Isabelle binder, it admits nested quantifications. For instance, $\exists! xy \,.\, P(x, y)$ abbreviates $\exists! x \,.\, \exists! y \,.\, P(x, y)$; note that this does not mean that there exists a unique pair (x, y) satisfying $P(x, y)$.

Some intuitionistic derived rules are shown in Fig. 16.3, again with their ML names. These include rules for the defined symbols $\neg$, $\leftrightarrow$ and $\exists!$. Natural deduction typically involves a combination of forward and backward reasoning, particularly with the destruction rules $(\land E)$, $(\to E)$, and $(\forall E)$. Isabelle's backward style handles these rules badly, so sequent-style rules are derived to eliminate conjunctions, implications, and universal quantifiers. Used with elim-resolution, `allE` eliminates a universal quantifier while `all_dupE` re-inserts the quantified formula for later use. The rules `conj_impE`, etc., support the intuitionistic proof procedure (see Sect. 16.3).

See the files `FOL/IFOL.thy`, `FOL/IFOL.ML` and `FOL/intprover.ML` for complete listings of the rules and derived rules.

16.2 Generic packages

`FOL` instantiates most of Isabelle's generic packages; see `FOL/ROOT.ML` for details.

- Because it includes a general substitution rule, `FOL` instantiates the tactic `hyp_subst_tac`, which substitutes for an equality throughout a subgoal and its hypotheses.
- It instantiates the simplifier. `IFOL_ss` is the simplification set for intuitionistic first-order logic, while `FOL_ss` is the simplification set for classical logic. Both equality ($=$) and the biconditional ($\leftrightarrow$) may be used for rewriting. See the file `FOL/simpdata.ML` for a complete listing of the simplification rules, and Sect. 13.5 for discussion.
- It instantiates the classical reasoner. See Sect. 16.4 for details.

16.3 Intuitionistic proof procedures

Implication elimination (the rules `mp` and `impE`) pose difficulties for automated proof. In intuitionistic logic, the assumption $P \to Q$ cannot be treated like $\neg P \vee Q$. Given $P \to Q$, we may use Q provided we can prove P; the proof of P may require repeated use of $P \to Q$. If the proof of P fails then the whole branch of the proof must be abandoned. Thus intuitionistic propositional logic requires backtracking.

For an elementary example, consider the intuitionistic proof of Q from $P \to Q$ and $(P \to Q) \to P$. The implication $P \to Q$ is needed twice:

$$\dfrac{P \to Q \qquad \dfrac{(P \to Q) \to P \quad P \to Q}{P}\,(\to E)}{Q}\,(\to E)$$

The theorem prover for intuitionistic logic does not use `impE`. Instead, it simplifies implications using derived rules (Fig. 16.3). It reduces the antecedents of implications to atoms and then uses Modus Ponens: from $P \to Q$ and P deduce Q. The rules `conj_impE` and `disj_impE` are straightforward: $(P \wedge Q) \to S$ is equivalent to $P \to (Q \to S)$, and $(P \vee Q) \to S$ is equivalent to the conjunction of $P \to S$ and $Q \to S$. The other `..._impE` rules are unsafe; the method requires backtracking. All the rules are derived in the same simple manner.

Dyckhoff has independently discovered similar rules, and (more importantly) has demonstrated their completeness for propositional logic [16]. How-

name	*meta-type*	*description*
Trueprop	$o \Rightarrow prop$	coercion to *prop*
Not	$o \Rightarrow o$	negation ($\neg$)
True	o	tautology ($\top$)
False	o	absurdity ($\bot$)

CONSTANTS

---

symbol	*name*	*meta-type*	*priority*	*description*
ALL	All	$(\alpha \Rightarrow o) \Rightarrow o$	10	universal quantifier ($\forall$)
EX	Ex	$(\alpha \Rightarrow o) \Rightarrow o$	10	existential quantifier ($\exists$)
EX!	Ex1	$(\alpha \Rightarrow o) \Rightarrow o$	10	unique existence ($\exists!$)

BINDERS

---

symbol	*meta-type*	*priority*	*description*
=	$[\alpha, \alpha] \Rightarrow o$	Left 50	equality ($=$)
&	$[o, o] \Rightarrow o$	Right 35	conjunction ($\wedge$)
|	$[o, o] \Rightarrow o$	Right 30	disjunction ($\vee$)
-->	$[o, o] \Rightarrow o$	Right 25	implication ($\rightarrow$)
<->	$[o, o] \Rightarrow o$	Right 25	biconditional ($\leftrightarrow$)

INFIXES

---

formula = expression of type o
| *term* = *term*
| *term* ~= *term*
| ~ *formula*
| *formula* & *formula*
| *formula* | *formula*
| *formula* --> *formula*
| *formula* <-> *formula*
| ALL *id id** . *formula*
| EX *id id** . *formula*
| EX! *id id** . *formula*

GRAMMAR

Fig. 16.1. Syntax of FOL

```
refl        a=a
subst       [| a=b;  P(a) |] ==> P(b)
```

EQUALITY RULES

```
conjI       [| P;  Q |] ==> P&Q
conjunct1   P&Q ==> P
conjunct2   P&Q ==> Q

disjI1      P ==> P|Q
disjI2      Q ==> P|Q
disjE       [| P|Q;  P ==> R;  Q ==> R |] ==> R

impI        (P ==> Q) ==> P-->Q
mp          [| P-->Q;  P |] ==> Q

FalseE      False ==> P
```

PROPOSITIONAL RULES

```
allI        (!!x. P(x))  ==> (ALL x.P(x))
spec        (ALL x.P(x)) ==> P(x)

exI         P(x) ==> (EX x.P(x))
exE         [| EX x.P(x);  !!x. P(x) ==> R |] ==> R
```

QUANTIFIER RULES

```
True_def    True         == False-->False
not_def     ~P           == P-->False
iff_def     P<->Q        == (P-->Q) & (Q-->P)
ex1_def     EX! x. P(x) == EX x. P(x) & (ALL y. P(y) --> y=x)
```

DEFINITIONS

Fig. 16.2. Rules of intuitionistic logic

```
sym         a=b ==> b=a
trans       [| a=b;  b=c |] ==> a=c
ssubst      [| b=a;  P(a) |] ==> P(b)
```

DERIVED EQUALITY RULES

```
TrueI       True

notI        (P ==> False) ==> ~P
notE        [| ~P;  P |] ==> R

iffI        [| P ==> Q;  Q ==> P |] ==> P<->Q
iffE        [| P <-> Q;  [| P-->Q; Q-->P |] ==> R |] ==> R
iffD1       [| P <-> Q;  P |] ==> Q
iffD2       [| P <-> Q;  Q |] ==> P

ex1I        [| P(a);  !!x. P(x) ==> x=a |]  ==>  EX! x. P(x)
ex1E        [| EX! x.P(x);  !!x.[| P(x);  ALL y. P(y) --> y=x |] ==> R
            |] ==> R
```

DERIVED RULES FOR $\top$, $\neg$, $\leftrightarrow$ AND $\exists!$

```
conjE       [| P&Q;  [| P; Q |] ==> R |] ==> R
impE        [| P-->Q;  P;  Q ==> R |] ==> R
allE        [| ALL x.P(x);  P(x) ==> R |] ==> R
all_dupE    [| ALL x.P(x);  [| P(x); ALL x.P(x) |] ==> R |] ==> R
```

SEQUENT-STYLE ELIMINATION RULES

```
conj_impE [| (P&Q)-->S;  P-->(Q-->S) ==> R |] ==> R
disj_impE [| (P|Q)-->S;  [| P-->S; Q-->S |] ==> R |] ==> R
imp_impE  [| (P-->Q)-->S;  [| P; Q-->S |] ==> Q;  S ==> R |] ==> R
not_impE  [| ~P --> S;  P ==> False;  S ==> R |] ==> R
iff_impE  [| (P<->Q)-->S; [| P; Q-->S |] ==> Q; [| Q; P-->S |] ==> P;
          S ==> R |] ==> R
all_impE  [| (ALL x.P(x))-->S;  !!x.P(x);  S ==> R |] ==> R
ex_impE   [| (EX x.P(x))-->S;  P(a)-->S ==> R |] ==> R
```

INTUITIONISTIC SIMPLIFICATION OF IMPLICATION

Fig. 16.3. Derived rules for intuitionistic logic

ever, the tactics given below are not complete for first-order logic because they discard universally quantified assumptions after a single use.

```
mp_tac            : int -> tactic
eq_mp_tac         : int -> tactic
Int.safe_step_tac : int -> tactic
Int.safe_tac      :        tactic
Int.inst_step_tac : int -> tactic
Int.step_tac      : int -> tactic
Int.fast_tac      : int -> tactic
Int.best_tac      : int -> tactic
```

Most of these belong to the structure `Int` and resemble the tactics of Isabelle's classical reasoner.

`mp_tac` i attempts to use `notE` or `impE` within the assumptions in subgoal i. For each assumption of the form $\neg P$ or $P \rightarrow Q$, it searches for another assumption unifiable with P. By contradiction with $\neg P$ it can solve the subgoal completely; by Modus Ponens it can replace the assumption $P \rightarrow Q$ by Q. The tactic can produce multiple outcomes, enumerating all suitable pairs of assumptions.

`eq_mp_tac` i is like `mp_tac` i, but may not instantiate unknowns — thus, it is safe.

`Int.safe_step_tac` i performs a safe step on subgoal i. This may include proof by assumption or Modus Ponens (taking care not to instantiate unknowns), or `hyp_subst_tac`.

`Int.safe_tac` repeatedly performs safe steps on all subgoals. It is deterministic, with at most one outcome.

`Int.inst_step_tac` i is like `safe_step_tac`, but allows unknowns to be instantiated.

`Int.step_tac` i tries `safe_tac` or `inst_step_tac`, or applies an unsafe rule. This is the basic step of the intuitionistic proof procedure.

`Int.fast_tac` i applies `step_tac`, using depth-first search, to solve subgoal i.

`Int.best_tac` i applies `step_tac`, using best-first search (guided by the size of the proof state) to solve subgoal i.

Here are some of the theorems that `Int.fast_tac` proves automatically. The latter three date from *Principia Mathematica* (*11.53, *11.55, *11.61) [59].

```
~~P & ~~(P --> Q) --> ~~Q
(ALL x y. P(x) --> Q(y)) <-> ((EX x. P(x)) --> (ALL y. Q(y)))
(EX x y. P(x) & Q(x,y)) <-> (EX x. P(x) & (EX y. Q(x,y)))
(EX y. ALL x. P(x) --> Q(x,y)) --> (ALL x. P(x) --> (EX y. Q(x,y)))
```

```
excluded_middle    ~P | P

disjCI     (~Q ==> P) ==> P|Q
exCI       (ALL x. ~P(x) ==> P(a)) ==> EX x.P(x)
impCE      [| P-->Q; ~P ==> R; Q ==> R |] ==> R
iffCE      [| P<->Q;  [| P; Q |] ==> R;  [| ~P; ~Q |] ==> R |] ==> R
notnotD    ~~P ==> P
swap       ~P ==> (~Q ==> P) ==> Q
```

Fig. 16.4. Derived rules for classical logic

16.4 Classical proof procedures

The classical theory, FOL, consists of intuitionistic logic plus the rule

$$\frac{\begin{array}{c}[\neg P]\\ \vdots\\ P\end{array}}{P} \quad (classical)$$

Natural deduction in classical logic is not really all that natural. FOL derives classical introduction rules for $\lor$ and $\exists$, as well as classical elimination rules for $\to$ and $\leftrightarrow$, and the swap rule (see Fig. 16.4).

The classical reasoner is set up for FOL, as the structure Cla. This structure is open, so ML identifiers such as step_tac, fast_tac, best_tac, etc., refer to it. Single-step proofs can be performed, using swap_res_tac to deal with negated assumptions.

FOL defines the following classical rule sets:

```
prop_cs    : claset
FOL_cs     : claset
FOL_dup_cs : claset
```

prop_cs contains the propositional rules, namely those for $\top$, $\bot$, $\land$, $\lor$, $\neg$, $\to$ and $\leftrightarrow$, along with the rule refl.

FOL_cs extends prop_cs with the safe rules allI and exE and the unsafe rules allE and exI, as well as rules for unique existence. Search using this is incomplete since quantified formulae are used at most once.

FOL_dup_cs extends prop_cs with the safe rules allI and exE and the unsafe rules all_dupE and exCI, as well as rules for unique existence. Search using this is complete — quantified formulae may be duplicated — but frequently fails to terminate. It is generally unsuitable for depth-first search.

See the file FOL/FOL.ML for derivations of the classical rules, and Chap. 14 for more discussion of classical proof methods.

16.5 An intuitionistic example

Here is a session similar to one in *Logic and Computation* [42, pages 222–3]. Isabelle treats quantifiers differently from LCF-based theorem provers such as HOL. The proof begins by entering the goal in intuitionistic logic, then applying the rule $(\to I)$.

```
goal IFOL.thy "(EX y. ALL x. Q(x,y)) -->  (ALL x. EX y. Q(x,y))";
  Level 0
  (EX y. ALL x. Q(x,y)) --> (ALL x. EX y. Q(x,y))
   1. (EX y. ALL x. Q(x,y)) --> (ALL x. EX y. Q(x,y))

by (resolve_tac [impI] 1);
  Level 1
  (EX y. ALL x. Q(x,y)) --> (ALL x. EX y. Q(x,y))
   1. EX y. ALL x. Q(x,y) ==> ALL x. EX y. Q(x,y)
```

In this example, we shall never have more than one subgoal. Applying $(\to I)$ replaces `-->` by `==>`, making $\exists y . \forall x . Q(x, y)$ an assumption. We have the choice of $(\exists E)$ and $(\forall I)$; let us try the latter.

```
by (resolve_tac [allI] 1);
  Level 2
  (EX y. ALL x. Q(x,y)) --> (ALL x. EX y. Q(x,y))
   1. !!x. EX y. ALL x. Q(x,y) ==> EX y. Q(x,y)
```

Applying $(\forall I)$ replaces the `ALL x` by `!!x`, changing the universal quantifier from object $(\forall)$ to meta $(\bigwedge)$. The bound variable is a **parameter** of the subgoal. We now must choose between $(\exists I)$ and $(\exists E)$. What happens if the wrong rule is chosen?

```
by (resolve_tac [exI] 1);
  Level 3
  (EX y. ALL x. Q(x,y)) --> (ALL x. EX y. Q(x,y))
   1. !!x. EX y. ALL x. Q(x,y) ==> Q(x,?y2(x))
```

The new subgoal 1 contains the function variable `?y2`. Instantiating `?y2` can replace `?y2(x)` by a term containing `x`, even though `x` is a bound variable. Now we analyse the assumption $\exists y . \forall x . Q(x, y)$ using elimination rules:

```
by (eresolve_tac [exE] 1);
  Level 4
  (EX y. ALL x. Q(x,y)) --> (ALL x. EX y. Q(x,y))
   1. !!x y. ALL x. Q(x,y) ==> Q(x,?y2(x))
```

Applying $(\exists E)$ has produced the parameter `y` and stripped the existential quantifier from the assumption. But the subgoal is unprovable: there is no way to unify `?y2(x)` with the bound variable `y`. Using `choplev` we can return to the critical point. This time we apply $(\exists E)$:

```
choplev 2;
  Level 2
  (EX y. ALL x. Q(x,y)) --> (ALL x. EX y. Q(x,y))
   1. !!x. EX y. ALL x. Q(x,y) ==> EX y. Q(x,y)
```

```
by (eresolve_tac [exE] 1);
  Level 3
  (EX y. ALL x. Q(x,y)) --> (ALL x. EX y. Q(x,y))
   1. !!x y. ALL x. Q(x,y) ==> EX y. Q(x,y)
```

We now have two parameters and no scheme variables. Applying $(\exists I)$ and $(\forall E)$ produces two scheme variables, which are applied to those parameters. Parameters should be produced early, as this example demonstrates.

```
by (resolve_tac [exI] 1);
  Level 4
  (EX y. ALL x. Q(x,y)) --> (ALL x. EX y. Q(x,y))
   1. !!x y. ALL x. Q(x,y) ==> Q(x,?y3(x,y))
by (eresolve_tac [allE] 1);
  Level 5
  (EX y. ALL x. Q(x,y)) --> (ALL x. EX y. Q(x,y))
   1. !!x y. Q(?x4(x,y),y) ==> Q(x,?y3(x,y))
```

The subgoal has variables `?y3` and `?x4` applied to both parameters. The obvious projection functions unify `?x4(x,y)` with `x` and `?y3(x,y)` with `y`.

```
by (assume_tac 1);
  Level 6
  (EX y. ALL x. Q(x,y)) --> (ALL x. EX y. Q(x,y))
  No subgoals!
```

The theorem was proved in six tactic steps, not counting the abandoned ones. But proof checking is tedious; `Int.fast_tac` proves the theorem in one step.

```
goal IFOL.thy "(EX y. ALL x. Q(x,y)) -->  (ALL x. EX y. Q(x,y))";
  Level 0
  (EX y. ALL x. Q(x,y)) --> (ALL x. EX y. Q(x,y))
   1. (EX y. ALL x. Q(x,y)) --> (ALL x. EX y. Q(x,y))
by (Int.fast_tac 1);
  Level 1
  (EX y. ALL x. Q(x,y)) --> (ALL x. EX y. Q(x,y))
  No subgoals!
```

16.6 An example of intuitionistic negation

The following example demonstrates the specialized forms of implication elimination. Even propositional formulae can be difficult to prove from the basic rules; the specialized rules help considerably.

Propositional examples are easy to invent. As Dummett notes [15, page 28], $\neg P$ is classically provable if and only if it is intuitionistically provable; therefore, P is classically provable if and only if $\neg\neg P$ is intuitionistically provable.[1] Proving $\neg\neg P$ intuitionistically is much harder than proving P classically.

Our example is the double negation of the classical tautology $(P \to Q) \vee (Q \to P)$. When stating the goal, we command Isabelle to expand negations to

[1] Of course this holds only for propositional logic, not if P is allowed to contain quantifiers.

implications using the definition $\neg P \equiv P \to \bot$. This allows use of the special implication rules.

```
goalw IFOL.thy [not_def] "~ ~ ((P-->Q) | (Q-->P))";
  Level 0
  ~ ~ ((P --> Q) | (Q --> P))
   1. ((P --> Q) | (Q --> P) --> False) --> False
```

The first step is trivial.

```
by (resolve_tac [impI] 1);
  Level 1
  ~ ~ ((P --> Q) | (Q --> P))
   1. (P --> Q) | (Q --> P) --> False ==> False
```

By $(\to E)$ it would suffice to prove $(P \to Q) \vee (Q \to P)$, but that formula is not a theorem of intuitionistic logic. Instead we apply the specialized implication rule `disj_impE`. It splits the assumption into two assumptions, one for each disjunct.

```
by (eresolve_tac [disj_impE] 1);
  Level 2
  ~ ~ ((P --> Q) | (Q --> P))
   1. [| (P --> Q) --> False; (Q --> P) --> False |] ==> False
```

We cannot hope to prove $P \to Q$ or $Q \to P$ separately, but their negations are inconsistent. Applying `imp_impE` breaks down the assumption $\neg(P \to Q)$, asking to show Q while providing new assumptions P and $\neg Q$.

```
by (eresolve_tac [imp_impE] 1);
  Level 3
  ~ ~ ((P --> Q) | (Q --> P))
   1. [| (Q --> P) --> False; P; Q --> False |] ==> Q
   2. [| (Q --> P) --> False; False |] ==> False
```

Subgoal 2 holds trivially; let us ignore it and continue working on subgoal 1. Thanks to the assumption P, we could prove $Q \to P$; applying `imp_impE` is simpler.

```
by (eresolve_tac [imp_impE] 1);
  Level 4
  ~ ~ ((P --> Q) | (Q --> P))
   1. [| P; Q --> False; Q; P --> False |] ==> P
   2. [| P; Q --> False; False |] ==> Q
   3. [| (Q --> P) --> False; False |] ==> False
```

The three subgoals are all trivial.

```
by (REPEAT (eresolve_tac [FalseE] 2));
  Level 5
  ~ ~ ((P --> Q) | (Q --> P))
   1. [| P; Q --> False; Q; P --> False |] ==> P
by (assume_tac 1);
  Level 6
  ~ ~ ((P --> Q) | (Q --> P))
  No subgoals!
```

This proof is also trivial for `Int.fast_tac`.

16.7 A classical example

To illustrate classical logic, we shall prove the theorem $\exists y . \forall x . P(y) \rightarrow P(x)$. Informally, the theorem can be proved as follows. Choose y such that $\neg P(y)$, if such exists; otherwise $\forall x . P(x)$ is true. Either way the theorem holds.

The formal proof does not conform in any obvious way to the sketch given above. The key inference is the first one, `exCI`; this classical version of $(\exists I)$ allows multiple instantiation of the quantifier.

```
goal FOL.thy "EX y. ALL x. P(y)-->P(x)";
  Level 0
  EX y. ALL x. P(y) --> P(x)
   1. EX y. ALL x. P(y) --> P(x)
by (resolve_tac [exCI] 1);
  Level 1
  EX y. ALL x. P(y) --> P(x)
   1. ALL y. ~ (ALL x. P(y) --> P(x)) ==> ALL x. P(?a) --> P(x)
```

We can either exhibit a term `?a` to satisfy the conclusion of subgoal 1, or produce a contradiction from the assumption. The next steps are routine.

```
by (resolve_tac [allI] 1);
  Level 2
  EX y. ALL x. P(y) --> P(x)
   1. !!x. ALL y. ~ (ALL x. P(y) --> P(x)) ==> P(?a) --> P(x)
by (resolve_tac [impI] 1);
  Level 3
  EX y. ALL x. P(y) --> P(x)
   1. !!x. [| ALL y. ~ (ALL x. P(y) --> P(x)); P(?a) |] ==> P(x)
```

By the duality between $\exists$ and $\forall$, applying $(\forall E)$ in effect applies $(\exists I)$ again.

```
by (eresolve_tac [allE] 1);
  Level 4
  EX y. ALL x. P(y) --> P(x)
   1. !!x. [| P(?a); ~ (ALL xa. P(?y3(x)) --> P(xa)) |] ==> P(x)
```

In classical logic, a negated assumption is equivalent to a conclusion. To get this effect, we create a swapped version of $(\forall I)$ and apply it using `eresolve_tac`; we could equivalently have applied $(\forall I)$ using `swap_res_tac`.

```
allI RSN (2,swap);
  val it = "[| ~ (ALL x. ?P1(x)); !!x. ~ ?Q ==> ?P1(x) |] ==> ?Q" : thm
by (eresolve_tac [it] 1);
  Level 5
  EX y. ALL x. P(y) --> P(x)
   1. !!x xa. [| P(?a); ~ P(x) |] ==> P(?y3(x)) --> P(xa)
```

The previous conclusion, `P(x)`, has become a negated assumption.

```
by (resolve_tac [impI] 1);
  Level 6
  EX y. ALL x. P(y) --> P(x)
   1. !!x xa. [| P(?a); ~ P(x); P(?y3(x)) |] ==> P(xa)
```

The subgoal has three assumptions. We produce a contradiction between the assumptions `~P(x)` and `P(?y3(x))`. The proof never instantiates the unknown `?a`.

```
by (eresolve_tac [notE] 1);
  Level 7
  EX y. ALL x. P(y) --> P(x)
   1. !!x xa. [| P(?a); P(?y3(x)) |] ==> P(x)
by (assume_tac 1);
  Level 8
  EX y. ALL x. P(y) --> P(x)
  No subgoals!
```

The civilised way to prove this theorem is through `best_tac`, supplying the classical version of $(\exists I)$:

```
goal FOL.thy "EX y. ALL x. P(y)-->P(x)";
  Level 0
  EX y. ALL x. P(y) --> P(x)
   1. EX y. ALL x. P(y) --> P(x)
by (best_tac FOL_dup_cs 1);
  Level 1
  EX y. ALL x. P(y) --> P(x)
  No subgoals!
```

If this theorem seems counterintuitive, then perhaps you are an intuitionist. In constructive logic, proving $\exists y \,.\, \forall x \,.\, P(y) \to P(x)$ requires exhibiting a particular term t such that $\forall x \,.\, P(t) \to P(x)$, which we cannot do without further knowledge about P.

16.8 Derived rules and the classical tactics

Classical first-order logic can be extended with the propositional connective *if*(P, Q, R), where

$$\textit{if}(P, Q, R) \equiv P \land Q \lor \neg P \land R. \qquad (\textit{if})$$

Theorems about *if* can be proved by treating this as an abbreviation, replacing *if*(P, Q, R) by $P \land Q \lor \neg P \land R$ in subgoals. But this duplicates P, causing an exponential blowup and an unreadable formula. Introducing further abbreviations makes the problem worse.

Natural deduction demands rules that introduce and eliminate *if*(P, Q, R) directly, without reference to its definition. The simple identity

$$\textit{if}(P, Q, R) \leftrightarrow (P \to Q) \land (\neg P \to R)$$

suggests that the *if*-introduction rule should be

$$\frac{\begin{array}{cc} [P] & [\neg P] \\ \vdots & \vdots \\ Q & R \end{array}}{\textit{if}(P, Q, R)} \; (\textit{if}\ I)$$

The *if*-elimination rule reflects the definition of $if(P,Q,R)$ and the elimination rules for $\vee$ and $\wedge$.

$$\frac{if(P,Q,R) \quad \begin{matrix}[P,Q]\\ \vdots \\ S\end{matrix} \quad \begin{matrix}[\neg P,R]\\ \vdots \\ S\end{matrix}}{S}\ (if\ E)$$

Having made these plans, we get down to work with Isabelle. The theory of classical logic, `FOL`, is extended with the constant $if :: [o,o,o] \Rightarrow o$. The axiom `if_def` asserts the equation (if).

```
If = FOL +
consts  if      :: "[o,o,o]=>o"
rules   if_def "if(P,Q,R) == P&Q | ~P&R"
end
```

The derivations of the introduction and elimination rules demonstrate the methods for rewriting with definitions. Classical reasoning is required, so we use `fast_tac`.

16.8.1 Deriving the introduction rule

The introduction rule, given the premises $P \Longrightarrow Q$ and $\neg P \Longrightarrow R$, concludes $if(P,Q,R)$. We propose the conclusion as the main goal using `goalw`, which uses `if_def` to rewrite occurrences of *if* in the subgoal.

```
val prems = goalw If.thy [if_def]
    "[| P ==> Q; ~ P ==> R |] ==> if(P,Q,R)";
  Level 0
  if(P,Q,R)
   1. P & Q | ~ P & R
```

The premises (bound to the ML variable `prems`) are passed as introduction rules to `fast_tac`:

```
by (fast_tac (FOL_cs addIs prems) 1);
  Level 1
  if(P,Q,R)
  No subgoals!
val ifI = result();
```

16.8.2 Deriving the elimination rule

The elimination rule has three premises, two of which are themselves rules. The conclusion is simply S.

```
val major::prems = goalw If.thy [if_def]
   "[| if(P,Q,R);  [| P; Q |] ==> S; [| ~ P; R |] ==> S |] ==> S";
  Level 0
  S
   1. S
```

The major premise contains an occurrence of *if*, but the version returned by `goalw` (and bound to the ML variable `major`) has the definition expanded. Now `cut_facts_tac` inserts `major` as an assumption in the subgoal, so that `fast_tac` can break it down.

```
by (cut_facts_tac [major] 1);
  Level 1
  S
   1. P & Q | ~ P & R ==> S
by (fast_tac (FOL_cs addIs prems) 1);
  Level 2
  S
  No subgoals!
val ifE = result();
```

As you may recall from Sect. 3.1.2, there are other ways of treating definitions when deriving a rule. We can start the proof using `goal`, which does not expand definitions, instead of `goalw`. We can use `rewrite_goals_tac` to expand definitions in the subgoals — perhaps after calling `cut_facts_tac` to insert the rule's premises. We can use `rewrite_rule`, which is a meta-inference rule, to expand definitions in the premises directly.

16.8.3 Using the derived rules

The rules just derived have been saved with the ML names `ifI` and `ifE`. They permit natural proofs of theorems such as the following:

$$\begin{aligned} \mathit{if}(P, \mathit{if}(Q, A, B), \mathit{if}(Q, C, D)) &\leftrightarrow \mathit{if}(Q, \mathit{if}(P, A, C), \mathit{if}(P, B, D)) \\ \mathit{if}(\mathit{if}(P, Q, R), A, B) &\leftrightarrow \mathit{if}(P, \mathit{if}(Q, A, B), \mathit{if}(R, A, B)) \end{aligned}$$

Proofs also require the classical reasoning rules and the $\leftrightarrow$ introduction rule (called `iffI`: do not confuse with `ifI`).

To display the *if*-rules in action, let us analyse a proof step by step.

```
goal If.thy
    "if(P, if(Q,A,B), if(Q,C,D)) <-> if(Q, if(P,A,C), if(P,B,D))";
  Level 0
  if(P,if(Q,A,B),if(Q,C,D)) <-> if(Q,if(P,A,C),if(P,B,D))
   1. if(P,if(Q,A,B),if(Q,C,D)) <-> if(Q,if(P,A,C),if(P,B,D))
by (resolve_tac [iffI] 1);
  Level 1
  if(P,if(Q,A,B),if(Q,C,D)) <-> if(Q,if(P,A,C),if(P,B,D))
   1. if(P,if(Q,A,B),if(Q,C,D)) ==> if(Q,if(P,A,C),if(P,B,D))
   2. if(Q,if(P,A,C),if(P,B,D)) ==> if(P,if(Q,A,B),if(Q,C,D))
```

The *if*-elimination rule can be applied twice in succession.

```
by (eresolve_tac [ifE] 1);
  Level 2
  if(P,if(Q,A,B),if(Q,C,D)) <-> if(Q,if(P,A,C),if(P,B,D))
   1. [| P; if(Q,A,B) |] ==> if(Q,if(P,A,C),if(P,B,D))
   2. [| ~ P; if(Q,C,D) |] ==> if(Q,if(P,A,C),if(P,B,D))
   3. if(Q,if(P,A,C),if(P,B,D)) ==> if(P,if(Q,A,B),if(Q,C,D))
```

```
by (eresolve_tac [ifE] 1);
  Level 3
  if(P,if(Q,A,B),if(Q,C,D)) <-> if(Q,if(P,A,C),if(P,B,D))
   1. [| P; Q; A |] ==> if(Q,if(P,A,C),if(P,B,D))
   2. [| P; ~ Q; B |] ==> if(Q,if(P,A,C),if(P,B,D))
   3. [| ~ P; if(Q,C,D) |] ==> if(Q,if(P,A,C),if(P,B,D))
   4. if(Q,if(P,A,C),if(P,B,D)) ==> if(P,if(Q,A,B),if(Q,C,D))
```

In the first two subgoals, all assumptions have been reduced to atoms. Now *if*-introduction can be applied. Observe how the *if*-rules break down occurrences of *if* when they become the outermost connective.

```
by (resolve_tac [ifI] 1);
  Level 4
  if(P,if(Q,A,B),if(Q,C,D)) <-> if(Q,if(P,A,C),if(P,B,D))
   1. [| P; Q; A; Q |] ==> if(P,A,C)
   2. [| P; Q; A; ~ Q |] ==> if(P,B,D)
   3. [| P; ~ Q; B |] ==> if(Q,if(P,A,C),if(P,B,D))
   4. [| ~ P; if(Q,C,D) |] ==> if(Q,if(P,A,C),if(P,B,D))
   5. if(Q,if(P,A,C),if(P,B,D)) ==> if(P,if(Q,A,B),if(Q,C,D))
by (resolve_tac [ifI] 1);
  Level 5
  if(P,if(Q,A,B),if(Q,C,D)) <-> if(Q,if(P,A,C),if(P,B,D))
   1. [| P; Q; A; Q; P |] ==> A
   2. [| P; Q; A; Q; ~ P |] ==> C
   3. [| P; Q; A; ~ Q |] ==> if(P,B,D)
   4. [| P; ~ Q; B |] ==> if(Q,if(P,A,C),if(P,B,D))
   5. [| ~ P; if(Q,C,D) |] ==> if(Q,if(P,A,C),if(P,B,D))
   6. if(Q,if(P,A,C),if(P,B,D)) ==> if(P,if(Q,A,B),if(Q,C,D))
```

Where do we stand? The first subgoal holds by assumption; the second and third, by contradiction. This is getting tedious. Let us revert to the initial proof state and let `fast_tac` solve it. The classical rule set `if_cs` contains the rules of `FOL` plus the derived rules for *if*.

```
choplev 0;
  Level 0
  if(P,if(Q,A,B),if(Q,C,D)) <-> if(Q,if(P,A,C),if(P,B,D))
   1. if(P,if(Q,A,B),if(Q,C,D)) <-> if(Q,if(P,A,C),if(P,B,D))
val if_cs = FOL_cs addSIs [ifI] addSEs[ifE];
by (fast_tac if_cs 1);
  Level 1
  if(P,if(Q,A,B),if(Q,C,D)) <-> if(Q,if(P,A,C),if(P,B,D))
  No subgoals!
```

This tactic also solves the other example.

```
goal If.thy "if(if(P,Q,R), A, B) <-> if(P, if(Q,A,B), if(R,A,B))";
  Level 0
  if(if(P,Q,R),A,B) <-> if(P,if(Q,A,B),if(R,A,B))
   1. if(if(P,Q,R),A,B) <-> if(P,if(Q,A,B),if(R,A,B))
by (fast_tac if_cs 1);
  Level 1
  if(if(P,Q,R),A,B) <-> if(P,if(Q,A,B),if(R,A,B))
  No subgoals!
```

16.8.4 Derived rules versus definitions

Dispensing with the derived rules, we can treat *if* as an abbreviation, and let `fast_tac` prove the expanded formula. Let us redo the previous proof:

```
choplev 0;
  Level 0
  if(if(P,Q,R),A,B) <-> if(P,if(Q,A,B),if(R,A,B))
   1. if(if(P,Q,R),A,B) <-> if(P,if(Q,A,B),if(R,A,B))
```

This time, simply unfold using the definition of *if*:

```
by (rewrite_goals_tac [if_def]);
  Level 1
  if(if(P,Q,R),A,B) <-> if(P,if(Q,A,B),if(R,A,B))
   1. (P & Q | ~ P & R) & A | ~ (P & Q | ~ P & R) & B <->
      P & (Q & A | ~ Q & B) | ~ P & (R & A | ~ R & B)
```

We are left with a subgoal in pure first-order logic:

```
by (fast_tac FOL_cs 1);
  Level 2
  if(if(P,Q,R),A,B) <-> if(P,if(Q,A,B),if(R,A,B))
  No subgoals!
```

Expanding definitions reduces the extended logic to the base logic. This approach has its merits — especially if the prover for the base logic is good — but can be slow. In these examples, proofs using `if_cs` (the derived rules) run about six times faster than proofs using `FOL_cs`.

Expanding definitions also complicates error diagnosis. Suppose we are having difficulties in proving some goal. If by expanding definitions we have made it unreadable, then we have little hope of diagnosing the problem.

Attempts at program verification often yield invalid assertions. Let us try to prove one:

```
goal If.thy "if(if(P,Q,R), A, B) <-> if(P, if(Q,A,B), if(R,B,A))";
  Level 0
  if(if(P,Q,R),A,B) <-> if(P,if(Q,A,B),if(R,B,A))
   1. if(if(P,Q,R),A,B) <-> if(P,if(Q,A,B),if(R,B,A))
by (fast_tac FOL_cs 1);
  by: tactic failed
```

This failure message is uninformative, but we can get a closer look at the situation by applying `step_tac`.

```
by (REPEAT (step_tac if_cs 1));
  Level 1
  if(if(P,Q,R),A,B) <-> if(P,if(Q,A,B),if(R,B,A))
   1. [| A; ~ P; R; ~ P; R |] ==> B
   2. [| B; ~ P; ~ R; ~ P; ~ R |] ==> A
   3. [| ~ P; R; B; ~ P; R |] ==> A
   4. [| ~ P; ~ R; A; ~ B; ~ P |] ==> R
```

Subgoal 1 is unprovable and yields a countermodel: P and B are false while R and A are true. This truth assignment reduces the main goal to $\mathit{true} \leftrightarrow \mathit{false}$, which is of course invalid.

We can repeat this analysis by expanding definitions, using just the rules of FOL:

```
choplev 0;
  Level 0
  if(if(P,Q,R),A,B) <-> if(P,if(Q,A,B),if(R,B,A))
   1. if(if(P,Q,R),A,B) <-> if(P,if(Q,A,B),if(R,B,A))
by (rewrite_goals_tac [if_def]);
  Level 1
  if(if(P,Q,R),A,B) <-> if(P,if(Q,A,B),if(R,B,A))
   1. (P & Q | ~ P & R) & A | ~ (P & Q | ~ P & R) & B <->
      P & (Q & A | ~ Q & B) | ~ P & (R & B | ~ R & A)
by (fast_tac FOL_cs 1);
  by: tactic failed
```

Again we apply `step_tac`:

```
by (REPEAT (step_tac FOL_cs 1));
  Level 2
  if(if(P,Q,R),A,B) <-> if(P,if(Q,A,B),if(R,B,A))
   1. [| A; ~ P; R; ~ P; R; ~ False |] ==> B
   2. [| A; ~ P; R; R; ~ False; ~ B; ~ B |] ==> Q
   3. [| B; ~ P; ~ R; ~ P; ~ A |] ==> R
   4. [| B; ~ P; ~ R; ~ Q; ~ A |] ==> R
   5. [| B; ~ R; ~ P; ~ A; ~ R; Q; ~ False |] ==> A
   6. [| ~ P; R; B; ~ P; R; ~ False |] ==> A
   7. [| ~ P; ~ R; A; ~ B; ~ R |] ==> P
   8. [| ~ P; ~ R; A; ~ B; ~ R |] ==> Q
```

Subgoal 1 yields the same countermodel as before. But each proof step has taken six times as long, and the final result contains twice as many subgoals.

Expanding definitions causes a great increase in complexity. This is why the classical prover has been designed to accept derived rules.

17. Zermelo-Fraenkel Set Theory

The theory `ZF` implements Zermelo-Fraenkel set theory [23, 56] as an extension of `FOL`, classical first-order logic. The theory includes a collection of derived natural deduction rules, for use with Isabelle's classical reasoner. Much of it is based on the work of Noël [38].

A tremendous amount of set theory has been formally developed, including the basic properties of relations, functions and ordinals. Significant results have been proved, such as the Schröder-Bernstein Theorem and a version of Ramsey's Theorem. General methods have been developed for solving recursion equations over monotonic functors; these have been applied to yield constructions of lists, trees, infinite lists, etc. The Recursion Theorem has been proved, admitting recursive definitions of functions over well-founded relations. Thus, we may even regard set theory as a computational logic, loosely inspired by Martin-Löf's Type Theory.

Because `ZF` is an extension of `FOL`, it provides the same packages, namely `hyp_subst_tac`, the simplifier, and the classical reasoner. The main simplification set is called `ZF_ss`. Several classical rule sets are defined, including `lemmas_cs`, `upair_cs` and `ZF_cs`.

`ZF` now has a flexible package for handling inductive definitions, such as inference systems, and datatype definitions, such as lists and trees. Moreover it also handles coinductive definitions, such as bisimulation relations, and codatatype definitions, such as streams. A recent paper describes the package [48].

Recent reports [49, 50] describe `ZF` less formally than this chapter. Isabelle employs a novel treatment of non-well-founded data structures within the standard ZF axioms including the Axiom of Foundation [51].

17.1 Which version of axiomatic set theory?

The two main axiom systems for set theory are Bernays-Gödel (BG) and Zermelo-Fraenkel (ZF). Resolution theorem provers can use BG because it is finite [3, 54]. ZF does not have a finite axiom system because of its Axiom Scheme of Replacement. This makes it awkward to use with many theorem provers, since instances of the axiom scheme have to be invoked explicitly. Since Isabelle has no difficulty with axiom schemes, we may adopt either axiom system.

These two theories differ in their treatment of **classes**, which are collections that are 'too big' to be sets. The class of all sets, V, cannot be a set without admitting Russell's Paradox. In BG, both classes and sets are individuals; $x \in V$ expresses that x is a set. In ZF, all variables denote sets; classes are identified with unary predicates. The two systems define essentially the same sets and classes, with similar properties. In particular, a class cannot belong to another class (let alone a set).

Modern set theorists tend to prefer ZF because they are mainly concerned with sets, rather than classes. BG requires tiresome proofs that various collections are sets; for instance, showing $x \in \{x\}$ requires showing that x is a set.

17.2 The syntax of set theory

The language of set theory, as studied by logicians, has no constants. The traditional axioms merely assert the existence of empty sets, unions, powersets, etc.; this would be intolerable for practical reasoning. The Isabelle theory declares constants for primitive sets. It also extends `FOL` with additional syntax for finite sets, ordered pairs, comprehension, general union/intersection, general sums/products, and bounded quantifiers. In most other respects, Isabelle implements precisely Zermelo-Fraenkel set theory.

Figure 17.1 lists the constants and infixes of `ZF`, while Figure 17.2 presents the syntax translations. Finally, Figure 17.3 presents the full grammar for set theory, including the constructs of `FOL`.

Set theory does not use polymorphism. All terms in `ZF` have type `i`, which is the type of individuals and lies in class `logic`. The type of first-order formulae, remember, is `o`.

Infix operators include binary union and intersection ($A \bigcup B$ and $A \bigcap B$), set difference ($A - B$), and the subset and membership relations. Note that a~$:b$ is translated to $\neg(a \in b)$. The union and intersection operators ($\bigcup A$ and $\bigcap A$) form the union or intersection of a set of sets; $\bigcup A$ means the same as $\bigcup_{x \in A} x$. Of these operators, only $\bigcup A$ is primitive.

The constant `Upair` constructs unordered pairs; thus `Upair(`A`,`B`)` denotes the set $\{A, B\}$ and `Upair(`A`,`A`)` denotes the singleton $\{A\}$. General union is used to define binary union. The Isabelle version goes on to define the constant `cons`:

$$\begin{aligned} A \cup B &\equiv \bigcup(\texttt{Upair}(A, B)) \\ \texttt{cons}(a, B) &\equiv \texttt{Upair}(a, a) \bigcup B \end{aligned}$$

The $\{\ldots\}$ notation abbreviates finite sets constructed in the obvious manner using `cons` and $\emptyset$ (the empty set):

$$\{a, b, c\} \equiv \texttt{cons}(a, \texttt{cons}(b, \texttt{cons}(c, \emptyset)))$$

The constant `Pair` constructs ordered pairs, as in `Pair(`a`,`b`)`. Ordered pairs may also be written within angle brackets, as `<`a`,`b`>`. The n-tuple

name	*meta-type*	*description*
0	i	empty set
cons	$[i, i] \Rightarrow i$	finite set constructor
Upair	$[i, i] \Rightarrow i$	unordered pairing
Pair	$[i, i] \Rightarrow i$	ordered pairing
Inf	i	infinite set
Pow	$i \Rightarrow i$	powerset
Union Inter	$i \Rightarrow i$	set union/intersection
split	$[[i, i] \Rightarrow i, i] \Rightarrow i$	generalized projection
fst snd	$i \Rightarrow i$	projections
converse	$i \Rightarrow i$	converse of a relation
succ	$i \Rightarrow i$	successor
Collect	$[i, i \Rightarrow o] \Rightarrow i$	separation
Replace	$[i, [i, i] \Rightarrow o] \Rightarrow i$	replacement
PrimReplace	$[i, [i, i] \Rightarrow o] \Rightarrow i$	primitive replacement
RepFun	$[i, i \Rightarrow i] \Rightarrow i$	functional replacement
Pi Sigma	$[i, i \Rightarrow i] \Rightarrow i$	general product/sum
domain	$i \Rightarrow i$	domain of a relation
range	$i \Rightarrow i$	range of a relation
field	$i \Rightarrow i$	field of a relation
Lambda	$[i, i \Rightarrow i] \Rightarrow i$	λ-abstraction
restrict	$[i, i] \Rightarrow i$	restriction of a function
The	$[i \Rightarrow o] \Rightarrow i$	definite description
if	$[o, i, i] \Rightarrow i$	conditional
Ball Bex	$[i, i \Rightarrow o] \Rightarrow o$	bounded quantifiers

CONSTANTS

symbol	*meta-type*	*priority*	*description*
``	$[i, i] \Rightarrow i$	Left 90	image
-``	$[i, i] \Rightarrow i$	Left 90	inverse image
`	$[i, i] \Rightarrow i$	Left 90	application
Int	$[i, i] \Rightarrow i$	Left 70	intersection ($\bigcap$)
Un	$[i, i] \Rightarrow i$	Left 65	union ($\bigcup$)
-	$[i, i] \Rightarrow i$	Left 65	set difference ($-$)
:	$[i, i] \Rightarrow o$	Left 50	membership ($\in$)
<=	$[i, i] \Rightarrow o$	Left 50	subset ($\subseteq$)

INFIXES

Fig. 17.1. Constants of ZF

external	*internal*	*description*
a ~: b	~(a : b)	negated membership
{a_1, ..., a_n}	cons(a_1,···,cons(a_n,0))	finite set
<a_1, ..., a_{n-1}, a_n>	Pair(a_1,...,Pair(a_{n-1},a_n)...)	ordered n-tuple
$\{x\!:\!A\,.\,P[x]\}$	Collect(A,$\lambda x\,.\,P[x]$)	separation
$\{y\,.\,x\!:\!A,\ Q[x,y]\}$	Replace(A,$\lambda x\,y\,.\,Q[x,y]$)	replacement
$\{b[x]\,.\,x\!:\!A\}$	RepFun(A,$\lambda x\,.\,b[x]$)	functional replacement
INT $x\!:\!A\,.\,B[x]$	Inter($\{B[x]\,.\,x\!:\!A\}$)	general intersection
UN $x\!:\!A\,.\,B[x]$	Union($\{B[x]\,.\,x\!:\!A\}$)	general union
PROD $x\!:\!A\,.\,B[x]$	Pi(A,$\lambda x\,.\,B[x]$)	general product
SUM $x\!:\!A\,.\,B[x]$	Sigma(A,$\lambda x\,.\,B[x]$)	general sum
A -> B	Pi(A,$\lambda x\,.\,B$)	function space
A * B	Sigma(A,$\lambda x\,.\,B$)	binary product
THE $x\,.\,P[x]$	The($\lambda x\,.\,P[x]$)	definite description
lam $x\!:\!A\,.\,b[x]$	Lambda(A,$\lambda x\,.\,b[x]$)	λ-abstraction
ALL $x\!:\!A\,.\,P[x]$	Ball(A,$\lambda x\,.\,P[x]$)	bounded $\forall$
EX $x\!:\!A\,.\,P[x]$	Bex(A,$\lambda x\,.\,P[x]$)	bounded $\exists$

Fig. 17.2. Translations for ZF

<a_1,...,a_{n-1},a_n> abbreviates the nest of pairs

Pair(a_1,...,Pair(a_{n-1},a_n)...).

In ZF, a function is a set of pairs. A ZF function f is simply an individual as far as Isabelle is concerned: its Isabelle type is i, not say $i \Rightarrow i$. The infix operator ‘ denotes the application of a function set to its argument; we must write f‘x, not $f(x)$. The syntax for image is f“A and that for inverse image is f−“A.

17.3 Binding operators

The constant Collect constructs sets by the principle of **separation**. The syntax for separation is $\{x\!:\!A\,.\,P[x]\}$, where $P[x]$ is a formula that may contain free occurrences of x. It abbreviates the set Collect(A,$\lambda x.P[x]$), which consists of all $x \in A$ that satisfy $P[x]$. Note that Collect is an unfortunate choice of name: some set theories adopt a set-formation principle, related to replacement, called collection.

The constant Replace constructs sets by the principle of **replacement**. The syntax $\{y\,.\,x\!:\!A, Q[x,y]\}$ denotes the set Replace(A,$\lambda x\,y\,.\,Q[x,y]$), which consists of all y such that there exists $x \in A$ satisfying $Q[x,y]$. The Replacement Axiom has the condition that Q must be single-valued over A: for all $x \in A$ there exists at most one y satisfying $Q[x,y]$. A single-valued binary predicate is also called a **class function**.

The constant RepFun expresses a special case of replacement, where $Q[x,y]$ has the form $y = b[x]$. Such a Q is trivially single-valued, since it is just the

```
term    = expression of type i
        | { term (,term)* }
        | < term (,term)* >
        | { id:term . formula }
        | { id . id:term, formula }
        | { term . id:term }
        | term `` term
        | term -`` term
        | term ` term
        | term * term
        | term Int term
        | term Un term
        | term - term
        | term -> term
        | THE id . formula
        | lam id:term . term
        | INT id:term . term
        | UN id:term . term
        | PROD id:term . term
        | SUM id:term . term

formula = expression of type o
        | term : term
        | term ~: term
        | term <= term
        | term = term
        | term ~= term
        | ~ formula
        | formula & formula
        | formula | formula
        | formula --> formula
        | formula <-> formula
        | ALL id:term . formula
        | EX id:term . formula
        | ALL id id* . formula
        | EX id id* . formula
        | EX! id id* . formula
```

Fig. 17.3. Full grammar for ZF

graph of the meta-level function $\lambda x . b[x]$. The resulting set consists of all $b[x]$ for $x \in A$. This is analogous to the ML functional `map`, since it applies a function to every element of a set. The syntax is $\{b[x].x:A\}$, which expands to `RepFun(`A`,`$\lambda x . b[x]$`)`.

General unions and intersections of indexed families of sets, namely $\bigcup_{x\in A} B[x]$ and $\bigcap_{x\in A} B[x]$, are written `UN` $x:A.B[x]$ and `INT` $x:A.B[x]$. Their meaning is expressed using `RepFun` as

$$\bigcup(\{B[x] . x \in A\}) \quad \text{and} \quad \bigcap(\{B[x] . x \in A\}).$$

General sums $\sum_{x\in A} B[x]$ and products $\prod_{x\in A} B[x]$ can be constructed in set theory, where $B[x]$ is a family of sets over A. They have as special cases $A \times B$ and $A \to B$, where B is simply a set. This is similar to the situation in Constructive Type Theory (set theory has 'dependent sets') and calls for similar syntactic conventions. The constants `Sigma` and `Pi` construct general sums and products. Instead of `Sigma(`A`,`B`)` and `Pi(`A`,`B`)` we may write `SUM` $x:A.B[x]$ and `PROD` $x:A.B[x]$. The special cases as A`*`B and A`->`B abbreviate general sums and products over a constant family.[1] Isabelle accepts these abbreviations in parsing and uses them whenever possible for printing.

As mentioned above, whenever the axioms assert the existence and uniqueness of a set, Isabelle's set theory declares a constant for that set. These constants can express the **definite description** operator $\iota x . P[x]$, which stands for the unique a satisfying $P[a]$, if such exists. Since all terms in `ZF` denote something, a description is always meaningful, but we do not know its value unless $P[x]$ defines it uniquely. Using the constant `The`, we may write descriptions as `The(`$\lambda x . P[x]$`)` or use the syntax `THE` $x.P[x]$.

Function sets may be written in λ-notation; $\lambda x \in A . b[x]$ stands for the set of all pairs $\langle x, b[x]\rangle$ for $x \in A$. In order for this to be a set, the function's domain A must be given. Using the constant `Lambda`, we may express function sets as `Lambda(`A`,`$\lambda x . b[x]$`)` or use the syntax `lam` $x:A.b[x]$.

Isabelle's set theory defines two **bounded quantifiers**:

$$\begin{array}{lll} \forall x \in A . P[x] & \text{abbreviates} & \forall x . x \in A \to P[x] \\ \exists x \in A . P[x] & \text{abbreviates} & \exists x . x \in A \land P[x] \end{array}$$

The constants `Ball` and `Bex` are defined accordingly. Instead of `Ball(`A`,`P`)` and `Bex(`A`,`P`)` we may write `ALL` $x:A.P[x]$ and `EX` $x:A.P[x]$.

17.4 The Zermelo-Fraenkel axioms

The axioms appear in Fig. 17.4. They resemble those presented by Suppes [56]. Most of the theory consists of definitions. In particular, bounded quantifiers

[1]Unlike normal infix operators, `*` and `->` merely define abbreviations; there are no constants `op *` and `op ->`.

```
Ball_def          Ball(A,P) == ALL x. x:A --> P(x)
Bex_def           Bex(A,P)  == EX x. x:A & P(x)

subset_def        A <= B  == ALL x:A. x:B
extension         A = B  <->  A <= B & B <= A

union_iff         A : Union(C) <-> (EX B:C. A:B)
power_set         A : Pow(B) <-> A <= B
foundation        A=0 | (EX x:A. ALL y:x. ~ y:A)

replacement       (ALL x:A. ALL y z. P(x,y) & P(x,z) --> y=z) ==>
                  b : PrimReplace(A,P) <-> (EX x:A. P(x,b))
```

THE ZERMELO-FRAENKEL AXIOMS

```
Replace_def  Replace(A,P) ==
                  PrimReplace(A, %x y. (EX!z.P(x,z)) & P(x,y))
RepFun_def   RepFun(A,f) == {y . x:A, y=f(x)}
the_def      The(P)      == Union({y . x:{0}, P(y)})
if_def       if(P,a,b)   == THE z. P & z=a | ~P & z=b
Collect_def  Collect(A,P) == {y . x:A, x=y & P(x)}
Upair_def    Upair(a,b)  ==
               {y. x:Pow(Pow(0)), (x=0 & y=a) | (x=Pow(0) & y=b)}
```

CONSEQUENCES OF REPLACEMENT

```
Inter_def    Inter(A) == { x:Union(A) . ALL y:A. x:y}
Un_def       A Un  B  == Union(Upair(A,B))
Int_def      A Int B  == Inter(Upair(A,B))
Diff_def     A - B    == { x:A . ~(x:B) }
```

UNION, INTERSECTION, DIFFERENCE

Fig. 17.4. Rules and axioms of ZF

```
cons_def      cons(a,A) == Upair(a,a) Un A
succ_def      succ(i) == cons(i,i)
infinity      0:Inf & (ALL y:Inf. succ(y): Inf)
```

FINITE AND INFINITE SETS

```
Pair_def      <a,b>        == {{a,a}, {a,b}}
split_def     split(c,p)   == THE y. EX a b. p=<a,b> & y=c(a,b)
fst_def       fst(A)       == split(%x y.x, p)
snd_def       snd(A)       == split(%x y.y, p)
Sigma_def     Sigma(A,B)   == UN x:A. UN y:B(x). {<x,y>}
```

ORDERED PAIRS AND CARTESIAN PRODUCTS

```
converse_def   converse(r) == {z. w:r, EX x y. w=<x,y> & z=<y,x>}
domain_def     domain(r)   == {x. w:r, EX y. w=<x,y>}
range_def      range(r)    == domain(converse(r))
field_def      field(r)    == domain(r) Un range(r)
image_def      r `` A      == {y : range(r) . EX x:A. <x,y> : r}
vimage_def     r -`` A     == converse(r)``A
```

OPERATIONS ON RELATIONS

```
lam_def     Lambda(A,b) == {<x,b(x)> . x:A}
apply_def   f`a         == THE y. <a,y> : f
Pi_def      Pi(A,B) == {f: Pow(Sigma(A,B)). ALL x:A. EX! y. <x,y>: f}
restrict_def   restrict(f,A) == lam x:A.f`x
```

FUNCTIONS AND GENERAL PRODUCT

Fig. 17.5. Further definitions of ZF

and the subset relation appear in other axioms. Object-level quantifiers and implications have been replaced by meta-level ones wherever possible, to simplify use of the axioms. See the file `ZF/ZF.thy` for details.

The traditional replacement axiom asserts

$$y \in \texttt{PrimReplace}(A, P) \leftrightarrow (\exists x \in A \,.\, P(x, y))$$

subject to the condition that $P(x, y)$ is single-valued for all $x \in A$. The Isabelle theory defines `Replace` to apply `PrimReplace` to the single-valued part of P, namely

$$(\exists! z \,.\, P(x, z)) \wedge P(x, y).$$

Thus $y \in \texttt{Replace}(A, P)$ if and only if there is some x such that $P(x, -)$ holds uniquely for y. Because the equivalence is unconditional, `Replace` is much easier to use than `PrimReplace`; it defines the same set, if $P(x, y)$ is single-valued. The nice syntax for replacement expands to `Replace`.

Other consequences of replacement include functional replacement (`RepFun`) and definite descriptions (`The`). Axioms for separation (`Collect`) and unordered pairs (`Upair`) are traditionally assumed, but they actually follow from replacement [56, pages 237–8].

The definitions of general intersection, etc., are straightforward. Note the definition of `cons`, which underlies the finite set notation. The axiom of infinity gives us a set that contains 0 and is closed under successor (`succ`). Although this set is not uniquely defined, the theory names it (`Inf`) in order to simplify the construction of the natural numbers.

Further definitions appear in Fig. 17.5. Ordered pairs are defined in the standard way, $\langle a, b\rangle \equiv \{\{a\}, \{a, b\}\}$. Recall that $\texttt{Sigma}(A, B)$ generalizes the Cartesian product of two sets. It is defined to be the union of all singleton sets $\{\langle x, y\rangle\}$, for $x \in A$ and $y \in B(x)$. This is a typical usage of general union.

The projections `fst` and `snd` are defined in terms of the generalized projection `split`. The latter has been borrowed from Martin-Löf's Type Theory, and is often easier to use than `fst` and `snd`.

Operations on relations include converse, domain, range, and image. The set $\texttt{Pi}(A, B)$ generalizes the space of functions between two sets. Note the simple definitions of λ-abstraction (using `RepFun`) and application (using a definite description). The function $\texttt{restrict}(f, A)$ has the same values as f, but only over the domain A.

17.5 From basic lemmas to function spaces

Faced with so many definitions, it is essential to prove lemmas. Even trivial theorems like $A \cap B = B \cap A$ would be difficult to prove from the definitions alone. Isabelle's set theory derives many rules using a natural deduction style. Ideally, a natural deduction rule should introduce or eliminate just one operator, but this is not always practical. For most operators, we may forget its definition and use its derived rules instead.

```
ballI       [| !!x. x:A ==> P(x) |] ==> ALL x:A. P(x)
bspec       [| ALL x:A. P(x);  x: A |] ==> P(x)
ballE       [| ALL x:A. P(x);  P(x) ==> Q;  ~ x:A ==> Q |] ==> Q

ball_cong   [| A=A';  !!x. x:A' ==> P(x) <-> P'(x) |] ==>
            (ALL x:A. P(x)) <-> (ALL x:A'. P'(x))

bexI        [| P(x);  x: A |] ==> EX x:A. P(x)
bexCI       [| ALL x:A. ~P(x) ==> P(a);  a: A |] ==> EX x:A.P(x)
bexE        [| EX x:A. P(x);  !!x. [| x:A; P(x) |] ==> Q |] ==> Q

bex_cong    [| A=A';  !!x. x:A' ==> P(x) <-> P'(x) |] ==>
            (EX x:A. P(x)) <-> (EX x:A'. P'(x))
```

BOUNDED QUANTIFIERS

```
subsetI       (!!x.x:A ==> x:B) ==> A <= B
subsetD       [| A <= B;  c:A |] ==> c:B
subsetCE      [| A <= B;  ~(c:A) ==> P;  c:B ==> P |] ==> P
subset_refl   A <= A
subset_trans  [| A<=B;  B<=C |] ==> A<=C

equalityI     [| A <= B;  B <= A |] ==> A = B
equalityD1    A = B ==> A<=B
equalityD2    A = B ==> B<=A
equalityE     [| A = B;  [| A<=B; B<=A |] ==> P |]  ==> P
```

SUBSETS AND EXTENSIONALITY

```
emptyE          a:0 ==> P
empty_subsetI   0 <= A
equals0I        [| !!y. y:A ==> False |] ==> A=0
equals0D        [| A=0;  a:A |] ==> P

PowI            A <= B ==> A : Pow(B)
PowD            A : Pow(B)  ==>  A<=B
```

THE EMPTY SET; POWER SETS

Fig. 17.6. Basic derived rules for ZF

17.5.1 Fundamental lemmas

Figure 17.6 presents the derived rules for the most basic operators. The rules for the bounded quantifiers resemble those for the ordinary quantifiers, but note that `ballE` uses a negated assumption in the style of Isabelle's classical reasoner. The congruence rules `ball_cong` and `bex_cong` are required by Isabelle's simplifier, but have few other uses. Congruence rules must be specially derived for all binding operators, and henceforth will not be shown.

Figure 17.6 also shows rules for the subset and equality relations (proof by extensionality), and rules about the empty set and the power set operator.

Figure 17.7 presents rules for replacement and separation. The rules for `Replace` and `RepFun` are much simpler than comparable rules for `PrimReplace` would be. The principle of separation is proved explicitly, although most proofs should use the natural deduction rules for `Collect`. The elimination rule `CollectE` is equivalent to the two destruction rules `CollectD1` and `CollectD2`, but each rule is suited to particular circumstances. Although too many rules can be confusing, there is no reason to aim for a minimal set of rules. See the file `ZF/ZF.ML` for a complete listing.

Figure 17.8 presents rules for general union and intersection. The empty intersection should be undefined. We cannot have $\bigcap(\emptyset) = V$ because V, the universal class, is not a set. All expressions denote something in ZF set theory; the definition of intersection implies $\bigcap(\emptyset) = \emptyset$, but this value is arbitrary. The rule `InterI` must have a premise to exclude the empty intersection. Some of the laws governing intersections require similar premises.

17.5.2 Unordered pairs and finite sets

Figure 17.9 presents the principle of unordered pairing, along with its derived rules. Binary union and intersection are defined in terms of ordered pairs (Fig. 17.10). Set difference is also included. The rule `UnCI` is useful for classical reasoning about unions, like `disjCI`; it supersedes `UnI1` and `UnI2`, but these rules are often easier to work with. For intersection and difference we have both elimination and destruction rules. Again, there is no reason to provide a minimal rule set.

Figure 17.11 is concerned with finite sets: it presents rules for `cons`, the finite set constructor, and rules for singleton sets. Figure 17.12 presents derived rules for the successor function, which is defined in terms of `cons`. The proof that `succ` is injective appears to require the Axiom of Foundation.

Definite descriptions (`THE`) are defined in terms of the singleton set $\{0\}$, but their derived rules fortunately hide this (Fig. 17.13). The rule `theI` is difficult to apply because of the two occurrences of $?P$. However, `the_equality` does not have this problem and the files contain many examples of its use.

Finally, the impossibility of having both $a \in b$ and $b \in a$ (`mem_anti_sym`) is proved by applying the Axiom of Foundation to the set $\{a, b\}$. The impossibility of $a \in a$ is a trivial consequence.

```
ReplaceI     [| x: A;  P(x,b);  !!y. P(x,y) ==> y=b |] ==>
             b : {y. x:A, P(x,y)}

ReplaceE     [| b : {y. x:A, P(x,y)};
                !!x. [| x: A;  P(x,b);  ALL y. P(x,y)-->y=b |] ==> R
             |] ==> R

RepFunI      [| a : A |] ==> f(a) : {f(x). x:A}
RepFunE      [| b : {f(x). x:A};
                !!x.[| x:A;  b=f(x) |] ==> P |] ==> P

separation    a : {x:A. P(x)} <-> a:A & P(a)
CollectI      [| a:A;  P(a) |] ==> a : {x:A. P(x)}
CollectE      [| a : {x:A. P(x)};  [| a:A; P(a) |] ==> R |] ==> R
CollectD1     a : {x:A. P(x)} ==> a:A
CollectD2     a : {x:A. P(x)} ==> P(a)
```

Fig. 17.7. Replacement and separation

```
UnionI     [| B: C;  A: B |] ==> A: Union(C)
UnionE     [| A : Union(C);  !!B.[| A: B;  B: C |] ==> R |] ==> R

InterI     [| !!x. x: C ==> A: x;  c:C |] ==> A : Inter(C)
InterD     [| A : Inter(C);  B : C |] ==> A : B
InterE     [| A : Inter(C);  A:B ==> R;  ~ B:C ==> R |] ==> R

UN_I       [| a: A;  b: B(a) |] ==> b: (UN x:A. B(x))
UN_E       [| b : (UN x:A. B(x));  !!x.[| x: A;  b: B(x) |] ==> R
           |] ==> R

INT_I      [| !!x. x: A ==> b: B(x);  a: A |] ==> b: (INT x:A. B(x))
INT_E      [| b : (INT x:A. B(x));  a: A |] ==> b : B(a)
```

Fig. 17.8. General union and intersection

```
pairing    a:Upair(b,c) <-> (a=b | a=c)
UpairI1    a : Upair(a,b)
UpairI2    b : Upair(a,b)
UpairE     [| a : Upair(b,c);  a = b ==> P;  a = c ==> P |] ==> P
```

Fig. 17.9. Unordered pairs

```
UnI1         c : A ==> c : A Un B
UnI2         c : B ==> c : A Un B
UnCI         (~c : B ==> c : A) ==> c : A Un B
UnE          [| c : A Un B;  c:A ==> P;  c:B ==> P |] ==> P

IntI         [| c : A;  c : B |] ==> c : A Int B
IntD1        c : A Int B ==> c : A
IntD2        c : A Int B ==> c : B
IntE         [| c : A Int B;  [| c:A; c:B |] ==> P |] ==> P

DiffI        [| c : A;  ~ c : B |] ==> c : A - B
DiffD1       c : A - B ==> c : A
DiffD2       [| c : A - B;  c : B |] ==> P
DiffE        [| c : A - B;  [| c:A; ~ c:B |] ==> P |] ==> P
```

Fig. 17.10. Union, intersection, difference

```
consI1       a : cons(a,B)
consI2       a : B ==> a : cons(b,B)
consCI       (~ a:B ==> a=b) ==> a: cons(b,B)
consE        [| a : cons(b,A);  a=b ==> P;  a:A ==> P |] ==> P

singletonI   a : {a}
singletonE   [| a : {b}; a=b ==> P |] ==> P
```

Fig. 17.11. Finite and singleton sets

```
succI1       i : succ(i)
succI2       i : j ==> i : succ(j)
succCI       (~ i:j ==> i=j) ==> i: succ(j)
succE        [| i : succ(j);  i=j ==> P;  i:j ==> P |] ==> P
succ_neq_0   [| succ(n)=0 |] ==> P
succ_inject  succ(m) = succ(n) ==> m=n
```

Fig. 17.12. The successor function

```
the_equality     [| P(a);  !!x. P(x) ==> x=a |] ==> (THE x. P(x)) = a
theI             EX! x. P(x) ==> P(THE x. P(x))

if_P             P ==> if(P,a,b) = a
if_not_P         ~P ==> if(P,a,b) = b

mem_anti_sym     [| a:b;  b:a |] ==> P
mem_anti_refl    a:a ==> P
```

Fig. 17.13. Descriptions; non-circularity

```
Union_upper       B:A ==> B <= Union(A)
Union_least       [| !!x. x:A ==> x<=C |] ==> Union(A) <= C

Inter_lower       B:A ==> Inter(A) <= B
Inter_greatest    [| a:A;  !!x. x:A ==> C<=x |] ==> C <= Inter(A)

Un_upper1         A <= A Un B
Un_upper2         B <= A Un B
Un_least          [| A<=C;  B<=C |] ==> A Un B <= C

Int_lower1        A Int B <= A
Int_lower2        A Int B <= B
Int_greatest      [| C<=A;  C<=B |] ==> C <= A Int B

Diff_subset       A-B <= A
Diff_contains     [| C<=A;  C Int B = 0 |] ==> C <= A-B

Collect_subset    Collect(A,P) <= A
```

Fig. 17.14. Subset and lattice properties

See the file ZF/upair.ML for full proofs of the rules discussed in this section.

17.5.3 Subset and lattice properties

The subset relation is a complete lattice. Unions form least upper bounds; non-empty intersections form greatest lower bounds. Figure 17.14 shows the corresponding rules. A few other laws involving subsets are included. Proofs are in the file ZF/subset.ML.

Reasoning directly about subsets often yields clearer proofs than reasoning about the membership relation. Section 17.9 below presents an example of this, proving the equation $\texttt{Pow}(A) \cap \texttt{Pow}(B) = \texttt{Pow}(A \cap B)$.

17.5.4 Ordered pairs

Figure 17.15 presents the rules governing ordered pairs, projections and general sums. File ZF/pair.ML contains the full (and tedious) proof that $\{\{a\},\{a,b\}\}$ functions as an ordered pair. This property is expressed as two destruction rules, Pair_inject1 and Pair_inject2, and equivalently as the elimination rule Pair_inject.

The rule Pair_neq_0 asserts $\langle a,b\rangle \neq \emptyset$. This is a property of $\{\{a\},\{a,b\}\}$, and need not hold for other encodings of ordered pairs. The non-standard ordered pairs mentioned below satisfy $\langle\emptyset;\emptyset\rangle = \emptyset$.

The natural deduction rules SigmaI and SigmaE assert that $\texttt{Sigma}(A,B)$ consists of all pairs of the form $\langle x,y\rangle$, for $x \in A$ and $y \in B(x)$. The rule SigmaE2 merely states that $\langle a,b\rangle \in \texttt{Sigma}(A,B)$ implies $a \in A$ and $b \in B(a)$.

```
Pair_inject1    <a,b> = <c,d> ==> a=c
Pair_inject2    <a,b> = <c,d> ==> b=d
Pair_inject     [| <a,b> = <c,d>;  [| a=c; b=d |] ==> P |] ==> P
Pair_neq_0      <a,b>=0 ==> P

fst_conv        fst(<a,b>) = a
snd_conv        snd(<a,b>) = b
split           split(%x y.c(x,y), <a,b>) = c(a,b)

SigmaI          [| a:A;  b:B(a) |] ==> <a,b> : Sigma(A,B)

SigmaE          [| c: Sigma(A,B);
                   !!x y.[| x:A; y:B(x); c=<x,y> |] ==> P |] ==> P

SigmaE2         [| <a,b> : Sigma(A,B);
                   [| a:A;  b:B(a) |] ==> P   |] ==> P
```

Fig. 17.15. Ordered pairs; projections; general sums

```
domainI        <a,b>: r ==> a : domain(r)
domainE        [| a : domain(r);  !!y. <a,y>: r ==> P |] ==> P
domain_subset  domain(Sigma(A,B)) <= A

rangeI         <a,b>: r ==> b : range(r)
rangeE         [| b : range(r);  !!x. <x,b>: r ==> P |] ==> P
range_subset   range(A*B) <= B

fieldI1        <a,b>: r ==> a : field(r)
fieldI2        <a,b>: r ==> b : field(r)
fieldCI        (~ <c,a>:r ==> <a,b>: r) ==> a : field(r)

fieldE         [| a : field(r);
                  !!x. <a,x>: r ==> P;
                  !!x. <x,a>: r ==> P
               |] ==> P

field_subset   field(A*A) <= A
```

Fig. 17.16. Domain, range and field of a relation

```
imageI         [| <a,b>: r;  a:A |] ==> b : r``A
imageE         [| b: r``A;  !!x.[| <x,b>: r;  x:A |] ==> P |] ==> P

vimageI        [| <a,b>: r;  b:B |] ==> a : r-``B
vimageE        [| a: r-``B;  !!x.[| <a,x>: r;  x:B |] ==> P |] ==> P
```

Fig. 17.17. Image and inverse image

```
fun_is_rel       f: Pi(A,B) ==> f <= Sigma(A,B)

apply_equality  [| <a,b>: f;  f: Pi(A,B) |] ==> f`a = b
apply_equality2 [| <a,b>: f;  <a,c>: f;  f: Pi(A,B) |] ==> b=c

apply_type      [| f: Pi(A,B);  a:A |] ==> f`a : B(a)
apply_Pair      [| f: Pi(A,B);  a:A |] ==> <a,f`a>: f
apply_iff       f: Pi(A,B) ==> <a,b>: f <-> a:A & f`a = b

fun_extension   [| f : Pi(A,B);  g: Pi(A,D);
                   !!x. x:A ==> f`x = g`x     |] ==> f=g

domain_type     [| <a,b> : f;  f: Pi(A,B) |] ==> a : A
range_type      [| <a,b> : f;  f: Pi(A,B) |] ==> b : B(a)

Pi_type         [| f: A->C;  !!x. x:A ==> f`x: B(x) |] ==> f: Pi(A,B)
domain_of_fun   f: Pi(A,B) ==> domain(f)=A
range_of_fun    f: Pi(A,B) ==> f: A->range(f)

restrict        a : A ==> restrict(f,A) ` a = f`a
restrict_type   [| !!x. x:A ==> f`x: B(x) |] ==>
                restrict(f,A) : Pi(A,B)
```

Fig. 17.18. Functions

17.5.5 Relations

Figure 17.16 presents rules involving relations, which are sets of ordered pairs. The converse of a relation r is the set of all pairs $\langle y, x\rangle$ such that $\langle x, y\rangle \in r$; if r is a function, then `converse`(r) is its inverse. The rules for the domain operation, namely `domainI` and `domainE`, assert that `domain`(r) consists of all x such that r contains some pair of the form $\langle x, y\rangle$. The range operation is similar, and the field of a relation is merely the union of its domain and range.

Figure 17.17 presents rules for images and inverse images. Note that these operations are generalisations of range and domain, respectively. See the file `ZF/domrange.ML` for derivations of the rules.

```
lamI        a:A ==> <a,b(a)> : (lam x:A. b(x))
lamE        [| p: (lam x:A. b(x));  !!x.[| x:A; p=<x,b(x)> |] ==> P
            |] ==>  P

lam_type    [| !!x. x:A ==> b(x): B(x) |] ==> (lam x:A.b(x)) : Pi(A,B)

beta        a : A ==> (lam x:A.b(x)) ` a = b(a)
eta         f : Pi(A,B) ==> (lam x:A. f`x) = f
```

Fig. 17.19. λ-abstraction

```
fun_empty            0: 0->0
fun_single           {<a,b>} : {a} -> {b}

fun_disjoint_Un      [| f: A->B;  g: C->D;  A Int C = 0  |] ==>
                     (f Un g) : (A Un C) -> (B Un D)

fun_disjoint_apply1  [| a:A;  f: A->B;  g: C->D;  A Int C = 0 |] ==>
                     (f Un g)`a = f`a

fun_disjoint_apply2  [| c:C;  f: A->B;  g: C->D;  A Int C = 0 |] ==>
                     (f Un g)`c = g`c
```

Fig. 17.20. Constructing functions from smaller sets

17.5.6 Functions

Functions, represented by graphs, are notoriously difficult to reason about. The file `ZF/func.ML` derives many rules, which overlap more than they ought. This section presents the more important rules.

Figure 17.18 presents the basic properties of `Pi`(A, B), the generalized function space. For example, if f is a function and $\langle a, b\rangle \in f$, then $f`a = b$ (`apply_equality`). Two functions are equal provided they have equal domains and deliver equals results (`fun_extension`).

By `Pi_type`, a function typing of the form $f \in A \to C$ can be refined to the dependent typing $f \in \prod_{x\in A} B(x)$, given a suitable family of sets $\{B(x)\}_{x\in A}$. Conversely, by `range_of_fun`, any dependent typing can be flattened to yield a function type of the form $A \to C$; here, $C =$ `range`(f).

Among the laws for λ-abstraction, `lamI` and `lamE` describe the graph of the generated function, while `beta` and `eta` are the standard conversions. We essentially have a dependently-typed λ-calculus (Fig. 17.19).

Figure 17.20 presents some rules that can be used to construct functions explicitly. We start with functions consisting of at most one pair, and may form the union of two functions provided their domains are disjoint.

17.6 Further developments

The next group of developments is complex and extensive, and only highlights can be covered here. It involves many theories and ML files of proofs.

Figure 17.21 presents commutative, associative, distributive, and idempotency laws of union and intersection, along with other equations. See file `ZF/equalities.ML`.

Theory `Bool` defines $\{0, 1\}$ as a set of booleans, with the usual operators including a conditional (Fig. 17.22). Although `ZF` is a first-order theory, you can obtain the effect of higher-order logic using `bool`-valued functions, for example. The constant `1` is translated to `succ(0)`.

```
Int_absorb          A Int A = A
Int_commute         A Int B = B Int A
Int_assoc           (A Int B) Int C  =  A Int (B Int C)
Int_Un_distrib      (A Un B) Int C  =  (A Int C) Un (B Int C)

Un_absorb           A Un A = A
Un_commute          A Un B = B Un A
Un_assoc            (A Un B) Un C  =  A Un (B Un C)
Un_Int_distrib      (A Int B) Un C  =  (A Un C) Int (B Un C)

Diff_cancel         A-A = 0
Diff_disjoint       A Int (B-A) = 0
Diff_partition      A<=B ==> A Un (B-A) = B
double_complement   [| A<=B; B<= C |] ==> (B - (C-A)) = A
Diff_Un             A - (B Un C) = (A-B) Int (A-C)
Diff_Int            A - (B Int C) = (A-B) Un (A-C)

Union_Un_distrib    Union(A Un B) = Union(A) Un Union(B)
Inter_Un_distrib    [| a:A;  b:B |] ==>
                    Inter(A Un B) = Inter(A) Int Inter(B)

Int_Union_RepFun    A Int Union(B) = (UN C:B. A Int C)

Un_Inter_RepFun     b:B ==>
                    A Un Inter(B) = (INT C:B. A Un C)

SUM_Un_distrib1     (SUM x:A Un B. C(x)) =
                    (SUM x:A. C(x)) Un (SUM x:B. C(x))

SUM_Un_distrib2     (SUM x:C. A(x) Un B(x)) =
                    (SUM x:C. A(x))  Un  (SUM x:C. B(x))

SUM_Int_distrib1    (SUM x:A Int B. C(x)) =
                    (SUM x:A. C(x)) Int (SUM x:B. C(x))

SUM_Int_distrib2    (SUM x:C. A(x) Int B(x)) =
                    (SUM x:C. A(x)) Int (SUM x:C. B(x))
```

Fig. 17.21. Equalities

```
bool_def        bool == {0,1}
cond_def        cond(b,c,d) == if(b=1,c,d)
not_def         not(b)  == cond(b,0,1)
and_def         a and b == cond(a,b,0)
or_def          a or b  == cond(a,1,b)
xor_def         a xor b == cond(a,not(b),b)

bool_1I         1 : bool
bool_0I         0 : bool
boolE           [| c: bool;  c=1 ==> P;  c=0 ==> P |] ==> P
cond_1          cond(1,c,d) = c
cond_0          cond(0,c,d) = d
```

Fig. 17.22. The booleans

symbol	*meta-type*	*priority*	*description*
`+`	$[i, i] \Rightarrow i$	Right 65	disjoint union operator
`Inl Inr`	$i \Rightarrow i$		injections
`case`	$[i \Rightarrow i, i \Rightarrow i, i] \Rightarrow i$		conditional for $A + B$

```
sum_def         A+B == {0}*A Un {1}*B
Inl_def         Inl(a) == <0,a>
Inr_def         Inr(b) == <1,b>
case_def        case(c,d,u) == split(%y z. cond(y, d(z), c(z)), u)

sum_InlI        a : A ==> Inl(a) : A+B
sum_InrI        b : B ==> Inr(b) : A+B

Inl_inject      Inl(a)=Inl(b) ==> a=b
Inr_inject      Inr(a)=Inr(b) ==> a=b
Inl_neq_Inr     Inl(a)=Inr(b) ==> P

sumE2   u: A+B ==> (EX x. x:A & u=Inl(x)) | (EX y. y:B & u=Inr(y))

case_Inl        case(c,d,Inl(a)) = c(a)
case_Inr        case(c,d,Inr(b)) = d(b)
```

Fig. 17.23. Disjoint unions

```
QPair_def       <a;b> == a+b
qsplit_def      qsplit(c,p)   == THE y. EX a b. p=<a;b> & y=c(a,b)
qfsplit_def     qfsplit(R,z) == EX x y. z=<x;y> & R(x,y)
qconverse_def   qconverse(r) == z. w:r, EX x y. w=<x;y> & z=<y;x>
QSigma_def      QSigma(A,B)  ==  UN x:A. UN y:B(x). <x;y>

qsum_def        A <+> B      == ({0} <*> A) Un ({1} <*> B)
QInl_def        QInl(a)      == <0;a>
QInr_def        QInr(b)      == <1;b>
qcase_def       qcase(c,d)   == qsplit(%y z. cond(y, d(z), c(z)))
```

Fig. 17.24. Non-standard pairs, products and sums

Theory `Sum` defines the disjoint union of two sets, with injections and a case analysis operator (Fig. 17.23). Disjoint unions play a role in datatype definitions, particularly when there is mutual recursion [50].

Theory `QPair` defines a notion of ordered pair that admits non-well-founded tupling (Fig. 17.24). Such pairs are written $<a;b>$. It also defines the eliminator `qsplit`, the converse operator `qconverse`, and the summation operator `QSigma`. These are completely analogous to the corresponding versions for standard ordered pairs. The theory goes on to define a non-standard notion of disjoint sum using non-standard pairs. All of these concepts satisfy the same properties as their standard counterparts; in addition, $<a;b>$ is continuous. The theory supports coinductive definitions, for example of infinite lists [51].

The Knaster-Tarski Theorem states that every monotone function over a complete lattice has a fixedpoint. Theory `Fixedpt` proves the Theorem only for a particular lattice, namely the lattice of subsets of a set (Fig. 17.25). The theory defines least and greatest fixedpoint operators with corresponding induction and coinduction rules. These are essential to many definitions that follow, including the natural numbers and the transitive closure operator. The (co)inductive definition package also uses the fixedpoint operators [48]. See Davey and Priestley [10] for more on the Knaster-Tarski Theorem and my paper [50] for discussion of the Isabelle proofs.

Monotonicity properties are proved for most of the set-forming operations: union, intersection, Cartesian product, image, domain, range, etc. These are useful for applying the Knaster-Tarski Fixedpoint Theorem. The proofs themselves are trivial applications of Isabelle's classical reasoner. See file `ZF/mono.ML`.

The theory `Perm` is concerned with permutations (bijections) and related concepts. These include composition of relations, the identity relation, and three specialized function spaces: injective, surjective and bijective. Figure 17.26 displays many of their properties that have been proved. These results are fundamental to a treatment of equipollence and cardinality.

Theory `Nat` defines the natural numbers and mathematical induction, along with a case analysis operator. The set of natural numbers, here called `nat`, is known in set theory as the ordinal ω.

```
bnd_mono_def   bnd_mono(D,h) ==
                 h(D)<=D & (ALL W X. W<=X --> X<=D --> h(W) <= h(X))

lfp_def        lfp(D,h) == Inter(X: Pow(D). h(X) <= X)
gfp_def        gfp(D,h) == Union(X: Pow(D). X <= h(X))


lfp_lowerbound [| h(A) <= A;  A<=D |] ==> lfp(D,h) <= A

lfp_subset     lfp(D,h) <= D

lfp_greatest   [| bnd_mono(D,h);
                  !!X. [| h(X) <= X;  X<=D |] ==> A<=X
               |] ==> A <= lfp(D,h)

lfp_Tarski     bnd_mono(D,h) ==> lfp(D,h) = h(lfp(D,h))

induct         [| a : lfp(D,h);  bnd_mono(D,h);
                  !!x. x : h(Collect(lfp(D,h),P)) ==> P(x)
               |] ==> P(a)

lfp_mono       [| bnd_mono(D,h);  bnd_mono(E,i);
                  !!X. X<=D ==> h(X) <= i(X)
               |] ==> lfp(D,h) <= lfp(E,i)

gfp_upperbound [| A <= h(A);  A<=D |] ==> A <= gfp(D,h)

gfp_subset     gfp(D,h) <= D

gfp_least      [| bnd_mono(D,h);
                  !!X. [| X <= h(X);  X<=D |] ==> X<=A
               |] ==> gfp(D,h) <= A

gfp_Tarski     bnd_mono(D,h) ==> gfp(D,h) = h(gfp(D,h))

coinduct       [| bnd_mono(D,h); a: X; X <= h(X Un gfp(D,h)); X <= D
               |] ==> a : gfp(D,h)

gfp_mono       [| bnd_mono(D,h);  D <= E;
                  !!X. X<=D ==> h(X) <= i(X)
               |] ==> gfp(D,h) <= gfp(E,i)
```

Fig. 17.25. Least and greatest fixedpoints

symbol	*meta-type*	*priority*	*description*
`O`	$[i,i] \Rightarrow i$	Right 60	composition ($\circ$)
`id`	$i \Rightarrow i$		identity function
`inj`	$[i,i] \Rightarrow i$		injective function space
`surj`	$[i,i] \Rightarrow i$		surjective function space
`bij`	$[i,i] \Rightarrow i$		bijective function space

```
comp_def  r O s      == {xz : domain(s)*range(r) .
                           EX x y z. xz=<x,z> & <x,y>:s & <y,z>:r}
id_def    id(A)      == (lam x:A. x)
inj_def   inj(A,B)   == { f: A->B. ALL w:A. ALL x:A. f`w=f`x --> w=x}
surj_def  surj(A,B)  == { f: A->B . ALL y:B. EX x:A. f`x=y}
bij_def   bij(A,B)   == inj(A,B) Int surj(A,B)

left_inverse      [| f: inj(A,B);  a: A |] ==> converse(f)`(f`a) = a
right_inverse     [| f: inj(A,B);  b: range(f) |] ==>
                  f`(converse(f)`b) = b

inj_converse_inj f: inj(A,B) ==> converse(f): inj(range(f), A)
bij_converse_bij f: bij(A,B) ==> converse(f): bij(B,A)

comp_type         [| s<=A*B;  r<=B*C |] ==> (r O s) <= A*C
comp_assoc        (r O s) O t = r O (s O t)

left_comp_id      r<=A*B ==> id(B) O r = r
right_comp_id     r<=A*B ==> r O id(A) = r

comp_func         [| g:A->B; f:B->C |] ==> (f O g):A->C
comp_func_apply   [| g:A->B; f:B->C; a:A |] ==> (f O g)`a = f`(g`a)

comp_inj          [| g:inj(A,B);  f:inj(B,C)  |] ==> (f O g):inj(A,C)
comp_surj         [| g:surj(A,B); f:surj(B,C) |] ==> (f O g):surj(A,C)
comp_bij          [| g:bij(A,B); f:bij(B,C) |] ==> (f O g):bij(A,C)

left_comp_inverse     f: inj(A,B) ==> converse(f) O f = id(A)
right_comp_inverse    f: surj(A,B) ==> f O converse(f) = id(B)

bij_disjoint_Un
    [| f: bij(A,B);  g: bij(C,D);  A Int C = 0;  B Int D = 0 |] ==>
    (f Un g) : bij(A Un C, B Un D)

restrict_bij  [| f:inj(A,B);  C<=A |] ==> restrict(f,C): bij(C, f``C)
```

Fig. 17.26. Permutations

symbol	*meta-type*	*priority*	*description*
nat	i		set of natural numbers
nat_case	$[i, i \Rightarrow i, i] \Rightarrow i$		conditional for *nat*
rec	$[i, i, [i, i] \Rightarrow i] \Rightarrow i$		recursor for *nat*
#*	$[i, i] \Rightarrow i$	Left 70	multiplication
div	$[i, i] \Rightarrow i$	Left 70	division
mod	$[i, i] \Rightarrow i$	Left 70	modulus
#+	$[i, i] \Rightarrow i$	Left 65	addition
#-	$[i, i] \Rightarrow i$	Left 65	subtraction

```
nat_def       nat == lfp(lam r: Pow(Inf). {0} Un {succ(x). x:r}

nat_case_def  nat_case(a,b,k) ==
              THE y. k=0 & y=a | (EX x. k=succ(x) & y=b(x))

rec_def       rec(k,a,b) ==
              transrec(k, %n f. nat_case(a, %m. b(m, f`m), n))

add_def       m#+n    == rec(m, n, %u v.succ(v))
diff_def      m#-n    == rec(n, m, %u v. rec(v, 0, %x y.x))
mult_def      m#*n    == rec(m, 0, %u v. n #+ v)
mod_def       m mod n == transrec(m, %j f. if(j:n, j, f`(j#-n)))
div_def       m div n == transrec(m, %j f. if(j:n, 0, succ(f`(j#-n))))


nat_0I        0 : nat
nat_succI     n : nat ==> succ(n) : nat

nat_induct
    [| n: nat;  P(0);  !!x. [| x: nat;  P(x) |] ==> P(succ(x))
    |] ==> P(n)

nat_case_0    nat_case(a,b,0) = a
nat_case_succ nat_case(a,b,succ(m)) = b(m)

rec_0         rec(0,a,b) = a
rec_succ      rec(succ(m),a,b) = b(m, rec(m,a,b))

mult_type      [| m:nat;  n:nat |] ==> m #* n : nat
mult_0         0 #* n = 0
mult_succ      succ(m) #* n = n #+ (m #* n)
mult_commute   [| m:nat;  n:nat |] ==> m #* n = n #* m
add_mult_dist
    [| m:nat;  k:nat |] ==> (m #+ n) #* k = (m #* k) #+ (n #* k)
mult_assoc
    [| m:nat;  n:nat;  k:nat |] ==> (m #* n) #* k = m #* (n #* k)
mod_quo_equality
    [| 0:n;  m:nat;  n:nat |] ==> (m div n)#*n #+ m mod n = m
```

Fig. 17.27. The natural numbers

```
Fin_0I          0 : Fin(A)
Fin_consI       [| a: A;  b: Fin(A) |] ==> cons(a,b) : Fin(A)

Fin_induct
    [| b: Fin(A);
       P(0);
       !!x y. [| x: A;  y: Fin(A);  x~:y;  P(y) |] ==> P(cons(x,y))
    |] ==> P(b)

Fin_mono        A<=B ==> Fin(A) <= Fin(B)
Fin_UnI         [| b: Fin(A);  c: Fin(A) |] ==> b Un c : Fin(A)
Fin_UnionI      C : Fin(Fin(A)) ==> Union(C) : Fin(A)
Fin_subset      [| c<=b;  b: Fin(A) |] ==> c: Fin(A)
```

Fig. 17.28. The finite set operator

Theory `Arith` defines primitive recursion and goes on to develop arithmetic on the natural numbers (Fig. 17.27). It defines addition, multiplication, subtraction, division, and remainder. Many of their properties are proved: commutative, associative and distributive laws, identity and cancellation laws, etc. The most interesting result is perhaps the theorem $a \bmod b + (a/b) \times b = a$. Division and remainder are defined by repeated subtraction, which requires well-founded rather than primitive recursion; the termination argument relies on the divisor's being non-zero.

Theory `Univ` defines a 'universe' $\texttt{univ}(A)$, for constructing datatypes such as trees. This set contains A and the natural numbers. Vitally, it is closed under finite products: $\texttt{univ}(A) \times \texttt{univ}(A) \subseteq \texttt{univ}(A)$. This theory also defines the cumulative hierarchy of axiomatic set theory, which traditionally is written V_α for an ordinal α. The 'universe' is a simple generalization of V_ω.

Theory `QUniv` defines a 'universe' $\texttt{quniv}(A)$, for constructing codatatypes such as streams. It is analogous to $\texttt{univ}(A)$ (and is defined in terms of it) but is closed under the non-standard product and sum.

Figure 17.28 presents the finite set operator; $\texttt{Fin}(A)$ is the set of all finite sets over A. The definition employs Isabelle's inductive definition package [48], which proves various rules automatically. The induction rule shown is stronger than the one proved by the package. See file `ZF/Fin.ML`.

Figure 17.29 presents the set of lists over A, $\texttt{list}(A)$. The definition employs Isabelle's datatype package, which defines the introduction and induction rules automatically, as well as the constructors and case operator (`list_case`). See file `ZF/List.ML`. The file `ZF/ListFn.thy` proceeds to define structural recursion and the usual list functions.

The constructions of the natural numbers and lists make use of a suite of operators for handling recursive function definitions. I have described the developments in detail elsewhere [50]. Here is a brief summary:

- Theory `Trancl` defines the transitive closure of a relation (as a least fixedpoint).

`list`	$i \Rightarrow i$		lists over some set
`list_case`	$[i, [i, i] \Rightarrow i, i] \Rightarrow i$		conditional for *list*(A)
`list_rec`	$[i, i, [i, i, i] \Rightarrow i] \Rightarrow i$		recursor for *list*(A)
`map`	$[i \Rightarrow i, i] \Rightarrow i$		mapping functional
`length`	$i \Rightarrow i$		length of a list
`rev`	$i \Rightarrow i$		reverse of a list
`@`	$[i, i] \Rightarrow i$	Right 60	append for lists
`flat`	$i \Rightarrow i$		append of list of lists

```
list_rec_def    list_rec(l,c,h) ==
                Vrec(l, %l g.list_case(c, %x xs. h(x, xs, g`xs), l))

map_def         map(f,l)  == list_rec(l,  0,  %x xs r. <f(x), r>)
length_def      length(l) == list_rec(l,  0,  %x xs r. succ(r))
app_def         xs@ys     == list_rec(xs, ys, %x xs r. <x,r>)
rev_def         rev(l)    == list_rec(l,  0,  %x xs r. r @ <x,0>)
flat_def        flat(ls)  == list_rec(ls, 0,  %l ls r. l @ r)


NilI            Nil : list(A)
ConsI           [| a: A;  l: list(A) |] ==> Cons(a,l) : list(A)

List.induct
    [| l: list(A);
       P(Nil);
       !!x y. [| x: A;  y: list(A);  P(y) |] ==> P(Cons(x,y))
    |] ==> P(l)

Cons_iff        Cons(a,l)=Cons(a',l') <-> a=a' & l=l'
Nil_Cons_iff    ~ Nil=Cons(a,l)

list_mono       A<=B ==> list(A) <= list(B)

list_rec_Nil    list_rec(Nil,c,h) = c
list_rec_Cons   list_rec(Cons(a,l), c, h) = h(a, l, list_rec(l,c,h))

map_ident       l: list(A) ==> map(%u.u, l) = l
map_compose     l: list(A) ==> map(h, map(j,l)) = map(%u.h(j(u)), l)
map_app_distrib xs: list(A) ==> map(h, xs@ys) = map(h,xs) @ map(h,ys)
map_type
    [| l: list(A);  !!x. x: A ==> h(x): B |] ==> map(h,l) : list(B)
map_flat
    ls: list(list(A)) ==> map(h, flat(ls)) = flat(map(map(h),ls))
```

Fig. 17.29. Lists

$$
\begin{aligned}
a \in \emptyset &\leftrightarrow \bot \\
a \in A \bigcup B &\leftrightarrow a \in A \lor a \in B \\
a \in A \bigcap B &\leftrightarrow a \in A \land a \in B \\
a \in A - B &\leftrightarrow a \in A \land \neg(a \in B) \\
\langle a, b\rangle \in \mathtt{Sigma}(A, B) &\leftrightarrow a \in A \land b \in B(a) \\
a \in \mathtt{Collect}(A, P) &\leftrightarrow a \in A \land P(a) \\
(\forall x \in \emptyset \,.\, P(x)) &\leftrightarrow \top \\
(\forall x \in A \,.\, \top) &\leftrightarrow \top
\end{aligned}
$$

Fig. 17.30. Rewrite rules for set theory

- Theory `WF` proves the Well-Founded Recursion Theorem, using an elegant approach of Tobias Nipkow. This theorem permits general recursive definitions within set theory.

- Theory `Ord` defines the notions of transitive set and ordinal number. It derives transfinite induction. A key definition is **less than**: $i < j$ if and only if i and j are both ordinals and $i \in j$. As a special case, it includes less than on the natural numbers.

- Theory `Epsilon` derives ϵ-induction and ϵ-recursion, which are generalisations of transfinite induction and recursion. It also defines $\mathtt{rank}(x)$, which is the least ordinal α such that x is constructed at stage α of the cumulative hierarchy (thus $x \in V_{\alpha+1}$).

17.7 Simplification rules

`ZF` does not merely inherit simplification from `FOL`, but modifies it extensively. File `ZF/simpdata.ML` contains the details.

The extraction of rewrite rules takes set theory primitives into account. It can strip bounded universal quantifiers from a formula; for example, $\forall x \in A \,.\, f(x) = g(x)$ yields the conditional rewrite rule $x \in A \Longrightarrow f(x) = g(x)$. Given $a \in \{x \in A \,.\, P(x)\}$ it extracts rewrite rules from $a \in A$ and $P(a)$. It can also break down $a \in A \cap B$ and $a \in A - B$.

The simplification set `ZF_ss` contains congruence rules for all the binding operators of `ZF`. It contains all the conversion rules, such as `fst` and `snd`, as well as the rewrites shown in Fig. 17.30.

17.8 The examples directory

The directory `ZF/ex` contains further developments in `ZF` set theory. Here is an overview; see the files themselves for more details. I describe much of this material in other publications [48, 49, 50].

`ZF/ex/misc.ML` contains miscellaneous examples such as Cantor's Theorem, the Schröder-Bernstein Theorem and the 'Composition of homomorphisms' challenge [3].

`ZF/ex/Ramsey.ML` proves the finite exponent 2 version of Ramsey's Theorem, following Basin and Kaufmann's presentation [2].

`ZF/ex/Equiv.ML` develops a theory of equivalence classes, not using the Axiom of Choice.

`ZF/ex/Integ.ML` develops a theory of the integers as equivalence classes of pairs of natural numbers.

`ZF/ex/Bin.ML` defines a datatype for two's complement binary integers. File `BinFn.ML` then develops rewrite rules for binary arithmetic. For instance, $1359 \times -2468 = -3354012$ takes under 14 seconds.

`ZF/ex/BT.ML` defines the recursive data structure `bt`(A), labelled binary trees.

`ZF/ex/Term.ML` and `TermFn.ML` define a recursive data structure for terms and term lists. These are simply finite branching trees.

`ZF/ex/TF.ML` and `TF_Fn.ML` define primitives for solving mutually recursive equations over sets. It constructs sets of trees and forests as an example, including induction and recursion rules that handle the mutual recursion.

`ZF/ex/Prop.ML` and `PropLog.ML` proves soundness and completeness of propositional logic [50]. This illustrates datatype definitions, inductive definitions, structural induction and rule induction.

`ZF/ex/ListN.ML` presents the inductive definition of the lists of n elements [40].

`ZF/ex/Acc.ML` presents the inductive definition of the accessible part of a relation [40].

`ZF/ex/Comb.ML` presents the datatype definition of combinators. The file `Contract0.ML` defines contraction, while file `ParContract.ML` defines parallel contraction and proves the Church-Rosser Theorem. This case study follows Camilleri and Melham [5].

`ZF/ex/LList.ML` and `LList_Eq.ML` develop lazy lists and a notion of coinduction for proving equations between them.

17.9 A proof about powersets

To demonstrate high-level reasoning about subsets, let us prove the equation $\texttt{Pow}(A) \cap \texttt{Pow}(B) = \texttt{Pow}(A \cap B)$. Compared with first-order logic, set theory involves a maze of rules, and theorems have many different proofs. Attempting other proofs of the theorem might be instructive. This proof exploits the lattice properties of intersection. It also uses the monotonicity of the powerset operation, from `ZF/mono.ML`:

```
Pow_mono        A<=B ==> Pow(A) <= Pow(B)
```

We enter the goal and make the first step, which breaks the equation into two inclusions by extensionality:

```
goal ZF.thy "Pow(A Int B) = Pow(A) Int Pow(B)";
  Level 0
  Pow(A Int B) = Pow(A) Int Pow(B)
   1. Pow(A Int B) = Pow(A) Int Pow(B)
by (resolve_tac [equalityI] 1);
  Level 1
  Pow(A Int B) = Pow(A) Int Pow(B)
   1. Pow(A Int B) <= Pow(A) Int Pow(B)
   2. Pow(A) Int Pow(B) <= Pow(A Int B)
```

Both inclusions could be tackled straightforwardly using `subsetI`. A shorter proof results from noting that intersection forms the greatest lower bound:

```
by (resolve_tac [Int_greatest] 1);
  Level 2
  Pow(A Int B) = Pow(A) Int Pow(B)
   1. Pow(A Int B) <= Pow(A)
   2. Pow(A Int B) <= Pow(B)
   3. Pow(A) Int Pow(B) <= Pow(A Int B)
```

Subgoal 1 follows by applying the monotonicity of `Pow` to $A \cap B \subseteq A$; subgoal 2 follows similarly:

```
by (resolve_tac [Int_lower1 RS Pow_mono] 1);
  Level 3
  Pow(A Int B) = Pow(A) Int Pow(B)
   1. Pow(A Int B) <= Pow(B)
   2. Pow(A) Int Pow(B) <= Pow(A Int B)
by (resolve_tac [Int_lower2 RS Pow_mono] 1);
  Level 4
  Pow(A Int B) = Pow(A) Int Pow(B)
   1. Pow(A) Int Pow(B) <= Pow(A Int B)
```

We are left with the opposite inclusion, which we tackle in the straightforward way:

```
by (resolve_tac [subsetI] 1);
  Level 5
  Pow(A Int B) = Pow(A) Int Pow(B)
   1. !!x. x : Pow(A) Int Pow(B) ==> x : Pow(A Int B)
```

The subgoal is to show $x \in$ `Pow`$(A \cap B)$ assuming $x \in$ `Pow`$(A) \cap$`Pow`(B); eliminating this assumption produces two subgoals. The rule `IntE` treats the intersection like a conjunction instead of unfolding its definition.

```
by (eresolve_tac [IntE] 1);
  Level 6
  Pow(A Int B) = Pow(A) Int Pow(B)
   1. !!x. [| x : Pow(A); x : Pow(B) |] ==> x : Pow(A Int B)
```

The next step replaces the `Pow` by the subset relation ($\subseteq$).

```
by (resolve_tac [PowI] 1);
  Level 7
  Pow(A Int B) = Pow(A) Int Pow(B)
   1. !!x. [| x : Pow(A); x : Pow(B) |] ==> x <= A Int B
```

We perform the same replacement in the assumptions. This is a good demonstration of the tactic `dresolve_tac`:

```
by (REPEAT (dresolve_tac [PowD] 1));
  Level 8
  Pow(A Int B) = Pow(A) Int Pow(B)
   1. !!x. [| x <= A; x <= B |] ==> x <= A Int B
```

The assumptions are that x is a lower bound of both A and B, but $A \cap B$ is the greatest lower bound:

```
by (resolve_tac [Int_greatest] 1);
  Level 9
  Pow(A Int B) = Pow(A) Int Pow(B)
   1. !!x. [| x <= A; x <= B |] ==> x <= A
   2. !!x. [| x <= A; x <= B |] ==> x <= B
```

To conclude the proof, we clear up the trivial subgoals:

```
by (REPEAT (assume_tac 1));
  Level 10
  Pow(A Int B) = Pow(A) Int Pow(B)
  No subgoals!
```

We could have performed this proof in one step by applying `fast_tac` with the classical rule set `ZF_cs`. Let us go back to the start:

```
choplev 0;
  Level 0
  Pow(A Int B) = Pow(A) Int Pow(B)
   1. Pow(A Int B) = Pow(A) Int Pow(B)
```

We must add `equalityI` to `ZF_cs` as an introduction rule. Extensionality is not used by default because many equalities can be proved by rewriting.

```
by (fast_tac (ZF_cs addIs [equalityI]) 1);
  Level 1
  Pow(A Int B) = Pow(A) Int Pow(B)
  No subgoals!
```

In the past this was regarded as a difficult proof, as indeed it is if all the symbols are replaced by their definitions.

17.10 Monotonicity of the union operator

For another example, we prove that general union is monotonic: $C \subseteq D$ implies $\bigcup(C) \subseteq \bigcup(D)$. To begin, we tackle the inclusion using subsetI:

```
val [prem] = goal ZF.thy "C<=D ==> Union(C) <= Union(D)";
  Level 0
  Union(C) <= Union(D)
   1. Union(C) <= Union(D)
  val prem = "C <= D  [C <= D]" : thm
by (resolve_tac [subsetI] 1);
  Level 1
  Union(C) <= Union(D)
   1. !!x. x : Union(C) ==> x : Union(D)
```

Big union is like an existential quantifier — the occurrence in the assumptions must be eliminated early, since it creates parameters.

```
by (eresolve_tac [UnionE] 1);
  Level 2
  Union(C) <= Union(D)
   1. !!x B. [| x : B; B : C |] ==> x : Union(D)
```

Now we may apply UnionI, which creates an unknown involving the parameters. To show $x \in \bigcup(D)$ it suffices to show that x belongs to some element, say $?B2(x, B)$, of D.

```
by (resolve_tac [UnionI] 1);
  Level 3
  Union(C) <= Union(D)
   1. !!x B. [| x : B; B : C |] ==> ?B2(x,B) : D
   2. !!x B. [| x : B; B : C |] ==> x : ?B2(x,B)
```

Combining subsetD with the premise $C \subseteq D$ yields $?a \in C \Longrightarrow ?a \in D$, which reduces subgoal 1:

```
by (resolve_tac [prem RS subsetD] 1);
  Level 4
  Union(C) <= Union(D)
   1. !!x B. [| x : B; B : C |] ==> ?B2(x,B) : C
   2. !!x B. [| x : B; B : C |] ==> x : ?B2(x,B)
```

The rest is routine. Note how $?B2(x, B)$ is instantiated.

```
by (assume_tac 1);
  Level 5
  Union(C) <= Union(D)
   1. !!x B. [| x : B; B : C |] ==> x : B
by (assume_tac 1);
  Level 6
  Union(C) <= Union(D)
  No subgoals!
```

Again, fast_tac with ZF_cs can do this proof in one step, provided we somehow supply it with prem. We can either add this premise to the assumptions using cut_facts_tac, or add prem RS subsetD to ZF_cs as an introduction rule.

The file ZF/equalities.ML has many similar proofs. Reasoning about general intersection can be difficult because of its anomalous behaviour on the empty set. However, fast_tac copes well with these. Here is a typical example, borrowed from Devlin [13, page 12]:

```
a:C ==> (INT x:C. A(x) Int B(x)) = (INT x:C.A(x)) Int (INT x:C.B(x))
```

In traditional notation this is

$$a \in C \Longrightarrow \bigcap_{x \in C} \Big(A(x) \cap B(x)\Big) = \Big(\bigcap_{x \in C} A(x)\Big) \cap \Big(\bigcap_{x \in C} B(x)\Big)$$

17.11 Low-level reasoning about functions

The derived rules lamI, lamE, lam_type, beta and eta support reasoning about functions in a λ-calculus style. This is generally easier than regarding functions as sets of ordered pairs. But sometimes we must look at the underlying representation, as in the following proof of fun_disjoint_apply1. This states that if f and g are functions with disjoint domains A and C, and if $a \in A$, then $(f \cup g)`a = f`a$:

```
val prems = goal ZF.thy
    "[| a:A;  f: A->B;  g: C->D;  A Int C = 0 |] ==>  \
\    (f Un g)`a = f`a";
  Level 0
  (f Un g) ` a = f ` a
   1. (f Un g) ` a = f ` a
```

Isabelle has produced the output above; the ML top-level now echoes the binding of prems.

```
val prems = ["a : A  [a : A]",
             "f : A -> B  [f : A -> B]",
             "g : C -> D  [g : C -> D]",
             "A Int C = 0  [A Int C = 0]"] : thm list
```

Using apply_equality, we reduce the equality to reasoning about ordered pairs. The second subgoal is to verify that $f \cup g$ is a function.

```
by (resolve_tac [apply_equality] 1);
  Level 1
  (f Un g) ` a = f ` a
   1. <a,f ` a> : f Un g
   2. f Un g : (PROD x:?A. ?B(x))
```

We must show that the pair belongs to f or g; by UnI1 we choose f:

```
by (resolve_tac [UnI1] 1);
  Level 2
  (f Un g) ` a = f ` a
   1. <a,f ` a> : f
   2. f Un g : (PROD x:?A. ?B(x))
```

To show $\langle a, f`a\rangle \in f$ we use apply_Pair, which is essentially the converse of apply_equality:

```
by (resolve_tac [apply_Pair] 1);
  Level 3
  (f Un g) ` a = f ` a
   1. f : (PROD x:?A2. ?B2(x))
   2. a : ?A2
   3. f Un g : (PROD x:?A. ?B(x))
```

Using the premises $f \in A \to B$ and $a \in A$, we solve the two subgoals from apply_Pair. Recall that a Π-set is merely a generalized function space, and observe that ?A2 is instantiated to A.

```
by (resolve_tac prems 1);
  Level 4
  (f Un g) ` a = f ` a
   1. a : A
   2. f Un g : (PROD x:?A. ?B(x))
by (resolve_tac prems 1);
  Level 5
  (f Un g) ` a = f ` a
   1. f Un g : (PROD x:?A. ?B(x))
```

To construct functions of the form $f \cup g$, we apply fun_disjoint_Un:

```
by (resolve_tac [fun_disjoint_Un] 1);
  Level 6
  (f Un g) ` a = f ` a
   1. f : ?A3 -> ?B3
   2. g : ?C3 -> ?D3
   3. ?A3 Int ?C3 = 0
```

The remaining subgoals are instances of the premises. Again, observe how unknowns are instantiated:

```
by (resolve_tac prems 1);
  Level 7
  (f Un g) ` a = f ` a
   1. g : ?C3 -> ?D3
   2. A Int ?C3 = 0
by (resolve_tac prems 1);
  Level 8
  (f Un g) ` a = f ` a
   1. A Int C = 0
by (resolve_tac prems 1);
  Level 9
  (f Un g) ` a = f ` a
  No subgoals!
```

See the files ZF/func.ML and ZF/WF.ML for more examples of reasoning about functions.

18. Higher-Order Logic

The theory HOL implements higher-order logic. It is based on Gordon's HOL system [22], which itself is based on Church's original paper [7]. Andrews's book [1] is a full description of higher-order logic. Experience with the HOL system has demonstrated that higher-order logic is useful for hardware verification; beyond this, it is widely applicable in many areas of mathematics. It is weaker than ZF set theory but for most applications this does not matter. If you prefer ML to Lisp, you will probably prefer HOL to ZF.

Previous releases of Isabelle included a different version of HOL, with explicit type inference rules [44]. This version no longer exists, but ZF supports a similar style of reasoning.

HOL has a distinct feel, compared with ZF and CTT. It identifies object-level types with meta-level types, taking advantage of Isabelle's built-in type checker. It identifies object-level functions with meta-level functions, so it uses Isabelle's operations for abstraction and application. There is no 'apply' operator: function applications are written as simply $f(a)$ rather than $f`a$.

These identifications allow Isabelle to support HOL particularly nicely, but they also mean that HOL requires more sophistication from the user — in particular, an understanding of Isabelle's type system. Beginners should work with `show_types` set to `true`. Gain experience by working in first-order logic before attempting to use higher-order logic. This chapter assumes familiarity with FOL.

18.1 Syntax

The type class of higher-order terms is called `term`. Type variables range over this class by default. The equality symbol and quantifiers are polymorphic over class `term`.

Class `ord` consists of all ordered types; the relations $<$ and $\leq$ are polymorphic over this class, as are the functions `mono`, `min` and `max`. Three other type classes — `plus`, `minus` and `times` — permit overloading of the operators `+`, `-` and `*`. In particular, `-` is overloaded for set difference and subtraction.

Figure 18.1 lists the constants (including infixes and binders), while Fig. 18.2 presents the grammar of higher-order logic. Note that `a~=b` is translated to $\neg(a = b)$.

name	*meta-type*	*description*
`Trueprop`	$bool \Rightarrow prop$	coercion to *prop*
`not`	$bool \Rightarrow bool$	negation ($\neg$)
`True`	$bool$	tautology ($\top$)
`False`	$bool$	absurdity ($\bot$)
`if`	$[bool, \alpha, \alpha] \Rightarrow \alpha :: term$	conditional
`Inv`	$(\alpha \Rightarrow \beta) \Rightarrow (\beta \Rightarrow \alpha)$	function inversion
`Let`	$[\alpha, \alpha \Rightarrow \beta] \Rightarrow \beta$	let binder

CONSTANTS

symbol	*name*	*meta-type*	*description*
`@`	`Eps`	$(\alpha \Rightarrow bool) \Rightarrow \alpha :: term$	Hilbert description (ϵ)
`!` or `ALL`	`All`	$(\alpha :: term \Rightarrow bool) \Rightarrow bool$	universal quantifier ($\forall$)
`?` or `EX`	`Ex`	$(\alpha :: term \Rightarrow bool) \Rightarrow bool$	existential quantifier ($\exists$)
`?!` or `EX!`	`Ex1`	$(\alpha :: term \Rightarrow bool) \Rightarrow bool$	unique existence ($\exists!$)

BINDERS

symbol	*meta-type*	*priority*	*description*
`o`	$[\beta \Rightarrow \gamma, \alpha \Rightarrow \beta] \Rightarrow (\alpha \Rightarrow \gamma)$	Right 50	composition ($\circ$)
`=`	$[\alpha :: term, \alpha] \Rightarrow bool$	Left 50	equality ($=$)
`<`	$[\alpha :: ord, \alpha] \Rightarrow bool$	Left 50	less than ($<$)
`<=`	$[\alpha :: ord, \alpha] \Rightarrow bool$	Left 50	less than or equals ($\leq$)
`&`	$[bool, bool] \Rightarrow bool$	Right 35	conjunction ($\wedge$)
`|`	$[bool, bool] \Rightarrow bool$	Right 30	disjunction ($\vee$)
`-->`	$[bool, bool] \Rightarrow bool$	Right 25	implication ($\rightarrow$)

INFIXES

Fig. 18.1. Syntax of `HOL`

```
term     =  expression of class term
         |  @ id id* . formula
         |  let id = term; ... ; id = term in term

formula  =  expression of type bool
         |  term = term
         |  term ~= term
         |  term < term
         |  term <= term
         |  ~ formula
         |  formula & formula
         |  formula | formula
         |  formula --> formula
         |  !   id id* . formula     |  ALL id id* . formula
         |  ?   id id* . formula     |  EX  id id* . formula
         |  ?!  id id* . formula     |  EX! id id* . formula
```

Fig. 18.2. Full grammar for HOL

! HOL has no if-and-only-if connective; logical equivalence is expressed using equality. But equality has a high priority, as befitting a relation, while if-and-only-if typically has the lowest priority. Thus, $\neg\neg P = P$ abbreviates $\neg\neg(P = P)$ and not $(\neg\neg P) = P$. When using $=$ to mean logical equivalence, enclose both operands in parentheses.

18.1.1 Types

The type of formulae, `bool`, belongs to class `term`; thus, formulae are terms. The built-in type `fun`, which constructs function types, is overloaded with arity `(term,term)term`. Thus, $\sigma \Rightarrow \tau$ belongs to class `term` if σ and τ do, allowing quantification over functions.

Types in HOL must be non-empty; otherwise the quantifier rules would be unsound. I have commented on this elsewhere [44, Sect. 7].

Gordon's HOL system supports **type definitions**. A type is defined by exhibiting an existing type σ, a predicate $P :: \sigma \Rightarrow bool$, and a theorem of the form $\exists x :: \sigma . P(x)$. Thus P specifies a non-empty subset of σ, and the new type denotes this subset. New function constants are generated to establish an isomorphism between the new type and the subset. If type σ involves type variables $\alpha_1, \ldots, \alpha_n$, then the type definition creates a type constructor $(\alpha_1, \ldots, \alpha_n)ty$ rather than a particular type. Melham [34] discusses type definitions at length, with examples.

Isabelle does not support type definitions at present. Instead, they are mimicked by explicit definitions of isomorphism functions. The definitions should be supported by theorems of the form $\exists x :: \sigma . P(x)$, but Isabelle cannot enforce this.

18.1.2 Binders

Hilbert's **description** operator $\epsilon x . P[x]$ stands for some a satisfying $P[a]$, if such exists. Since all terms in HOL denote something, a description is always meaningful, but we do not know its value unless $P[x]$ defines it uniquely. We may write descriptions as `Eps`(P) or use the syntax `@`$x . P[x]$.

Existential quantification is defined by

$$\exists x . P(x) \equiv P(\epsilon x . P(x)).$$

The unique existence quantifier, $\exists! x . P[x]$, is defined in terms of $\exists$ and $\forall$. An Isabelle binder, it admits nested quantifications. For instance, $\exists! xy . P(x, y)$ abbreviates $\exists! x . \exists! y . P(x, y)$; note that this does not mean that there exists a unique pair (x, y) satisfying $P(x, y)$.

Quantifiers have two notations. As in Gordon's HOL system, `HOL` uses `!` and `?` to stand for $\forall$ and $\exists$. The existential quantifier must be followed by a space; thus `?x` is an unknown, while `? x.f(x)=y` is a quantification. Isabelle's usual notation for quantifiers, `ALL` and `EX`, is also available. Both notations are accepted for input. The ML reference `HOL_quantifiers` governs the output notation. If set to `true`, then `!` and `?` are displayed; this is the default. If set to `false`, then `ALL` and `EX` are displayed.

All these binders have priority 10.

18.1.3 The `let` and `case` constructions

Local abbreviations can be introduced by a `let` construct whose syntax appears in Fig. 18.2. Internally it is translated into the constant `Let`. It can be expanded by rewriting with its definition, `Let_def`.

`HOL` also defines the basic syntax

$$\texttt{case}\ e\ \texttt{of}\ c_1\ \texttt{=>}\ e_1\ |\ \ldots\ |\ c_n\ \texttt{=>}\ e_n$$

as a uniform means of expressing `case` constructs. Therefore `case` and `of` are reserved words. However, so far this is mere syntax and has no logical meaning. By declaring translations, you can cause instances of the `case` construct to denote applications of particular case operators. The patterns supplied for $c_1, \ldots, c_n$ distinguish among the different case operators. For an example, see the case construct for lists on page 257 below.

```
refl           t = t::'a
subst          [| s=t; P(s) |] ==> P(t::'a)
ext            (!!x::'a. f(x)::'b = g(x)) ==> (%x.f(x)) = (%x.g(x))
impI           (P ==> Q) ==> P-->Q
mp             [| P-->Q;  P |] ==> Q
iff            (P-->Q) --> (Q-->P) --> (P=Q)
selectI        P(x::'a) ==> P(@x.P(x))
True_or_False  (P=True) | (P=False)
```

Fig. 18.3. The HOL rules

```
True_def   True  == ((%x.x)=(%x.x))
All_def    All   == (%P. P = (%x.True))
Ex_def     Ex    == (%P. P(@x.P(x)))
False_def  False == (!P.P)
not_def    not   == (%P. P-->False)
and_def    op &  == (%P Q. !R. (P-->Q-->R) --> R)
or_def     op |  == (%P Q. !R. (P-->R) --> (Q-->R) --> R)
Ex1_def    Ex1   == (%P. ? x. P(x) & (! y. P(y) --> y=x))

Inv_def    Inv   == (%(f::'a=>'b) y. @x. f(x)=y)
o_def      op o  == (%(f::'b=>'c) g (x::'a). f(g(x)))
if_def     if    == (%P x y.@z::'a.(P=True --> z=x) & (P=False --> z=y))
Let_def    Let(s,f) == f(s)
```

Fig. 18.4. The HOL definitions

18.2 Rules of inference

Figure 18.3 shows the inference rules of HOL, with their ML names. Some of the rules deserve additional comments:

`ext` expresses extensionality of functions.

`iff` asserts that logically equivalent formulae are equal.

`selectI` gives the defining property of the Hilbert ϵ-operator. It is a form of the Axiom of Choice. The derived rule `select_equality` (see below) is often easier to use.

`True_or_False` makes the logic classical.[1]

HOL follows standard practice in higher-order logic: only a few connectives are taken as primitive, with the remainder defined obscurely (Fig. 18.4). Gordon's HOL system expresses the corresponding definitions [22, page 270] using object-equality (=), which is possible because equality in higher-order logic may equate formulae and even functions over formulae. But theory HOL, like all other Isabelle theories, uses meta-equality (==) for definitions.

Some of the rules mention type variables; for example, `refl` mentions the type variable `'a`. This allows you to instantiate type variables explicitly by calling `res_inst_tac`. By default, explicit type variables have class `term`.

Include type constraints whenever you state a polymorphic goal. Type inference may otherwise make the goal more polymorphic than you intended, with confusing results.

! If resolution fails for no obvious reason, try setting `show_types` to `true`, causing Isabelle to display types of terms. Possibly set `show_sorts` to `true` as well, causing Isabelle to display sorts.

Where function types are involved, Isabelle's unification code does not guarantee to find instantiations for type variables automatically. Be prepared to use `res_inst_tac` instead of `resolve_tac`, possibly instantiating type variables. Setting `Unify.trace_types` to `true` causes Isabelle to report omitted search paths during unification.

Some derived rules are shown in Figures 18.5 and 18.6, with their ML names. These include natural rules for the logical connectives, as well as sequent-style elimination rules for conjunctions, implications, and universal quantifiers.

Note the equality rules: `ssubst` performs substitution in backward proofs, while `box_equals` supports reasoning by simplifying both sides of an equation.

---

[1] In fact, the ϵ-operator already makes the logic classical, as shown by Diaconescu; see Paulson [44] for details.

```
sym          s=t ==> t=s
trans        [| r=s; s=t |] ==> r=t
ssubst       [| t=s; P(s) |] ==> P(t::'a)
box_equals   [| a=b;  a=c;  b=d |] ==> c=d
arg_cong     s=t ==> f(s)=f(t)
fun_cong     s::'a=>'b = t ==> s(x)=t(x)
```

EQUALITY

```
TrueI        True
FalseE       False ==> P

conjI        [| P; Q |] ==> P&Q
conjunct1    [| P&Q |] ==> P
conjunct2    [| P&Q |] ==> Q
conjE        [| P&Q;  [| P; Q |] ==> R |] ==> R

disjI1       P ==> P|Q
disjI2       Q ==> P|Q
disjE        [| P | Q; P ==> R; Q ==> R |] ==> R

notI         (P ==> False) ==> ~ P
notE         [| ~ P;  P |] ==> R
impE         [| P-->Q;  P;  Q ==> R |] ==> R
```

PROPOSITIONAL LOGIC

```
iffI         [| P ==> Q;  Q ==> P |] ==> P=Q
iffD1        [| P=Q; P |] ==> Q
iffD2        [| P=Q; Q |] ==> P
iffE         [| P=Q; [| P --> Q; Q --> P |] ==> R |] ==> R

eqTrueI      P ==> P=True
eqTrueE      P=True ==> P
```

LOGICAL EQUIVALENCE

Fig. 18.5. Derived rules for HOL

```
allI      (!!x::'a. P(x)) ==> !x. P(x)
spec      !x::'a.P(x) ==> P(x)
allE      [| !x.P(x);  P(x) ==> R |] ==> R
all_dupE  [| !x.P(x);  [| P(x); !x.P(x) |] ==> R |] ==> R

exI       P(x) ==> ? x::'a.P(x)
exE       [| ? x::'a.P(x); !!x. P(x) ==> Q |] ==> Q

ex1I      [| P(a);  !!x. P(x) ==> x=a |] ==> ?! x. P(x)
ex1E      [| ?! x.P(x);  !!x. [| P(x);  ! y. P(y) --> y=x |] ==> R
          |] ==> R

select_equality [| P(a);  !!x. P(x) ==> x=a |] ==> (@x.P(x)) = a
```

QUANTIFIERS AND DESCRIPTIONS

```
ccontr           (~P ==> False) ==> P
classical        (~P ==> P) ==> P
excluded_middle ~P | P

disjCI           (~Q ==> P) ==> P|Q
exCI             (! x. ~ P(x) ==> P(a)) ==> ? x.P(x)
impCE            [| P-->Q; ~ P ==> R; Q ==> R |] ==> R
iffCE            [| P=Q;  [| P;Q |] ==> R;  [| ~P; ~Q |] ==> R |] ==> R
notnotD          ~~P ==> P
swap             ~P ==> (~Q ==> P) ==> Q
```

CLASSICAL LOGIC

```
if_True          if(True,x,y) = x
if_False         if(False,x,y) = y
if_P             P ==> if(P,x,y) = x
if_not_P         ~ P ==> if(P,x,y) = y
expand_if        P(if(Q,x,y)) = ((Q --> P(x)) & (~Q --> P(y)))
```

CONDITIONALS

Fig. 18.6. More derived rules

name	*meta-type*	*description*
{}	$\alpha\, set$	the empty set
insert	$[\alpha, \alpha\, set] \Rightarrow \alpha\, set$	insertion of element
Collect	$(\alpha \Rightarrow bool) \Rightarrow \alpha\, set$	comprehension
Compl	$(\alpha\, set) \Rightarrow \alpha\, set$	complement
INTER	$[\alpha\, set, \alpha \Rightarrow \beta\, set] \Rightarrow \beta\, set$	intersection over a set
UNION	$[\alpha\, set, \alpha \Rightarrow \beta\, set] \Rightarrow \beta\, set$	union over a set
Inter	$((\alpha\, set) set) \Rightarrow \alpha\, set$	set of sets intersection
Union	$((\alpha\, set) set) \Rightarrow \alpha\, set$	set of sets union
range	$(\alpha \Rightarrow \beta) \Rightarrow \beta\, set$	range of a function
Ball Bex	$[\alpha\, set, \alpha \Rightarrow bool] \Rightarrow bool$	bounded quantifiers
mono	$(\alpha\, set \Rightarrow \beta\, set) \Rightarrow bool$	monotonicity
inj surj	$(\alpha \Rightarrow \beta) \Rightarrow bool$	injective/surjective
inj_onto	$[\alpha \Rightarrow \beta, \alpha\, set] \Rightarrow bool$	injective over subset

CONSTANTS

symbol	*name*	*meta-type*	*priority*	*description*
INT	INTER1	$(\alpha \Rightarrow \beta\, set) \Rightarrow \beta\, set$	10	intersection over a type
UN	UNION1	$(\alpha \Rightarrow \beta\, set) \Rightarrow \beta\, set$	10	union over a type

BINDERS

symbol	*meta-type*	*priority*	*description*
``	$[\alpha \Rightarrow \beta, \alpha\, set] \Rightarrow (\beta\, set)$	Left 90	image
Int	$[\alpha\, set, \alpha\, set] \Rightarrow \alpha\, set$	Left 70	intersection ($\bigcap$)
Un	$[\alpha\, set, \alpha\, set] \Rightarrow \alpha\, set$	Left 65	union ($\bigcup$)
:	$[\alpha, \alpha\, set] \Rightarrow bool$	Left 50	membership ($\in$)
<=	$[\alpha\, set, \alpha\, set] \Rightarrow bool$	Left 50	subset ($\subseteq$)

INFIXES

Fig. 18.7. Syntax of the theory Set

external	*internal*	*description*
`a ~: b`	`~(a : b)`	non-membership
$\{a_1, \ldots\}$	`insert(`a_1`, ...{})`	finite set
$\{x . P[x]\}$	`Collect(`$\lambda x . P[x]$`)`	comprehension
`INT` $x{:}A.B[x]$	`INTER(`$A, \lambda x . B[x]$`)`	intersection
`UN` $x{:}A.B[x]$	`UNION(`$A, \lambda x . B[x]$`)`	union
`!` $x{:}A.P[x]$ or `ALL` $x{:}A.P[x]$	`Ball(`$A, \lambda x . P[x]$`)`	bounded $\forall$
`?` $x{:}A.P[x]$ or `EX` $x{:}A.P[x]$	`Bex(`$A, \lambda x . P[x]$`)`	bounded $\exists$

TRANSLATIONS

```
term    =  other terms...
        |  {}
        |  { term (,term)* }
        |  { id . formula }
        |  term `` term
        |  term Int term
        |  term Un term
        |  INT id:term . term
        |  UN  id:term . term
        |  INT id id* . term
        |  UN  id id* . term

formula =  other formulae...
        |  term : term
        |  term ~: term
        |  term <= term
        |  ! id:term . formula   |  ALL id:term . formula
        |  ? id:term . formula   |  EX  id:term . formula
```

FULL GRAMMAR

Fig. 18.8. Syntax of the theory `Set` (continued)

18.3 A formulation of set theory

Historically, higher-order logic gives a foundation for Russell and Whitehead's theory of classes. Let us use modern terminology and call them **sets**, but note that these sets are distinct from those of `ZF` set theory, and behave more like `ZF` classes.

- Sets are given by predicates over some type σ. Types serve to define universes for sets, but type checking is still significant.
- There is a universal set (for each type). Thus, sets have complements, and may be defined by absolute comprehension.
- Although sets may contain other sets as elements, the containing set must have a more complex type.

Finite unions and intersections have the same behaviour in `HOL` as they do in `ZF`. In `HOL` the intersection of the empty set is well-defined, denoting the universal set for the given type.

18.3.1 Syntax of set theory

`HOL`'s set theory is called `Set`. The type $\alpha\, set$ is essentially the same as $\alpha \Rightarrow bool$. The new type is defined for clarity and to avoid complications involving function types in unification. Since Isabelle does not support type definitions (as mentioned in Sect. 18.1.1), the isomorphisms between the two types are declared explicitly. Here they are natural: `Collect` maps $\alpha \Rightarrow bool$ to $\alpha\, set$, while `op :` maps in the other direction (ignoring argument order).

Figure 18.7 lists the constants, infixes, and syntax translations. Figure 18.8 presents the grammar of the new constructs. Infix operators include union and intersection ($A \bigcup B$ and $A \bigcap B$), the subset and membership relations, and the image operator `''`. Note that a`~`$:b$ is translated to $\neg(a \in b)$.

The $\{\ldots\}$ notation abbreviates finite sets constructed in the obvious manner using `insert` and $\{\}$:

$$\{a_1, \ldots, a_n\} \quad \equiv \quad \texttt{insert}(a_1, \ldots, \texttt{insert}(a_n, \{\}))$$

The set $\{x \,.\, P[x]\}$ consists of all x (of suitable type) that satisfy $P[x]$, where $P[x]$ is a formula that may contain free occurrences of x. This syntax expands to $\texttt{Collect}(\lambda x \,.\, P[x])$. It defines sets by absolute comprehension, which is impossible in `ZF`; the type of x implicitly restricts the comprehension.

The set theory defines two **bounded quantifiers**:

$$\begin{array}{lll} \forall x \in A \,.\, P[x] & \text{abbreviates} & \forall x \,.\, x \in A \to P[x] \\ \exists x \in A \,.\, P[x] & \text{abbreviates} & \exists x \,.\, x \in A \land P[x] \end{array}$$

The constants `Ball` and `Bex` are defined accordingly. Instead of `Ball(`A`,`P`)` and `Bex(`A`,`P`)` we may write `!` $x\!:\!A.P[x]$ and `?` $x\!:\!A.P[x]$. Isabelle's usual

```
mem_Collect_eq      (a : {x.P(x)}) = P(a)
Collect_mem_eq      {x.x:A} = A

empty_def           {}             == {x.x=False}
insert_def          insert(a,B) == {x.x=a} Un B
Ball_def            Ball(A,P)      == ! x. x:A --> P(x)
Bex_def             Bex(A,P)       == ? x. x:A & P(x)
subset_def          A <= B         == ! x:A. x:B
Un_def              A Un B         == {x.x:A | x:B}
Int_def             A Int B        == {x.x:A & x:B}
set_diff_def        A - B          == {x.x:A & x~:B}
Compl_def           Compl(A)       == {x. ~ x:A}
INTER_def           INTER(A,B)     == {y. ! x:A. y: B(x)}
UNION_def           UNION(A,B)     == {y. ? x:A. y: B(x)}
INTER1_def          INTER1(B)      == INTER({x.True}, B)
UNION1_def          UNION1(B)      == UNION({x.True}, B)
Inter_def           Inter(S)       == (INT x:S. x)
Union_def           Union(S)       ==  (UN x:S. x)
image_def           f``A           == {y. ? x:A. y=f(x)}
range_def           range(f)       == {y. ? x. y=f(x)}
mono_def            mono(f)        == !A B. A <= B --> f(A) <= f(B)
inj_def             inj(f)         == ! x y. f(x)=f(y) --> x=y
surj_def            surj(f)        == ! y. ? x. y=f(x)
inj_onto_def        inj_onto(f,A) == !x:A. !y:A. f(x)=f(y) --> x=y
```

Fig. 18.9. Rules of the theory `Set`

quantifier symbols, `ALL` and `EX`, are also accepted for input. As with the primitive quantifiers, the ML reference `HOL_quantifiers` specifies which notation to use for output.

Unions and intersections over sets, namely $\bigcup_{x\in A} B[x]$ and $\bigcap_{x\in A} B[x]$, are written `UN` $x : A . B[x]$ and `INT` $x : A . B[x]$.

Unions and intersections over types, namely $\bigcup_x B[x]$ and $\bigcap_x B[x]$, are written `UN` $x . B[x]$ and `INT` $x . B[x]$. They are equivalent to the previous union and intersection operators when A is the universal set.

The operators $\bigcup A$ and $\bigcap A$ act upon sets of sets. They are not binders, but are equal to $\bigcup_{x\in A} x$ and $\bigcap_{x\in A} x$, respectively.

18.3.2 Axioms and rules of set theory

Figure 18.9 presents the rules of theory `Set`. The axioms `mem_Collect_eq` and `Collect_mem_eq` assert that the functions `Collect` and `op :` are isomorphisms. Of course, `op :` also serves as the membership relation.

All the other axioms are definitions. They include the empty set, bounded quantifiers, unions, intersections, complements and the subset relation. They also include straightforward properties of functions: image (` `` `) and `range`, and predicates concerning monotonicity, injectiveness and surjectiveness.

```
CollectI        [| P(a) |] ==> a : {x.P(x)}
CollectD        [| a : {x.P(x)} |] ==> P(a)
CollectE        [| a : {x.P(x)};  P(a) ==> W |] ==> W

ballI           [| !!x. x:A ==> P(x) |] ==> ! x:A. P(x)
bspec           [| ! x:A. P(x);  x:A |] ==> P(x)
ballE           [| ! x:A. P(x);  P(x) ==> Q;  ~ x:A ==> Q |] ==> Q

bexI            [| P(x);  x:A |] ==> ? x:A. P(x)
bexCI           [| ! x:A. ~ P(x) ==> P(a);  a:A |] ==> ? x:A.P(x)
bexE            [| ? x:A. P(x);  !!x. [| x:A; P(x) |] ==> Q  |] ==> Q
```

COMPREHENSION AND BOUNDED QUANTIFIERS

```
subsetI         (!!x.x:A ==> x:B) ==> A <= B
subsetD         [| A <= B;  c:A |] ==> c:B
subsetCE        [| A <= B;  ~ (c:A) ==> P;  c:B ==> P |] ==> P

subset_refl     A <= A
subset_antisym  [| A <= B;  B <= A |] ==> A = B
subset_trans    [| A<=B;  B<=C |] ==> A<=C

set_ext         [| !!x. (x:A) = (x:B) |] ==> A = B
equalityD1      A = B ==> A<=B
equalityD2      A = B ==> B<=A
equalityE       [| A = B;  [| A<=B; B<=A |] ==> P |]  ==>  P

equalityCE      [| A = B;  [| c:A; c:B |] ==> P;
                           [| ~ c:A; ~ c:B |] ==> P
                |]  ==>  P
```

THE SUBSET AND EQUALITY RELATIONS

Fig. 18.10. Derived rules for set theory

```
emptyE    a : {} ==> P

insertI1 a : insert(a,B)
insertI2 a : B ==> a : insert(b,B)
insertE  [| a : insert(b,A);  a=b ==> P;  a:A ==> P |] ==> P

ComplI   [| c:A ==> False |] ==> c : Compl(A)
ComplD   [| c : Compl(A) |] ==> ~ c:A

UnI1     c:A ==> c : A Un B
UnI2     c:B ==> c : A Un B
UnCI     (~c:B ==> c:A) ==> c : A Un B
UnE      [| c : A Un B;  c:A ==> P;  c:B ==> P |] ==> P

IntI     [| c:A;  c:B |] ==> c : A Int B
IntD1    c : A Int B ==> c:A
IntD2    c : A Int B ==> c:B
IntE     [| c : A Int B;  [| c:A; c:B |] ==> P |] ==> P

UN_I     [| a:A;  b: B(a) |] ==> b: (UN x:A. B(x))
UN_E     [| b: (UN x:A. B(x));  !!x.[| x:A;  b:B(x) |] ==> R |] ==> R

INT_I    (!!x. x:A ==> b: B(x)) ==> b : (INT x:A. B(x))
INT_D    [| b: (INT x:A. B(x));  a:A |] ==> b: B(a)
INT_E    [| b: (INT x:A. B(x));  b: B(a) ==> R;  ~ a:A ==> R |] ==> R

UnionI   [| X:C;  A:X |] ==> A : Union(C)
UnionE   [| A : Union(C);  !!X.[| A:X;  X:C |] ==> R |] ==> R

InterI   [| !!X. X:C ==> A:X |] ==> A : Inter(C)
InterD   [| A : Inter(C);  X:C |] ==> A:X
InterE   [| A : Inter(C);  A:X ==> R;  ~ X:C ==> R |] ==> R
```

Fig. 18.11. Further derived rules for set theory

```
Inv_f_f     inj(f) ==> Inv(f,f(x)) = x
f_Inv_f     y : range(f) ==> f(Inv(f,y)) = y

imageI      [| x:A |] ==> f(x) : f``A
imageE      [| b : f``A;  !!x.[| b=f(x);  x:A |] ==> P |] ==> P

rangeI      f(x) : range(f)
rangeE      [| b : range(f);  !!x.[| b=f(x) |] ==> P |] ==> P

monoI       [| !!A B. A <= B ==> f(A) <= f(B) |] ==> mono(f)
monoD       [| mono(f);  A <= B |] ==> f(A) <= f(B)

injI        [| !! x y. f(x) = f(y) ==> x=y |] ==> inj(f)
inj_inverseI                 (!!x. g(f(x)) = x) ==> inj(f)
injD        [| inj(f); f(x) = f(y) |] ==> x=y

inj_ontoI  (!!x y. [| f(x)=f(y); x:A; y:A |] ==> x=y) ==> inj_onto(f,A)
inj_ontoD  [| inj_onto(f,A);  f(x)=f(y);  x:A;  y:A |] ==> x=y

inj_onto_inverseI
    (!!x. x:A ==> g(f(x)) = x) ==> inj_onto(f,A)
inj_onto_contraD
    [| inj_onto(f,A);  x~=y;  x:A;  y:A |] ==> ~ f(x)=f(y)
```

Fig. 18.12. Derived rules involving functions

The predicate `inj_onto` is used for simulating type definitions. The statement `inj_onto`(f, A) asserts that f is injective on the set A, which specifies a subset of its domain type. In a type definition, f is the abstraction function and A is the set of valid representations; we should not expect f to be injective outside of A.

Figures 18.10 and 18.11 present derived rules. Most are obvious and resemble rules of Isabelle's `ZF` set theory. Certain rules, such as `subsetCE`, `bexCI` and `UnCI`, are designed for classical reasoning; the rules `subsetD`, `bexI`, `Un1` and `Un2` are not strictly necessary but yield more natural proofs. Similarly, `equalityCE` supports classical reasoning about extensionality, after the fashion of `iffCE`. See the file `HOL/Set.ML` for proofs pertaining to set theory.

Figure 18.12 presents derived inference rules involving functions. They also include rules for `Inv`, which is defined in theory `HOL`; note that `Inv`(f) applies the Axiom of Choice to yield an inverse of f. They also include natural deduction rules for the image and range operators, and for the predicates `inj` and `inj_onto`. Reasoning about function composition (the operator `o`) and the predicate `surj` is done simply by expanding the definitions. See the file `HOL/fun.ML` for a complete listing of the derived rules.

Figure 18.13 presents lattice properties of the subset relation. Unions form least upper bounds; non-empty intersections form greatest lower bounds. Reasoning directly about subsets often yields clearer proofs than reasoning about the membership relation. See the file `HOL/subset.ML`.

```
Union_upper      B:A ==> B <= Union(A)
Union_least      [| !!X. X:A ==> X<=C |] ==> Union(A) <= C

Inter_lower      B:A ==> Inter(A) <= B
Inter_greatest   [| !!X. X:A ==> C<=X |] ==> C <= Inter(A)

Un_upper1        A <= A Un B
Un_upper2        B <= A Un B
Un_least         [| A<=C;  B<=C |] ==> A Un B <= C

Int_lower1       A Int B <= A
Int_lower2       A Int B <= B
Int_greatest     [| C<=A;  C<=B |] ==> C <= A Int B
```

Fig. 18.13. Derived rules involving subsets

```
Int_absorb          A Int A = A
Int_commute         A Int B = B Int A
Int_assoc           (A Int B) Int C  =  A Int (B Int C)
Int_Un_distrib      (A Un B)  Int C  =  (A Int C) Un (B Int C)

Un_absorb           A Un A = A
Un_commute          A Un B = B Un A
Un_assoc            (A Un B)  Un C  =  A Un (B Un C)
Un_Int_distrib      (A Int B) Un C  =  (A Un C) Int (B Un C)

Compl_disjoint      A Int Compl(A) = {x.False}
Compl_partition     A Un  Compl(A) = {x.True}
double_complement   Compl(Compl(A)) = A
Compl_Un            Compl(A Un B)  = Compl(A) Int Compl(B)
Compl_Int           Compl(A Int B) = Compl(A) Un Compl(B)

Union_Un_distrib    Union(A Un B) = Union(A) Un Union(B)
Int_Union           A Int Union(B) = (UN C:B. A Int C)
Un_Union_image      (UN x:C. A(x) Un B(x)) = Union(A``C) Un Union(B``C)

Inter_Un_distrib    Inter(A Un B) = Inter(A) Int Inter(B)
Un_Inter            A Un Inter(B) = (INT C:B. A Un C)
Int_Inter_image     (INT x:C. A(x) Int B(x)) = Inter(A``C) Int Inter(B``C)
```

Fig. 18.14. Set equalities

symbol	*meta-type*	*description*
`Pair`	$[\alpha, \beta] \Rightarrow \alpha \times \beta$	ordered pairs $\langle a, b \rangle$
`fst`	$\alpha \times \beta \Rightarrow \alpha$	first projection
`snd`	$\alpha \times \beta \Rightarrow \beta$	second projection
`split`	$[\alpha \times \beta, [\alpha, \beta] \Rightarrow \gamma] \Rightarrow \gamma$	generalized projection
`Sigma`	$[\alpha\, set, \alpha \Rightarrow \beta\, set] \Rightarrow (\alpha \times \beta) set$	general sum of sets

```
fst_def      fst(p)      == @a. ? b. p = <a,b>
snd_def      snd(p)      == @b. ? a. p = <a,b>
split_def    split(p,c) == c(fst(p),snd(p))
Sigma_def    Sigma(A,B) == UN x:A. UN y:B(x). {<x,y>}

Pair_inject  [| <a, b> = <a',b'>;  [| a=a';  b=b' |] ==> R |] ==> R
fst_conv     fst(<a,b>) = a
snd_conv     snd(<a,b>) = b
split        split(<a,b>, c) = c(a,b)

surjective_pairing  p = <fst(p),snd(p)>

SigmaI       [| a:A;  b:B(a) |] ==> <a,b> : Sigma(A,B)

SigmaE       [| c: Sigma(A,B);
                !!x y.[| x:A; y:B(x); c=<x,y> |] ==> P |] ==> P
```

Fig. 18.15. Type $\alpha \times \beta$

Figure 18.14 presents many common set equalities. They include commutative, associative and distributive laws involving unions, intersections and complements. The proofs are mostly trivial, using the classical reasoner; see file `HOL/equalities.ML`.

18.4 Generic packages and classical reasoning

`HOL` instantiates most of Isabelle's generic packages; see `HOL/ROOT.ML` for details.

- Because it includes a general substitution rule, `HOL` instantiates the tactic `hyp_subst_tac`, which substitutes for an equality throughout a subgoal and its hypotheses.
- It instantiates the simplifier, defining `HOL_ss` as the simplification set for higher-order logic. Equality (=), which also expresses logical equivalence, may be used for rewriting. See the file `HOL/simpdata.ML` for a complete listing of the simplification rules.
- It instantiates the classical reasoner, as described below.

symbol	*meta-type*	*description*
`Inl`	$\alpha \Rightarrow \alpha + \beta$	first injection
`Inr`	$\beta \Rightarrow \alpha + \beta$	second injection
`sum_case`	$[\alpha + \beta, \alpha \Rightarrow \gamma, \beta \Rightarrow \gamma] \Rightarrow \gamma$	conditional

```
sum_case_def   sum_case == (%p f g. @z. (!x. p=Inl(x) --> z=f(x)) &
                                        (!y. p=Inr(y) --> z=g(y)))

Inl_not_Inr    ~ Inl(a)=Inr(b)

inj_Inl        inj(Inl)
inj_Inr        inj(Inr)

sumE           [| !!x::'a. P(Inl(x));  !!y::'b. P(Inr(y)) |] ==> P(s)

sum_case_Inl   sum_case(Inl(x), f, g) = f(x)
sum_case_Inr   sum_case(Inr(x), f, g) = g(x)

surjective_sum sum_case(s, %x::'a. f(Inl(x)), %y::'b. f(Inr(y))) = f(s)
```

Fig. 18.16. Type $\alpha + \beta$

HOL derives classical introduction rules for $\vee$ and $\exists$, as well as classical elimination rules for $\rightarrow$ and $\leftrightarrow$, and the swap rule; recall Fig. 18.6 above.

The classical reasoner is set up as the structure `Classical`. This structure is open, so ML identifiers such as `step_tac`, `fast_tac`, `best_tac`, etc., refer to it. HOL defines the following classical rule sets:

```
prop_cs    : claset
HOL_cs     : claset
HOL_dup_cs : claset
set_cs     : claset
```

`prop_cs` contains the propositional rules, namely those for $\top$, $\bot$, $\wedge$, $\vee$, $\neg$, $\rightarrow$ and $\leftrightarrow$, along with the rule `refl`.

`HOL_cs` extends `prop_cs` with the safe rules `allI` and `exE` and the unsafe rules `allE` and `exI`, as well as rules for unique existence. Search using this classical set is incomplete: quantified formulae are used at most once.

`HOL_dup_cs` extends `prop_cs` with the safe rules `allI` and `exE` and the unsafe rules `all_dupE` and `exCI`, as well as rules for unique existence. Search using this is complete — quantified formulae may be duplicated — but frequently fails to terminate. It is generally unsuitable for depth-first search.

`set_cs` extends `HOL_cs` with rules for the bounded quantifiers, subsets, comprehensions, unions and intersections, complements, finite sets, images and ranges.

See Chap. 14 for more discussion of classical proof methods.

18.5 Types

The basic higher-order logic is augmented with a tremendous amount of material, including support for recursive function and type definitions. A detailed discussion appears elsewhere [47]. The simpler definitions are the same as those used the HOL system, but my treatment of recursive types differs from Melham's [34]. The present section describes product, sum, natural number and list types.

18.5.1 Product and sum types

Theory `Prod` defines the product type $\alpha\times\beta$, with the ordered pair syntax `<`a`,`b`>`. Theory `Sum` defines the sum type $\alpha+\beta$. These use fairly standard constructions; see Figs. 18.15 and 18.16. Because Isabelle does not support abstract type definitions, the isomorphisms between these types and their representations are made explicitly.

Most of the definitions are suppressed, but observe that the projections and conditionals are defined as descriptions. Their properties are easily proved using `select_equality`.

18.5.2 The type of natural numbers, nat

The theory `Nat` defines the natural numbers in a roundabout but traditional way. The axiom of infinity postulates an type `ind` of individuals, which is non-empty and closed under an injective operation. The natural numbers are inductively generated by choosing an arbitrary individual for 0 and using the injective operation to take successors. As usual, the isomorphisms between `nat` and its representation are made explicitly.

The definition makes use of a least fixed point operator `lfp`, defined using the Knaster-Tarski theorem. This is used to define the operator `trancl`, for taking the transitive closure of a relation. Primitive recursion makes use of `wfrec`, an operator for recursion along arbitrary well-founded relations. The corresponding theories are called `Lfp`, `Trancl` and `WF`. Elsewhere I have described similar constructions in the context of set theory [50].

Type `nat` is postulated to belong to class `ord`, which overloads $<$ and $\leq$ on the natural numbers. As of this writing, Isabelle provides no means of verifying that such overloading is sensible; there is no means of specifying the operators' properties and verifying that instances of the operators satisfy those properties. To be safe, the `HOL` theory includes no polymorphic axioms asserting general properties of $<$ and $\leq$.

symbol	meta-type	priority	description
0	nat		zero
Suc	$nat \Rightarrow nat$		successor function
nat_case	$[nat, \alpha, nat \Rightarrow \alpha] \Rightarrow \alpha$		conditional
nat_rec	$[nat, \alpha, [nat, \alpha] \Rightarrow \alpha] \Rightarrow \alpha$		primitive recursor
pred_nat	$(nat \times nat)set$		predecessor relation
*	$[nat, nat] \Rightarrow nat$	Left 70	multiplication
div	$[nat, nat] \Rightarrow nat$	Left 70	division
mod	$[nat, nat] \Rightarrow nat$	Left 70	modulus
+	$[nat, nat] \Rightarrow nat$	Left 65	addition
-	$[nat, nat] \Rightarrow nat$	Left 65	subtraction

CONSTANTS AND INFIXES

```
nat_case_def  nat_case == (%n a f. @z.  (n=0 --> z=a) &
                                        (!x. n=Suc(x) --> z=f(x)))
pred_nat_def  pred_nat == {p. ? n. p = <n, Suc(n)>}
less_def      m<n      == <m,n>:pred_nat^+
nat_rec_def   nat_rec(n,c,d) ==
                wfrec(pred_nat, n, %l g.nat_case(l, c, %m.d(m,g(m))))

add_def   m+n     == nat_rec(m, n, %u v.Suc(v))
diff_def  m-n     == nat_rec(n, m, %u v. nat_rec(v, 0, %x y.x))
mult_def  m*n     == nat_rec(m, 0, %u v. n + v)
mod_def   m mod n == wfrec(trancl(pred_nat), m, %j f. if(j<n,j,f(j-n)))
quo_def   m div n == wfrec(trancl(pred_nat),
                           m, %j f. if(j<n,0,Suc(f(j-n))))
```

DEFINITIONS

Fig. 18.17. Defining nat, the type of natural numbers

```
nat_induct      [| P(0); !!k. [| P(k) |] ==> P(Suc(k)) |]  ==> P(n)

Suc_not_Zero    Suc(m) ~= 0
inj_Suc         inj(Suc)
n_not_Suc_n     n~=Suc(n)
```

BASIC PROPERTIES

```
pred_natI       <n, Suc(n)> : pred_nat
pred_natE
    [| p : pred_nat;  !!x n. [| p = <n, Suc(n)> |] ==> R |] ==> R

nat_case_0      nat_case(0, a, f) = a
nat_case_Suc    nat_case(Suc(k), a, f) = f(k)

wf_pred_nat     wf(pred_nat)
nat_rec_0       nat_rec(0,c,h) = c
nat_rec_Suc     nat_rec(Suc(n), c, h) = h(n, nat_rec(n,c,h))
```

CASE ANALYSIS AND PRIMITIVE RECURSION

```
less_trans      [| i<j;  j<k |] ==> i<k
lessI           n < Suc(n)
zero_less_Suc   0 < Suc(n)

less_not_sym    n<m --> ~ m<n
less_not_refl   ~ n<n
not_less0       ~ n<0

Suc_less_eq     (Suc(m) < Suc(n)) = (m<n)
less_induct     [| !!n. [| ! m. m<n --> P(m) |] ==> P(n) |]  ==>  P(n)

less_linear     m<n | m=n | n<m
```

THE LESS-THAN RELATION

Fig. 18.18. Derived rules for nat

Theory `Arith` develops arithmetic on the natural numbers. It defines addition, multiplication, subtraction, division, and remainder. Many of their properties are proved: commutative, associative and distributive laws, identity and cancellation laws, etc. The most interesting result is perhaps the theorem $a \bmod b + (a/b) \times b = a$. Division and remainder are defined by repeated subtraction, which requires well-founded rather than primitive recursion. See Figs. 18.17 and 18.18.

The predecessor relation, `pred_nat`, is shown to be well-founded. Recursion along this relation resembles primitive recursion, but is stronger because we are in higher-order logic; using primitive recursion to define a higher-order function, we can easily Ackermann's function, which is not primitive recursive [58, page 104]. The transitive closure of `pred_nat` is $<$. Many functions on the natural numbers are most easily expressed using recursion along $<$.

The tactic `nat_ind_tac "`n`"` i performs induction over the variable n in subgoal i.

18.5.3 The type constructor for lists, `list`

`HOL`'s definition of lists is an example of an experimental method for handling recursive data types. Figure 18.19 presents the theory `List`: the basic list operations with their types and properties.

The `case` construct is defined by the following translation:

$$\begin{array}{l} \texttt{case}\ e\ \texttt{of}\ \texttt{[]}\ \ \texttt{=>}\ a \\ \qquad\qquad \texttt{|}\ x\texttt{\#}xs\ \texttt{=>}\ b \end{array} \equiv \texttt{list\_case}(e, a, \lambda x\ xs\ .\ b)$$

The theory includes `list_rec`, a primitive recursion operator for lists. It is derived from well-founded recursion, a general principle that can express arbitrary total recursive functions.

The simpset `list_ss` contains, along with additional useful lemmas, the basic rewrite rules that appear in Fig. 18.20.

The tactic `list_ind_tac "`xs`"` i performs induction over the variable xs in subgoal i.

18.5.4 The type constructor for lazy lists, `llist`

The definition of lazy lists demonstrates methods for handling infinite data structures and coinduction in higher-order logic. Theory `LList` defines an operator for corecursion on lazy lists, which is used to define a few simple functions such as map and append. Corecursion cannot easily define operations such as filter, which can compute indefinitely before yielding the next element (if any!) of the lazy list. A coinduction principle is defined for proving equations on lazy lists.

I have written a paper discussing the treatment of lazy lists; it also covers finite lists [47].

symbol	*meta-type*	*priority*	*description*
`Nil`	$\alpha list$		empty list
`#`	$[\alpha, \alpha list] \Rightarrow \alpha list$	Right 65	list constructor
`null`	$\alpha list \Rightarrow bool$		emptiness test
`hd`	$\alpha list \Rightarrow \alpha$		head
`tl`	$\alpha list \Rightarrow \alpha list$		tail
`ttl`	$\alpha list \Rightarrow \alpha list$		total tail
`@`	$[\alpha list, \alpha list] \Rightarrow \alpha list$	Left 65	append
`mem`	$[\alpha, \alpha list] \Rightarrow bool$	Left 55	membership
`map`	$(\alpha \Rightarrow \beta) \Rightarrow (\alpha list \Rightarrow \beta list)$		mapping functional
`filter`	$(\alpha \Rightarrow bool) \Rightarrow (\alpha list \Rightarrow \alpha list)$		filter functional
`list_all`	$(\alpha \Rightarrow bool) \Rightarrow (\alpha list \Rightarrow bool)$		forall functional
`list_rec`	$[\alpha list, \beta, [\alpha, \alpha list, \beta] \Rightarrow \beta] \Rightarrow \beta$		list recursor

CONSTANTS AND INFIXES

external	*internal*	*description*
`[]`	`Nil`	empty list
`[`x_1`, ..., `x_n`]`	x_1 `#` $\cdots$ `#` x_n `# []`	finite list
`[`x`:`l`. `P`]`	`filter(`$\lambda x.P$`, `l`)`	list comprehension

TRANSLATIONS

```
list_induct    [| P([]);  !!x xs. [| P(xs) |] ==> P(x#xs)) |]  ==> P(l)

Cons_not_Nil   (x # xs) ~= []
Cons_Cons_eq   ((x # xs) = (y # ys)) = (x=y & xs=ys)
```

INDUCTION AND FREENESS

Fig. 18.19. The theory `List`

```
list_rec_Nil        list_rec([],c,h) = c
list_rec_Cons       list_rec(a # l, c, h) = h(a, l, list_rec(l,c,h))

list_case_Nil       list_case([],c,h) = c
list_case_Cons      list_case(x # xs, c, h) = h(x, xs)

map_Nil             map(f,[]) = []
map_Cons            map(f, x # xs) = f(x) # map(f,xs)

null_Nil            null([]) = True
null_Cons           null(x # xs) = False

hd_Cons             hd(x # xs) = x
tl_Cons             tl(x # xs) = xs

ttl_Nil             ttl([]) = []
ttl_Cons            ttl(x # xs) = xs

append_Nil          [] @ ys = ys
append_Cons         (x # xs) @ ys = x # xs @ ys

mem_Nil             x mem [] = False
mem_Cons            x mem y # ys = if(y = x, True, x mem ys)

filter_Nil          filter(P, []) = []
filter_Cons         filter(P,x#xs) = if(P(x),x#filter(P,xs),filter(P,xs))

list_all_Nil        list_all(P,[]) = True
list_all_Cons       list_all(P, x # xs) = (P(x) & list_all(P, xs))
```

Fig. 18.20. Rewrite rules for lists

18.6 The examples directories

Directory `HOL/Subst` contains Martin Coen's mechanisation of a theory of substitutions and unifiers. It is based on Paulson's previous mechanisation in LCF [41] of Manna and Waldinger's theory [31].

Directory `HOL/ex` contains other examples and experimental proofs in `HOL`. Here is an overview of the more interesting files.

`HOL/ex/cla.ML` demonstrates the classical reasoner on over sixty predicate calculus theorems, ranging from simple tautologies to moderately difficult problems involving equality and quantifiers.

`HOL/ex/meson.ML` contains an experimental implementation of the MESON proof procedure, inspired by Plaisted [53]. It is much more powerful than Isabelle's classical reasoner. But it is less useful in practice because it works only for pure logic; it does not accept derived rules for the set theory primitives, for example.

`HOL/ex/mesontest.ML` contains test data for the MESON proof procedure. These are mostly taken from Pelletier [52].

`HOL/ex/set.ML` proves Cantor's Theorem, which is presented in Sect. 18.7 below, and the Schröder-Bernstein Theorem.

`HOL/ex/InSort.ML` and `HOL/ex/Qsort.ML` contain correctness proofs about insertion sort and quick sort.

`HOL/ex/PL.ML` proves the soundness and completeness of classical propositional logic, given a truth table semantics. The only connective is $\rightarrow$. A Hilbert-style axiom system is specified, and its set of theorems defined inductively. A similar proof in `ZF` is described elsewhere [50].

`HOL/ex/Term.ML` contains proofs about an experimental recursive type definition; the recursion goes through the type constructor `list`.

`HOL/ex/Simult.ML` defines primitives for solving mutually recursive equations over sets. It constructs sets of trees and forests as an example, including induction and recursion rules that handle the mutual recursion.

`HOL/ex/MT.ML` contains Jacob Frost's formalization [19] of Milner and Tofte's coinduction example [36]. This substantial proof concerns the soundness of a type system for a simple functional language. The semantics of recursion is given by a cyclic environment, which makes a coinductive argument appropriate.

18.7 Example: Cantor's Theorem

Cantor's Theorem states that every set has more subsets than it has elements. It has become a favourite example in higher-order logic since it is so easily expressed:

$$\forall f :: [\alpha, \alpha] \Rightarrow bool \,.\, \exists S :: \alpha \Rightarrow bool \,.\, \forall x :: \alpha \,.\, f(x) \neq S$$

Viewing types as sets, $\alpha \Rightarrow bool$ represents the powerset of α. This version states that for every function from α to its powerset, some subset is outside its range.

The Isabelle proof uses `HOL`'s set theory, with the type α *set* and the operator `range`. The set S is given as an unknown instead of a quantified variable so that we may inspect the subset found by the proof.

```
goal Set.thy "~ ?S : range(f :: 'a=>'a set)";
Level 0
~ ?S : range(f)
 1. ~ ?S : range(f)
```

The first two steps are routine. The rule `rangeE` replaces $?S \in$ `range`(f) by $?S = f(x)$ for some x.

```
by (resolve_tac [notI] 1);
Level 1
~ ?S : range(f)
 1. ?S : range(f) ==> False
by (eresolve_tac [rangeE] 1);
Level 2
~ ?S : range(f)
 1. !!x. ?S = f(x) ==> False
```

Next, we apply `equalityCE`, reasoning that since $?S = f(x)$, we have $?c \in ?S$ if and only if $?c \in f(x)$ for any $?c$.

```
by (eresolve_tac [equalityCE] 1);
Level 3
~ ?S : range(f)
 1. !!x. [| ?c3(x) : ?S; ?c3(x) : f(x) |] ==> False
 2. !!x. [| ~ ?c3(x) : ?S; ~ ?c3(x) : f(x) |] ==> False
```

Now we use a bit of creativity. Suppose that $?S$ has the form of a comprehension. Then $?c \in \{x \,.\, ?P(x)\}$ implies $?P(?c)$. Destruct-resolution using `CollectD` instantiates $?S$ and creates the new assumption.

```
by (dresolve_tac [CollectD] 1);
Level 4
~ {x. ?P7(x)} : range(f)
 1. !!x. [| ?c3(x) : f(x); ?P7(?c3(x)) |] ==> False
 2. !!x. [| ~ ?c3(x) : {x. ?P7(x)}; ~ ?c3(x) : f(x) |] ==> False
```

Forcing a contradiction between the two assumptions of subgoal 1 completes the instantiation of S. It is now the set $\{x \,.\, x \notin f(x)\}$, which is the standard diagonal construction.

```
by (contr_tac 1);
  Level 5
  ~ {x. ~ x : f(x)} : range(f)
   1. !!x. [| ~ x : {x. ~ x : f(x)}; ~ x : f(x) |] ==> False
```

The rest should be easy. To apply `CollectI` to the negated assumption, we employ `swap_res_tac`:

```
by (swap_res_tac [CollectI] 1);
  Level 6
  ~ {x. ~ x : f(x)} : range(f)
   1. !!x. [| ~ x : f(x); ~ False |] ==> ~ x : f(x)
by (assume_tac 1);
  Level 7
  ~ {x. ~ x : f(x)} : range(f)
  No subgoals!
```

How much creativity is required? As it happens, Isabelle can prove this theorem automatically. The classical set `set_cs` contains rules for most of the constructs of `HOL`'s set theory. We must augment it with `equalityCE` to break up set equalities, and then apply best-first search. Depth-first search would diverge, but best-first search successfully navigates through the large search space.

```
choplev 0;
  Level 0
  ~ ?S : range(f)
   1. ~ ?S : range(f)
by (best_tac (set_cs addSEs [equalityCE]) 1);
  Level 1
  ~ {x. ~ x : f(x)} : range(f)
  No subgoals!
```

19. First-Order Sequent Calculus

The theory `LK` implements classical first-order logic through Gentzen's sequent calculus (see Gallier [21] or Takeuti [57]). Resembling the method of semantic tableaux, the calculus is well suited for backwards proof. Assertions have the form $\Gamma \vdash \Delta$, where Γ and Δ are lists of formulae. Associative unification, simulated by higher-order unification, handles lists.

The logic is many-sorted, using Isabelle's type classes. The class of first-order terms is called `term`. No types of individuals are provided, but extensions can define types such as `nat::term` and type constructors such as `list::(term)term`. Below, the type variable α ranges over class `term`; the equality symbol and quantifiers are polymorphic (many-sorted). The type of formulae is `o`, which belongs to class `logic`.

No generic packages are instantiated, since Isabelle does not provide packages for sequent calculi at present. `LK` implements a classical logic theorem prover that is as powerful as the generic classical reasoner, except that it does not perform equality reasoning.

19.1 Unification for lists

Higher-order unification includes associative unification as a special case, by an encoding that involves function composition [28, page 37]. To represent lists, let C be a new constant. The empty list is $\lambda x . x$, while $[t_1, t_2, \ldots, t_n]$ is represented by

$$\lambda x . C(t_1, C(t_2, \ldots, C(t_n, x))).$$

The unifiers of this with $\lambda x . ?f(?g(x))$ give all the ways of expressing $[t_1, t_2, \ldots, t_n]$ as the concatenation of two lists.

Unlike orthodox associative unification, this technique can represent certain infinite sets of unifiers by flex-flex equations. But note that the term $\lambda x . C(t, ?a)$ does not represent any list. Flex-flex constraints containing such garbage terms may accumulate during a proof.

This technique lets Isabelle formalize sequent calculus rules, where the comma is the associative operator:

$$\frac{\Gamma, P, Q, \Delta \vdash \Theta}{\Gamma, P \wedge Q, \Delta \vdash \Theta} \ (\wedge\text{-left})$$

name	*meta-type*	*description*
`Trueprop`	$[sobj \Rightarrow sobj, sobj \Rightarrow sobj] \Rightarrow prop$	coercion to *prop*
`Seqof`	$[o, sobj] \Rightarrow sobj$	singleton sequence
`Not`	$o \Rightarrow o$	negation ($\neg$)
`True`	o	tautology ($\top$)
`False`	o	absurdity ($\bot$)

CONSTANTS

symbol	*name*	*meta-type*	*priority*	*description*
`ALL`	`All`	$(\alpha \Rightarrow o) \Rightarrow o$	10	universal quantifier ($\forall$)
`EX`	`Ex`	$(\alpha \Rightarrow o) \Rightarrow o$	10	existential quantifier ($\exists$)
`THE`	`The`	$(\alpha \Rightarrow o) \Rightarrow \alpha$	10	definite description (ι)

BINDERS

symbol	*meta-type*	*priority*	*description*
`=`	$[\alpha, \alpha] \Rightarrow o$	Left 50	equality ($=$)
`&`	$[o, o] \Rightarrow o$	Right 35	conjunction ($\wedge$)
`|`	$[o, o] \Rightarrow o$	Right 30	disjunction ($\vee$)
`-->`	$[o, o] \Rightarrow o$	Right 25	implication ($\rightarrow$)
`<->`	$[o, o] \Rightarrow o$	Right 25	biconditional ($\leftrightarrow$)

INFIXES

external	*internal*	*description*
Γ `|-` Δ	`Trueprop(`Γ`,` Δ`)`	sequent $\Gamma \vdash \Delta$

TRANSLATIONS

Fig. 19.1. Syntax of LK

```
    prop  =  sequence |- sequence

sequence  =  elem (, elem)*
          |  empty

    elem  =  $ id
          |  $ var
          |  formula

 formula  =  expression of type o
          |  term = term
          |  ~ formula
          |  formula & formula
          |  formula | formula
          |  formula --> formula
          |  formula <-> formula
          |  ALL id id* . formula
          |  EX  id id* . formula
          |  THE id . formula
```

Fig. 19.2. Grammar of LK

Multiple unifiers occur whenever this is resolved against a goal containing more than one conjunction on the left.

LK exploits this representation of lists. As an alternative, the sequent calculus can be formalized using an ordinary representation of lists, with a logic program for removing a formula from a list. Amy Felty has applied this technique using the language λProlog [17].

Explicit formalization of sequents can be tiresome. But it gives precise control over contraction and weakening, and is essential to handle relevant and linear logics.

19.2 Syntax and rules of inference

Figure 19.1 gives the syntax for LK, which is complicated by the representation of sequents. Type $sobj \Rightarrow sobj$ represents a list of formulae.

The **definite description** operator $\iota x \,.\, P[x]$ stands for some a satisfying $P[a]$, if one exists and is unique. Since all terms in LK denote something, a description is always meaningful, but we do not know its value unless $P[x]$ defines it uniquely. The Isabelle notation is THE $x \,.\, P[x]$. The corresponding rule (Fig. 19.3) does not entail the Axiom of Choice because it requires uniqueness.

Figure 19.2 presents the grammar of LK. Traditionally, Γ and Δ are meta-

```
basic      $H, P, $G |- $E, P, $F
thinR      $H |- $E, $F ==> $H |- $E, P, $F
thinL      $H, $G |- $E ==> $H, P, $G |- $E
cut        [| $H |- $E, P;  $H, P |- $E |] ==> $H |- $E
```

STRUCTURAL RULES

```
refl       $H |- $E, a=a, $F
sym        $H |- $E, a=b, $F ==> $H |- $E, b=a, $F
trans      [| $H|- $E, a=b, $F;  $H|- $E, b=c, $F |] ==>
           $H|- $E, a=c, $F
```

EQUALITY RULES

```
True_def   True  == False-->False
iff_def    P<->Q == (P-->Q) & (Q-->P)

conjR   [| $H|- $E, P, $F;  $H|- $E, Q, $F |] ==> $H|- $E, P&Q, $F
conjL   $H, P, Q, $G |- $E ==> $H, P & Q, $G |- $E

disjR   $H |- $E, P, Q, $F ==> $H |- $E, P|Q, $F
disjL   [| $H, P, $G |- $E;  $H, Q, $G |- $E |] ==> $H, P|Q, $G |- $E

impR    $H, P |- $E, Q, $F ==> $H |- $E, P-->Q, $
impL    [| $H,$G |- $E,P;  $H, Q, $G |- $E |] ==> $H, P-->Q, $G |- $E

notR    $H, P |- $E, $F ==> $H |- $E, ~P, $F
notL    $H, $G |- $E, P ==> $H, ~P, $G |- $E

FalseL  $H, False, $G |- $E

allR    (!!x.$H|- $E, P(x), $F) ==> $H|- $E, ALL x.P(x), $F
allL    $H, P(x), $G, ALL x.P(x) |- $E ==> $H, ALL x.P(x), $G|- $E

exR     $H|- $E, P(x), $F, EX x.P(x) ==> $H|- $E, EX x.P(x), $F
exL     (!!x.$H, P(x), $G|- $E) ==> $H, EX x.P(x), $G|- $E

The     [| $H |- $E, P(a), $F;  !!x.$H, P(x) |- $E, x=a, $F |] ==>
        $H |- $E, P(THE x.P(x)), $F
```

LOGICAL RULES

Fig. 19.3. Rules of LK

```
conR        $H |- $E, P, $F, P ==> $H |- $E, P, $F
conL        $H, P, $G, P |- $E ==> $H, P, $G |- $E

symL        $H, $G, B = A |- $E ==> $H, A = B, $G |- $E

TrueR       $H |- $E, True, $F

iffR        [| $H, P |- $E, Q, $F;  $H, Q |- $E, P, $F |] ==>
            $H |- $E, P<->Q, $F

iffL        [| $H, $G |- $E, P, Q;  $H, Q, P, $G |- $E |] ==>
            $H, P<->Q, $G |- $E

allL_thin   $H, P(x), $G |- $E ==> $H, ALL x.P(x), $G |- $E
exR_thin    $H |- $E, P(x), $F ==> $H |- $E, EX x.P(x), $F
```

Fig. 19.4. Derived rules for LK

variables for sequences. In Isabelle's notation, the prefix $ on a variable makes it range over sequences. In a sequent, anything not prefixed by $ is taken as a formula.

Figure 19.3 presents the rules of theory LK. The connective $\leftrightarrow$ is defined using $\wedge$ and $\rightarrow$. The axiom for basic sequents is expressed in a form that provides automatic thinning: redundant formulae are simply ignored. The other rules are expressed in the form most suitable for backward proof — they do not require exchange or contraction rules. The contraction rules are actually derivable (via cut) in this formulation.

Figure 19.4 presents derived rules, including rules for $\leftrightarrow$. The weakened quantifier rules discard each quantification after a single use; in an automatic proof procedure, they guarantee termination, but are incomplete. Multiple use of a quantifier can be obtained by a contraction rule, which in backward proof duplicates a formula. The tactic `res_inst_tac` can instantiate the variable `?P` in these rules, specifying the formula to duplicate.

See the files `LK/LK.thy` and `LK/LK.ML` for complete listings of the rules and derived rules.

19.3 Tactics for the cut rule

According to the cut-elimination theorem, the cut rule can be eliminated from proofs of sequents. But the rule is still essential. It can be used to structure a proof into lemmas, avoiding repeated proofs of the same formula. More importantly, the cut rule can not be eliminated from derivations of rules. For example, there is a trivial cut-free proof of the sequent $P \wedge Q \vdash Q \wedge P$. Noting this, we might want to derive a rule for swapping the conjuncts in a right-hand formula:

$$\frac{\Gamma \vdash \Delta, P \wedge Q}{\Gamma \vdash \Delta, Q \wedge P}$$

The cut rule must be used, for $P \wedge Q$ is not a subformula of $Q \wedge P$. Most cuts directly involve a premise of the rule being derived (a meta-assumption). In a few cases, the cut formula is not part of any premise, but serves as a bridge between the premises and the conclusion. In such proofs, the cut formula is specified by calling an appropriate tactic.

```
cutR_tac : string -> int -> tactic
cutL_tac : string -> int -> tactic
```

These tactics refine a subgoal into two by applying the cut rule. The cut formula is given as a string, and replaces some other formula in the sequent.

`cutR_tac` P i reads an `LK` formula P, and applies the cut rule to subgoal i. It then deletes some formula from the right side of subgoal i, replacing that formula by P.

`cutL_tac` P i reads an `LK` formula P, and applies the cut rule to subgoal i. It then deletes some formula from the left side of the new subgoal $i + 1$, replacing that formula by P.

All the structural rules — cut, contraction, and thinning — can be applied to particular formulae using `res_inst_tac`.

19.4 Tactics for sequents

```
forms_of_seq         : term -> term list
could_res            : term * term -> bool
could_resolve_seq    : term * term -> bool
filseq_resolve_tac : thm list -> int -> int -> tactic
```

Associative unification is not as efficient as it might be, in part because the representation of lists defeats some of Isabelle's internal optimisations. The following operations implement faster rule application, and may have other uses.

`forms_of_seq` t returns the list of all formulae in the sequent t, removing sequence variables.

`could_res` (t, u) tests whether two formula lists could be resolved. List t is from a premise or subgoal, while u is from the conclusion of an object-rule. Assuming that each formula in u is surrounded by sequence variables, it checks that each conclusion formula is unifiable (using `could_unify`) with some subgoal formula.

`could_resolve_seq` (t, u) tests whether two sequents could be resolved. Sequent t is a premise or subgoal, while u is the conclusion of an object-rule. It simply calls `could_res` twice to check that both the left and the right sides of the sequents are compatible.

`filseq_resolve_tac` *thms maxr i* uses `filter_thms could_resolve` to extract the *thms* that are applicable to subgoal *i*. If more than *maxr* theorems are applicable then the tactic fails. Otherwise it calls `resolve_tac`. Thus, it is the sequent calculus analogue of `filt_resolve_tac`.

19.5 Packaging sequent rules

Section 15.2 described the distinction between safe and unsafe rules. An unsafe rule may reduce a provable goal to an unprovable set of subgoals, and should only be used as a last resort. Typical examples are the weakened quantifier rules `allL_thin` and `exR_thin`.

A **pack** is a pair whose first component is a list of safe rules and whose second is a list of unsafe rules. Packs can be extended in an obvious way to allow reasoning with various collections of rules. For clarity, LK declares `pack` as an ML datatype, although is essentially a type synonym:

```
datatype pack = Pack of thm list * thm list;
```

Pattern-matching using constructor `Pack` can inspect a pack's contents. Packs support the following operations:

```
empty_pack  : pack
prop_pack   : pack
LK_pack     : pack
LK_dup_pack : pack
add_safes   : pack * thm list -> pack                      infix 4
add_unsafes : pack * thm list -> pack                      infix 4
```

`empty_pack` is the empty pack.

`prop_pack` contains the propositional rules, namely those for $\wedge$, $\vee$, $\neg$, $\rightarrow$ and $\leftrightarrow$, along with the rules `basic` and `refl`. These are all safe.

`LK_pack` extends `prop_pack` with the safe rules `allR` and `exL` and the unsafe rules `allL_thin` and `exR_thin`. Search using this is incomplete since quantified formulae are used at most once.

`LK_dup_pack` extends `prop_pack` with the safe rules `allR` and `exL` and the unsafe rules `allL` and `exR`. Search using this is complete, since quantified formulae may be reused, but frequently fails to terminate. It is generally unsuitable for depth-first search.

pack `add_safes` *rules* adds some safe *rules* to the pack *pack*.

pack `add_unsafes` *rules* adds some unsafe *rules* to the pack *pack*.

19.6 Proof procedures

The LK proof procedure is similar to the classical reasoner described in Chap. 14. In fact it is simpler, since it works directly with sequents rather than simulating them. There is no need to distinguish introduction rules from elimination rules, and of course there is no swap rule. As always, Isabelle's classical proof procedures are less powerful than resolution theorem provers. But they are more natural and flexible, working with an open-ended set of rules.

Backtracking over the choice of a safe rule accomplishes nothing: applying them in any order leads to essentially the same result. Backtracking may be necessary over basic sequents when they perform unification. Suppose that 0, 1, 2, 3 are constants in the subgoals

$$P(0), P(1), P(2) \vdash P(?a)$$
$$P(0), P(2), P(3) \vdash P(?a)$$
$$P(1), P(3), P(2) \vdash P(?a)$$

The only assignment that satisfies all three subgoals is $?a \mapsto 2$, and this can only be discovered by search. The tactics given below permit backtracking only over axioms, such as `basic` and `refl`; otherwise they are deterministic.

19.6.1 Method A

```
reresolve_tac    : thm list -> int -> tactic
repeat_goal_tac : pack -> int -> tactic
pc_tac           : pack -> int -> tactic
```

These tactics use a method developed by Philippe de Groote. A subgoal is refined and the resulting subgoals are attempted in reverse order. For some reason, this is much faster than attempting the subgoals in order. The method is inherently depth-first.

At present, these tactics only work for rules that have no more than two premises. They fail — return no next state — if they can do nothing.

`reresolve_tac` *thms* *i* repeatedly applies the *thms* to subgoal *i* and the resulting subgoals.

`repeat_goal_tac` *pack* *i* applies the safe rules in the pack to a goal and the resulting subgoals. If no safe rule is applicable then it applies an unsafe rule and continues.

`pc_tac` *pack* *i* applies `repeat_goal_tac` using depth-first search to solve subgoal *i*.

19.6.2 Method B

```
safe_goal_tac : pack -> int -> tactic
step_tac      : pack -> int -> tactic
fast_tac      : pack -> int -> tactic
best_tac      : pack -> int -> tactic
```

These tactics are precisely analogous to those of the generic classical reasoner. They use 'Method A' only on safe rules. They fail if they can do nothing.

`safe_goal_tac` *pack i* applies the safe rules in the pack to a goal and the resulting subgoals. It ignores the unsafe rules.

`step_tac` *pack i* either applies safe rules (using `safe_goal_tac`) or applies one unsafe rule.

`fast_tac` *pack i* applies `step_tac` using depth-first search to solve subgoal i. Despite its name, it is frequently slower than `pc_tac`.

`best_tac` *pack i* applies `step_tac` using best-first search to solve subgoal i. It is particularly useful for quantifier duplication (using `LK_dup_pack`).

19.7 A simple example of classical reasoning

The theorem $\vdash \exists y . \forall x . P(y) \rightarrow P(x)$ is a standard example of the classical treatment of the existential quantifier. Classical reasoning is easy using `LK`, as you can see by comparing this proof with the one given in Sect. 16.7. From a logical point of view, the proofs are essentially the same; the key step here is to use `exR` rather than the weaker `exR_thin`.

```
goal LK.thy "|- EX y. ALL x. P(y)-->P(x)";
  Level 0
   |- EX y. ALL x. P(y) --> P(x)
   1.  |- EX y. ALL x. P(y) --> P(x)
by (resolve_tac [exR] 1);
  Level 1
   |- EX y. ALL x. P(y) --> P(x)
   1.  |- ALL x. P(?x) --> P(x), EX x. ALL xa. P(x) --> P(xa)
```

There are now two formulae on the right side. Keeping the existential one in reserve, we break down the universal one.

```
by (resolve_tac [allR] 1);
  Level 2
   |- EX y. ALL x. P(y) --> P(x)
   1. !!x.  |- P(?x) --> P(x), EX x. ALL xa. P(x) --> P(xa)
by (resolve_tac [impR] 1);
  Level 3
   |- EX y. ALL x. P(y) --> P(x)
   1. !!x. P(?x) |- P(x), EX x. ALL xa. P(x) --> P(xa)
```

Because LK is a sequent calculus, the formula $P(?x)$ does not become an assumption; instead, it moves to the left side. The resulting subgoal cannot be instantiated to a basic sequent: the bound variable x is not unifiable with the unknown $?x$.

```
by (resolve_tac [basic] 1);
  by: tactic failed
```

We reuse the existential formula using `exR_thin`, which discards it; we shall not need it a third time. We again break down the resulting formula.

```
by (resolve_tac [exR_thin] 1);
  Level 4
   |- EX y. ALL x. P(y) --> P(x)
   1. !!x. P(?x) |- P(x), ALL xa. P(?x7(x)) --> P(xa)
by (resolve_tac [allR] 1);
  Level 5
   |- EX y. ALL x. P(y) --> P(x)
   1. !!x xa. P(?x) |- P(x), P(?x7(x)) --> P(xa)
by (resolve_tac [impR] 1);
  Level 6
   |- EX y. ALL x. P(y) --> P(x)
   1. !!x xa. P(?x), P(?x7(x)) |- P(x), P(xa)
```

Subgoal 1 seems to offer lots of possibilities. Actually the only useful step is instantiating $?x_7$ to $\lambda x\,.\,x$, transforming $?x_7(x)$ into x.

```
by (resolve_tac [basic] 1);
  Level 7
   |- EX y. ALL x. P(y) --> P(x)
  No subgoals!
```

This theorem can be proved automatically. Because it involves quantifier duplication, we employ best-first search:

```
goal LK.thy "|- EX y. ALL x. P(y)-->P(x)";
  Level 0
   |- EX y. ALL x. P(y) --> P(x)
   1.  |- EX y. ALL x. P(y) --> P(x)
by (best_tac LK_dup_pack 1);
  Level 1
   |- EX y. ALL x. P(y) --> P(x)
  No subgoals!
```

19.8 A more complex proof

Many of Pelletier's test problems for theorem provers [52] can be solved automatically. Problem 39 concerns set theory, asserting that there is no Russell set — a set consisting of those sets that are not members of themselves:

$$\vdash \neg(\exists x\,.\,\forall y\,.\,y \in x \leftrightarrow y \notin y)$$

This does not require special properties of membership; we may generalize $x \in y$ to an arbitrary predicate $F(x, y)$. The theorem has a short manual proof. See the directory LK/ex for many more examples.

We set the main goal and move the negated formula to the left.

```
goal LK.thy "|- ~ (EX x. ALL y. F(y,x) <-> ~F(y,y))";
  Level 0
   |- ~ (EX x. ALL y. F(y,x) <-> ~ F(y,y))
   1.  |- ~ (EX x. ALL y. F(y,x) <-> ~ F(y,y))
by (resolve_tac [notR] 1);
  Level 1
   |- ~ (EX x. ALL y. F(y,x) <-> ~ F(y,y))
   1. EX x. ALL y. F(y,x) <-> ~ F(y,y) |-
```

The right side is empty; we strip both quantifiers from the formula on the left.

```
by (resolve_tac [exL] 1);
  Level 2
   |- ~ (EX x. ALL y. F(y,x) <-> ~ F(y,y))
   1. !!x. ALL y. F(y,x) <-> ~ F(y,y) |-
by (resolve_tac [allL_thin] 1);
  Level 3
   |- ~ (EX x. ALL y. F(y,x) <-> ~ F(y,y))
   1. !!x. F(?x2(x),x) <-> ~ F(?x2(x),?x2(x)) |-
```

The rule iffL says, if $P \leftrightarrow Q$ then P and Q are either both true or both false. It yields two subgoals.

```
by (resolve_tac [iffL] 1);
  Level 4
   |- ~ (EX x. ALL y. F(y,x) <-> ~ F(y,y))
   1. !!x.  |- F(?x2(x),x), ~ F(?x2(x),?x2(x))
   2. !!x. ~ F(?x2(x),?x2(x)), F(?x2(x),x) |-
```

We must instantiate $?x_2$, the shared unknown, to satisfy both subgoals. Beginning with subgoal 2, we move a negated formula to the left and create a basic sequent.

```
by (resolve_tac [notL] 2);
  Level 5
   |- ~ (EX x. ALL y. F(y,x) <-> ~ F(y,y))
   1. !!x.  |- F(?x2(x),x), ~ F(?x2(x),?x2(x))
   2. !!x. F(?x2(x),x) |- F(?x2(x),?x2(x))
by (resolve_tac [basic] 2);
  Level 6
   |- ~ (EX x. ALL y. F(y,x) <-> ~ F(y,y))
   1. !!x.  |- F(x,x), ~ F(x,x)
```

Thanks to the instantiation of $?x_2$, subgoal 1 is obviously true.

```
by (resolve_tac [notR] 1);
  Level 7
   |- ~ (EX x. ALL y. F(y,x) <-> ~ F(y,y))
   1. !!x. F(x,x) |- F(x,x)
by (resolve_tac [basic] 1);
  Level 8
   |- ~ (EX x. ALL y. F(y,x) <-> ~ F(y,y))
  No subgoals!
```

20. Constructive Type Theory

Martin-Löf's Constructive Type Theory [33, 39] can be viewed at many different levels. It is a formal system that embodies the principles of intuitionistic mathematics; it embodies the interpretation of propositions as types; it is a vehicle for deriving programs from proofs.

Thompson's book [58] gives a readable and thorough account of Type Theory. Nuprl is an elaborate implementation [9]. ALF is a more recent tool that allows proof terms to be edited directly [30].

Isabelle's original formulation of Type Theory was a kind of sequent calculus, following Martin-Löf [33]. It included rules for building the context, namely variable bindings with their types. A typical judgement was

$$a(x_1,\ldots,x_n) \in A(x_1,\ldots,x_n)\ [x_1 \in A_1, x_2 \in A_2(x_1), \ldots, x_n \in A_n(x_1,\ldots,x_{n-1})]$$

This sequent calculus was not satisfactory because assumptions like 'suppose A is a type' or 'suppose $B(x)$ is a type for all x in A' could not be formalized.

The theory CTT implements Constructive Type Theory, using natural deduction. The judgement above is expressed using $\bigwedge$ and $\Longrightarrow$:

$$\begin{array}{c} \bigwedge x_1 \ldots x_n . [\![x_1 \in A_1; x_2 \in A_2(x_1); \cdots\ x_n \in A_n(x_1,\ldots,x_{n-1})]\!] \Longrightarrow \\ a(x_1,\ldots,x_n) \in A(x_1,\ldots,x_n) \end{array}$$

Assumptions can use all the judgement forms, for instance to express that B is a family of types over A:

$$\bigwedge x \,.\, x \in A \Longrightarrow B(x)\ \text{type}$$

To justify the CTT formulation it is probably best to appeal directly to the semantic explanations of the rules [33], rather than to the rules themselves. The order of assumptions no longer matters, unlike in standard Type Theory. Contexts, which are typical of many modern type theories, are difficult to represent in Isabelle. In particular, it is difficult to enforce that all the variables in a context are distinct.

The theory does not use polymorphism. Terms in CTT have type `i`, the type of individuals. Types in CTT have type `t`.

CTT supports all of Type Theory apart from list types, well-ordering types, and universes. Universes could be introduced *à la Tarski*, adding new constants as names for types. The formulation *à la Russell*, where types denote themselves, is only possible if we identify the meta-types `i` and `t`. Most published

name	*meta-type*	*description*
Type	$t \to prop$	judgement form
Eqtype	$[t, t] \to prop$	judgement form
Elem	$[i, t] \to prop$	judgement form
Eqelem	$[i, i, t] \to prop$	judgement form
Reduce	$[i, i] \to prop$	extra judgement form
N	t	natural numbers type
0	i	constructor
succ	$i \to i$	constructor
rec	$[i, i, [i, i] \to i] \to i$	eliminator
Prod	$[t, i \to t] \to t$	general product type
lambda	$(i \to i) \to i$	constructor
Sum	$[t, i \to t] \to t$	general sum type
pair	$[i, i] \to i$	constructor
split	$[i, [i, i] \to i] \to i$	eliminator
fst snd	$i \to i$	projections
inl inr	$i \to i$	constructors for +
when	$[i, i \to i, i \to i] \to i$	eliminator for +
Eq	$[t, i, i] \to t$	equality type
eq	i	constructor
F	t	empty type
contr	$i \to i$	eliminator
T	t	singleton type
tt	i	constructor

Fig. 20.1. The constants of CTT

formulations of well-ordering types have difficulties involving extensionality of functions; I suggest that you use some other method for defining recursive types. List types are easy to introduce by declaring new rules.

CTT uses the 1982 version of Type Theory, with extensional equality. The computation $a = b \in A$ and the equality $c \in Eq(A, a, b)$ are interchangeable. Its rewriting tactics prove theorems of the form $a = b \in A$. It could be modified to have intensional equality, but rewriting tactics would have to prove theorems of the form $c \in Eq(A, a, b)$ and the computation rules might require a separate simplifier.

20.1 Syntax

The constants are shown in Fig. 20.1. The infixes include the function application operator (sometimes called 'apply'), and the 2-place type operators. Note that meta-level abstraction and application, $\lambda x . b$ and $f(a)$, differ from object-level abstraction and application, `lam` $x . b$ and $b`a$. A CTT function f is simply an individual as far as Isabelle is concerned: its Isabelle type is i, not say $i \Rightarrow i$.

The notation for CTT (Fig. 20.2) is based on that of Nordström et al. [39]. The empty type is called F and the one-element type is T; other finite types are built as $T + T + T$, etc.

Quantification is expressed using general sums $\sum_{x \in A} B[x]$ and products $\prod_{x \in A} B[x]$. Instead of `Sum(`A`,`B`)` and `Prod(`A`,`B`)` we may write `SUM` $x : A . B[x]$ and `PROD` $x : A . B[x]$. For example, we may write

```
SUM y:B. PROD x:A. C(x,y)    for    Sum(B, %y. Prod(A, %x. C(x,y)))
```

The special cases as A`*`B and A`-->`B abbreviate general sums and products over a constant family.[1] Isabelle accepts these abbreviations in parsing and uses them whenever possible for printing.

20.2 Rules of inference

The rules obey the following naming conventions. Type formation rules have the suffix `F`. Introduction rules have the suffix `I`. Elimination rules have the suffix `E`. Computation rules, which describe the reduction of eliminators, have the suffix `C`. The equality versions of the rules (which permit reductions on subterms) are called **long** rules; their names have the suffix `L`. Introduction and computation rules are often further suffixed with constructor names.

Figure 20.3 presents the equality rules. Most of them are straightforward: reflexivity, symmetry, transitivity and substitution. The judgement `Reduce` does not belong to Type Theory proper; it has been added to implement rewriting. The judgement `Reduce`(a, b) holds when $a = b : A$ holds. It also holds

[1] Unlike normal infix operators, `*` and `-->` merely define abbreviations; there are no constants `op *` and `op -->`.

symbol	*name*	*meta-type*	*priority*	*description*
`lam`	`lambda`	$(i \Rightarrow o) \Rightarrow i$	10	λ-abstraction

BINDERS

symbol	*meta-type*	*priority*	*description*
`	$[i, i] \to i$	Left 55	function application
+	$[t, t] \to t$	Right 30	sum of two types

INFIXES

external	*internal*	*standard notation*
`PROD` $x{:}A\ .\ B[x]$	`Prod(`A`,` $\lambda x\,.\,B[x]$`)`	product $\prod_{x\in A} B[x]$
`SUM` $x{:}A\ .\ B[x]$	`Sum(`A`,` $\lambda x\,.\,B[x]$`)`	sum $\sum_{x\in A} B[x]$
A `-->` B	`Prod(`A`,` $\lambda x\,.\,B$`)`	function space $A \to B$
A `*` B	`Sum(`A`,` $\lambda x\,.\,B$`)`	binary product $A \times B$

TRANSLATIONS

prop = *type* `type`
  | *type* `=` *type*
  | *term* `:` *type*
  | *term* `=` *term* `:` *type*

type = expression of type t
  | `PROD` *id* `:` *type* `.` *type*
  | `SUM` *id* `:` *type* `.` *type*

term = expression of type i
  | `lam` *id id** `.` *term*
  | `<` *term* `,` *term* `>`

GRAMMAR

Fig. 20.2. Syntax of CTT

```
refl_type        A type ==> A = A
refl_elem        a : A ==> a = a : A

sym_type         A = B ==> B = A
sym_elem         a = b : A ==> b = a : A

trans_type       [| A = B;  B = C |] ==> A = C
trans_elem       [| a = b : A;  b = c : A |] ==> a = c : A

equal_types      [| a : A;  A = B |] ==> a : B
equal_typesL     [| a = b : A;  A = B |] ==> a = b : B

subst_type       [| a : A;  !!z. z:A ==> B(z) type |] ==> B(a) type
subst_typeL      [| a = c : A;  !!z. z:A ==> B(z) = D(z)
                 |] ==> B(a) = D(c)

subst_elem       [| a : A;  !!z. z:A ==> b(z):B(z) |] ==> b(a):B(a)
subst_elemL      [| a = c : A;  !!z. z:A ==> b(z) = d(z) : B(z)
                 |] ==> b(a) = d(c) : B(a)

refl_red         Reduce(a,a)
red_if_equal     a = b : A ==> Reduce(a,b)
trans_red        [| a = b : A;  Reduce(b,c) |] ==> a = c : A
```

Fig. 20.3. General equality rules

when a and b are syntactically identical, even if they are ill-typed, because rule `refl_red` does not verify that a belongs to A.

The `Reduce` rules do not give rise to new theorems about the standard judgements. The only rule with `Reduce` in a premise is `trans_red`, whose other premise ensures that a and b (and thus c) are well-typed.

Figure 20.4 presents the rules for N, the type of natural numbers. They include `zero_ne_succ`, which asserts $0 \neq n+1$. This is the fourth Peano axiom and cannot be derived without universes [33, page 91].

The constant `rec` constructs proof terms when mathematical induction, rule `NE`, is applied. It can also express primitive recursion. Since `rec` can be applied to higher-order functions, it can even express Ackermann's function, which is not primitive recursive [58, page 104].

Figure 20.5 shows the rules for general product types, which include function types as a special case. The rules correspond to the predicate calculus rules for universal quantifiers and implication. They also permit reasoning about functions, with the rules of a typed λ-calculus.

Figure 20.6 shows the rules for general sum types, which include binary product types as a special case. The rules correspond to the predicate calculus rules for existential quantifiers and conjunction. They also permit reasoning about ordered pairs, with the projections `fst` and `snd`.

Figure 20.7 shows the rules for binary sum types. They correspond to the predicate calculus rules for disjunction. They also permit reasoning about dis-

```
NF          N type

NI0         0 : N
NI_succ     a : N ==> succ(a) : N
NI_succL    a = b : N ==> succ(a) = succ(b) : N

NE          [| p: N;  a: C(0);
               !!u v. [| u: N; v: C(u) |] ==> b(u,v): C(succ(u))
            |] ==> rec(p, a, %u v.b(u,v)) : C(p)

NEL         [| p = q : N;  a = c : C(0);
               !!u v. [| u: N; v: C(u) |] ==> b(u,v)=d(u,v): C(succ(u))
            |] ==> rec(p, a, %u v.b(u,v)) = rec(q,c,d) : C(p)

NC0         [| a: C(0);
               !!u v. [| u: N; v: C(u) |] ==> b(u,v): C(succ(u))
            |] ==> rec(0, a, %u v.b(u,v)) = a : C(0)

NC_succ     [| p: N;  a: C(0);
               !!u v. [| u: N; v: C(u) |] ==> b(u,v): C(succ(u))
            |] ==> rec(succ(p), a, %u v.b(u,v)) =
                   b(p, rec(p, a, %u v.b(u,v))) : C(succ(p))

zero_ne_succ      [| a: N;  0 = succ(a) : N |] ==> 0: F
```

Fig. 20.4. Rules for type N

```
ProdF       [| A type; !!x. x:A ==> B(x) type |] ==> PROD x:A.B(x) type
ProdFL      [| A = C;  !!x. x:A ==> B(x) = D(x) |] ==>
          PROD x:A.B(x) = PROD x:C.D(x)

ProdI       [| A type;  !!x. x:A ==> b(x):B(x)
            |] ==> lam x.b(x) : PROD x:A.B(x)
ProdIL      [| A type;  !!x. x:A ==> b(x) = c(x) : B(x)
            |] ==> lam x.b(x) = lam x.c(x) : PROD x:A.B(x)

ProdE       [| p : PROD x:A.B(x);  a : A |] ==> p`a : B(a)
ProdEL      [| p=q: PROD x:A.B(x);  a=b : A |] ==> p`a = q`b : B(a)

ProdC       [| a : A;  !!x. x:A ==> b(x) : B(x)
            |] ==> (lam x.b(x)) ` a = b(a) : B(a)

ProdC2      p : PROD x:A.B(x) ==> (lam x. p`x) = p : PROD x:A.B(x)
```

Fig. 20.5. Rules for the product type $\prod_{x \in A} B[x]$

```
SumF     [| A type;  !!x. x:A ==> B(x) type |] ==> SUM x:A.B(x) type
SumFL    [| A = C;  !!x. x:A ==> B(x) = D(x)
         |] ==> SUM x:A.B(x) = SUM x:C.D(x)

SumI     [| a : A;  b : B(a) |] ==> <a,b> : SUM x:A.B(x)
SumIL    [| a=c:A;  b=d:B(a) |] ==> <a,b> = <c,d> : SUM x:A.B(x)

SumE     [| p: SUM x:A.B(x);
            !!x y. [| x:A; y:B(x) |] ==> c(x,y): C(<x,y>)
         |] ==> split(p, %x y.c(x,y)) : C(p)

SumEL    [| p=q : SUM x:A.B(x);
            !!x y. [| x:A; y:B(x) |] ==> c(x,y)=d(x,y): C(<x,y>)
         |] ==> split(p, %x y.c(x,y)) = split(q, %x y.d(x,y)) : C(p)

SumC     [| a: A;  b: B(a);
            !!x y. [| x:A; y:B(x) |] ==> c(x,y): C(<x,y>)
         |] ==> split(<a,b>, %x y.c(x,y)) = c(a,b) : C(<a,b>)

fst_def  fst(a) == split(a, %x y.x)
snd_def  snd(a) == split(a, %x y.y)
```

Fig. 20.6. Rules for the sum type $\sum_{x \in A} B[x]$

joint sums, with the injections `inl` and `inr` and case analysis operator `when`.

Figure 20.8 shows the rules for the empty and unit types, F and T. They correspond to the predicate calculus rules for absurdity and truth.

Figure 20.9 shows the rules for equality types. If $a = b \in A$ is provable then `eq` is a canonical element of the type $Eq(A, a, b)$, and vice versa. These rules define extensional equality; the most recent versions of Type Theory use intensional equality [39].

Figure 20.10 presents the derived rules. The rule `subst_prodE` is derived from `prodE`, and is easier to use in backwards proof. The rules `SumE_fst` and `SumE_snd` express the typing of `fst` and `snd`; together, they are roughly equivalent to `SumE` with the advantage of creating no parameters. Section 20.12 below demonstrates these rules in a proof of the Axiom of Choice.

All the rules are given in η-expanded form. For instance, every occurrence of $\lambda u\, v\,.\, b(u, v)$ could be abbreviated to b in the rules for N. The expanded form permits Isabelle to preserve bound variable names during backward proof. Names of bound variables in the conclusion (here, u and v) are matched with corresponding bound variables in the premises.

20.3 Rule lists

The Type Theory tactics provide rewriting, type inference, and logical reasoning. Many proof procedures work by repeatedly resolving certain Type Theory

```
PlusF       [| A type;  B type |] ==> A+B type
PlusFL      [| A = C;  B = D |] ==> A+B = C+D

PlusI_inl   [| a : A;  B type |] ==> inl(a) : A+B
PlusI_inlL  [| a = c : A;  B type |] ==> inl(a) = inl(c) : A+B

PlusI_inr   [| A type;  b : B |] ==> inr(b) : A+B
PlusI_inrL  [| A type;  b = d : B |] ==> inr(b) = inr(d) : A+B

PlusE      [| p: A+B;
              !!x. x:A ==> c(x): C(inl(x));
              !!y. y:B ==> d(y): C(inr(y))
           |] ==> when(p, %x.c(x), %y.d(y)) : C(p)

PlusEL     [| p = q : A+B;
              !!x. x: A ==> c(x) = e(x) : C(inl(x));
              !!y. y: B ==> d(y) = f(y) : C(inr(y))
           |] ==> when(p, %x.c(x), %y.d(y)) =
                  when(q, %x.e(x), %y.f(y)) : C(p)

PlusC_inl [| a: A;
              !!x. x:A ==> c(x): C(inl(x));
              !!y. y:B ==> d(y): C(inr(y))
           |] ==> when(inl(a), %x.c(x), %y.d(y)) = c(a) : C(inl(a))

PlusC_inr [| b: B;
              !!x. x:A ==> c(x): C(inl(x));
              !!y. y:B ==> d(y): C(inr(y))
           |] ==> when(inr(b), %x.c(x), %y.d(y)) = d(b) : C(inr(b))
```

Fig. 20.7. Rules for the binary sum type $A + B$

```
FF          F type
FE          [| p: F;  C type |] ==> contr(p) : C
FEL         [| p = q : F;  C type |] ==> contr(p) = contr(q) : C

TF          T type
TI          tt : T
TE          [| p : T;  c : C(tt) |] ==> c : C(p)
TEL         [| p = q : T;  c = d : C(tt) |] ==> c = d : C(p)
TC          p : T ==> p = tt : T)
```

Fig. 20.8. Rules for types F and T

```
EqF         [| A type;  a : A;  b : A |] ==> Eq(A,a,b) type
EqFL        [| A=B;  a=c: A;  b=d : A |] ==> Eq(A,a,b) = Eq(B,c,d)
EqI         a = b : A ==> eq : Eq(A,a,b)
EqE         p : Eq(A,a,b) ==> a = b : A
EqC         p : Eq(A,a,b) ==> p = eq : Eq(A,a,b)
```

Fig. 20.9. Rules for the equality type $Eq(A, a, b)$

```
replace_type     [| B = A;  a : A |] ==> a : B
subst_eqtyparg   [| a=c : A;  !!z. z:A ==> B(z) type |] ==> B(a)=B(c)

subst_prodE      [| p: Prod(A,B);  a: A;  !!z. z: B(a) ==> c(z): C(z)
                 |] ==> c(p`a): C(p`a)

SumIL2      [| c=a : A;  d=b : B(a) |] ==> <c,d> = <a,b> : Sum(A,B)

SumE_fst  p : Sum(A,B) ==> fst(p) : A

SumE_snd  [| p: Sum(A,B);  A type;  !!x. x:A ==> B(x) type
          |] ==> snd(p) : B(fst(p))
```

Fig. 20.10. Derived rules for CTT

rules against a proof state. CTT defines lists — each with type `thm list` — of related rules.

`form_rls` contains formation rules for the types N, Π, Σ, $+$, Eq, F, and T.

`formL_rls` contains long formation rules for Π, Σ, $+$, and Eq. (For other types use `refl_type`.)

`intr_rls` contains introduction rules for the types N, Π, Σ, $+$, and T.

`intrL_rls` contains long introduction rules for N, Π, Σ, and $+$. (For T use `refl_elem`.)

`elim_rls` contains elimination rules for the types N, Π, Σ, $+$, and F. The rules for Eq and T are omitted because they involve no eliminator.

`elimL_rls` contains long elimination rules for N, Π, Σ, $+$, and F.

`comp_rls` contains computation rules for the types N, Π, Σ, and $+$. Those for Eq and T involve no eliminator.

`basic_defs` contains the definitions of `fst` and `snd`.

20.4 Tactics for subgoal reordering

```
test_assume_tac : int -> tactic
typechk_tac     : thm list -> tactic
equal_tac       : thm list -> tactic
intr_tac        : thm list -> tactic
```

Blind application of CTT rules seldom leads to a proof. The elimination rules, especially, create subgoals containing new unknowns. These subgoals unify with anything, creating a huge search space. The standard tactic `filt_resolve_tac` (see Sect. 6.4.2) fails for goals that are too flexible; so does the CTT tactic `test_assume_tac`. Used with the tactical `REPEAT_FIRST` they achieve a simple kind of subgoal reordering: the less flexible subgoals are attempted first. Do some single step proofs, or study the examples below, to see why this is necessary.

`test_assume_tac` *i* uses `assume_tac` to solve the subgoal by assumption, but only if subgoal *i* has the form $a \in A$ and the head of *a* is not an unknown. Otherwise, it fails.

`typechk_tac` *thms* uses *thms* with formation, introduction, and elimination rules to check the typing of constructions. It is designed to solve goals of the form $a \in ?A$, where *a* is rigid and $?A$ is flexible. Thus it performs Hindley-Milner type inference. The tactic can also solve goals of the form A type.

`equal_tac` *thms* uses *thms* with the long introduction and elimination rules to solve goals of the form $a = b \in A$, where *a* is rigid. It is intended for deriving the long rules for defined constants such as the arithmetic operators. The tactic can also perform type checking.

`intr_tac` *thms* uses *thms* with the introduction rules to break down a type. It is designed for goals like $?a \in A$ where $?a$ is flexible and *A* rigid. These typically arise when trying to prove a proposition *A*, expressed as a type.

20.5 Rewriting tactics

```
rew_tac     : thm list -> tactic
hyp_rew_tac : thm list -> tactic
```

Object-level simplification is accomplished through proof, using the CTT equality rules and the built-in rewriting functor `TSimpFun`.[2] The rewrites include the computation rules and other equations. The long versions of the other rules permit rewriting of subterms and subtypes. Also used are transitivity and the ex-

[2]This should not be confused with Isabelle's main simplifier; `TSimpFun` is only useful for CTT and similar logics with type inference rules. At present it is undocumented.

tra judgement form `Reduce`. Meta-level simplification handles only definitional equality.

`rew_tac` *thms* applies *thms* and the computation rules as left-to-right rewrites. It solves the goal $a = b \in A$ by rewriting a to b. If b is an unknown then it is assigned the rewritten form of a. All subgoals are rewritten.

`hyp_rew_tac` *thms* is like `rew_tac`, but includes as rewrites any equations present in the assumptions.

20.6 Tactics for logical reasoning

Interpreting propositions as types lets `CTT` express statements of intuitionistic logic. However, Constructive Type Theory is not just another syntax for first-order logic. There are fundamental differences.

Can assumptions be deleted after use? Not every occurrence of a type represents a proposition, and Type Theory assumptions declare variables. In first-order logic, $\vee$-elimination with the assumption $P \vee Q$ creates one subgoal assuming P and another assuming Q, and $P \vee Q$ can be deleted safely. In Type Theory, $+$-elimination with the assumption $z \in A + B$ creates one subgoal assuming $x \in A$ and another assuming $y \in B$ (for arbitrary x and y). Deleting $z \in A+B$ when other assumptions refer to z may render the subgoal unprovable: arguably, meaningless.

Isabelle provides several tactics for predicate calculus reasoning in `CTT`:

```
mp_tac       : int -> tactic
add_mp_tac   : int -> tactic
safestep_tac : thm list -> int -> tactic
safe_tac     : thm list -> int -> tactic
step_tac     : thm list -> int -> tactic
pc_tac       : thm list -> int -> tactic
```

These are loosely based on the intuitionistic proof procedures of `FOL`. For the reasons discussed above, a rule that is safe for propositional reasoning may be unsafe for type checking; thus, some of the 'safe' tactics are misnamed.

`mp_tac` i searches in subgoal i for assumptions of the form $f \in \Pi(A, B)$ and $a \in A$, where A may be found by unification. It replaces $f \in \Pi(A, B)$ by $z \in B(a)$, where z is a new parameter. The tactic can produce multiple outcomes for each suitable pair of assumptions. In short, `mp_tac` performs Modus Ponens among the assumptions.

`add_mp_tac` i is like `mp_tac` i but retains the assumption $f \in \Pi(A, B)$. It avoids information loss but obviously loops if repeated.

`safestep_tac` *thms* i attacks subgoal i using formation rules and certain other 'safe' rules (`FE`, `ProdI`, `SumE`, `PlusE`), calling `mp_tac` when appropriate. It also uses *thms*, which are typically premises of the rule being derived.

`safe_tac` *thms* i attempts to solve subgoal i by means of backtracking, using `safestep_tac`.

`step_tac` *thms* i tries to reduce subgoal i using `safestep_tac`, then tries unsafe rules. It may produce multiple outcomes.

`pc_tac` *thms* i tries to solve subgoal i by backtracking, using `step_tac`.

20.7 A theory of arithmetic

`Arith` is a theory of elementary arithmetic. It proves the properties of addition, multiplication, subtraction, division, and remainder, culminating in the theorem

$$a \bmod b + (a/b) \times b = a.$$

Figure 20.11 presents the definitions and some of the key theorems, including commutative, distributive, and associative laws.

The operators `#+`, `-`, `|-|`, `#*`, `mod` and `div` stand for sum, difference, absolute difference, product, remainder and quotient, respectively. Since Type Theory has only primitive recursion, some of their definitions may be obscure.

The difference $a - b$ is computed by taking b predecessors of a, where the predecessor function is $\lambda v\,.\,\mathtt{rec}(v, 0, \lambda x\; y\,.\,x)$.

The remainder $a \bmod b$ counts up to a in a cyclic fashion, using 0 as the successor of $b - 1$. Absolute difference is used to test the equality $succ(v) = b$.

The quotient a/b is computed by adding one for every number x such that $0 \leq x \leq a$ and $x \bmod b = 0$.

20.8 The examples directory

This directory contains examples and experimental proofs in `CTT`.

`CTT/ex/typechk.ML` contains simple examples of type checking and type deduction.

`CTT/ex/elim.ML` contains some examples from Martin-Löf [33], proved using `pc_tac`.

`CTT/ex/equal.ML` contains simple examples of rewriting.

`CTT/ex/synth.ML` demonstrates the use of unknowns with some trivial examples of program synthesis.

symbol	*meta-type*	*priority*	*description*
`#*`	$[i, i] \Rightarrow i$	Left 70	multiplication
`div`	$[i, i] \Rightarrow i$	Left 70	division
`mod`	$[i, i] \Rightarrow i$	Left 70	modulus
`#+`	$[i, i] \Rightarrow i$	Left 65	addition
`-`	$[i, i] \Rightarrow i$	Left 65	subtraction
`|-|`	$[i, i] \Rightarrow i$	Left 65	absolute difference

```
add_def          a#+b  == rec(a, b, %u v.succ(v))
diff_def         a-b   == rec(b, a, %u v.rec(v, 0, %x y.x))
absdiff_def      a|-|b == (a-b) #+ (b-a)
mult_def         a#*b  == rec(a, 0, %u v. b #+ v)

mod_def          a mod b ==
                 rec(a, 0, %u v. rec(succ(v) |-| b, 0, %x y.succ(v)))

div_def          a div b ==
                 rec(a, 0, %u v. rec(succ(u) mod b, succ(v), %x y.v))

add_typing       [| a:N;  b:N |] ==> a #+ b : N
addC0            b:N ==> 0 #+ b = b : N
addC_succ        [| a:N;  b:N |] ==> succ(a) #+ b = succ(a #+ b) : N

add_assoc        [| a:N;  b:N;  c:N |] ==>
                 (a #+ b) #+ c = a #+ (b #+ c) : N

add_commute      [| a:N;  b:N |] ==> a #+ b = b #+ a : N

mult_typing      [| a:N;  b:N |] ==> a #* b : N
multC0           b:N ==> 0 #* b = 0 : N
multC_succ       [| a:N;  b:N |] ==> succ(a) #* b = b #+ (a#*b) : N
mult_commute     [| a:N;  b:N |] ==> a #* b = b #* a : N

add_mult_dist    [| a:N;  b:N;  c:N |] ==>
                 (a #+ b) #* c = (a #* c) #+ (b #* c) : N

mult_assoc       [| a:N;  b:N;  c:N |] ==>
                 (a #* b) #* c = a #* (b #* c) : N

diff_typing      [| a:N;  b:N |] ==> a - b : N
diffC0           a:N ==> a - 0 = a : N
diff_0_eq_0      b:N ==> 0 - b = 0 : N
diff_succ_succ   [| a:N;  b:N |] ==> succ(a) - succ(b) = a - b : N
diff_self_eq_0   a:N ==> a - a = 0 : N
add_inverse_diff [| a:N;  b:N;  b-a=0 : N |] ==> b #+ (a-b) = a : N
```

Fig. 20.11. The theory of arithmetic

20.9 Example: type inference

Type inference involves proving a goal of the form $a \in ?A$, where a is a term and $?A$ is an unknown standing for its type. The type, initially unknown, takes shape in the course of the proof. Our example is the predecessor function on the natural numbers.

```
goal CTT.thy "lam n. rec(n, 0, %x y.x) : ?A";
  Level 0
  lam n. rec(n,0,%x y. x) : ?A
   1. lam n. rec(n,0,%x y. x) : ?A
```

Since the term is a Constructive Type Theory λ-abstraction (not to be confused with a meta-level abstraction), we apply the rule ProdI, for Π-introduction. This instantiates $?A$ to a product type of unknown domain and range.

```
by (resolve_tac [ProdI] 1);
  Level 1
  lam n. rec(n,0,%x y. x) : PROD x:?A1. ?B1(x)
   1. ?A1 type
   2. !!n. n : ?A1 ==> rec(n,0,%x y. x) : ?B1(n)
```

Subgoal 1 is too flexible. It can be solved by instantiating $?A_1$ to any type, but most instantiations will invalidate subgoal 2. We therefore tackle the latter subgoal. It asks the type of a term beginning with `rec`, which can be found by N-elimination.

```
by (eresolve_tac [NE] 2);
  Level 2
  lam n. rec(n,0,%x y. x) : PROD x:N. ?C2(x,x)
   1. N type
   2. !!n. 0 : ?C2(n,0)
   3. !!n x y. [| x : N; y : ?C2(n,x) |] ==> x : ?C2(n,succ(x))
```

Subgoal 1 is no longer flexible: we now know $?A_1$ is the type of natural numbers. However, let us continue proving nontrivial subgoals. Subgoal 2 asks, what is the type of 0?

```
by (resolve_tac [NIO] 2);
  Level 3
  lam n. rec(n,0,%x y. x) : N --> N
   1. N type
   2. !!n x y. [| x : N; y : N |] ==> x : N
```

The type $?A$ is now fully determined. It is the product type $\prod_{x \in N} N$, which is shown as the function type $N \to N$ because there is no dependence on x. But we must prove all the subgoals to show that the original term is validly typed. Subgoal 2 is provable by assumption and the remaining subgoal falls by N-formation.

```
by (assume_tac 2);
  Level 4
  lam n. rec(n,0,%x y. x) : N --> N
   1. N type
```

```
by (resolve_tac [NF] 1);
  Level 5
  lam n. rec(n,0,%x y. x) : N --> N
  No subgoals!
```

Calling `typechk_tac` can prove this theorem in one step.

Even if the original term is ill-typed, one can infer a type for it, but unprovable subgoals will be left. As an exercise, try to prove the following invalid goal:

```
goal CTT.thy "lam n. rec(n, 0, %x y.tt) : ?A";
```

20.10 An example of logical reasoning

Logical reasoning in Type Theory involves proving a goal of the form $?a \in A$, where type A expresses a proposition and $?a$ stands for its proof term, a value of type A. The proof term is initially unknown and takes shape during the proof.

Our example expresses a theorem about quantifiers in a sorted logic:

$$\frac{\exists x \in A \,.\, P(x) \vee Q(x)}{(\exists x \in A \,.\, P(x)) \vee (\exists x \in A \,.\, Q(x))}$$

By the propositions-as-types principle, this is encoded using Σ and $+$ types. A special case of it expresses a distributive law of Type Theory:

$$\frac{A \times (B + C)}{(A \times B) + (A \times C)}$$

Generalizing this from $\times$ to Σ, and making the typing conditions explicit, yields the rule we must derive:

$$\frac{A \text{ type} \quad \begin{matrix}[x \in A] \\ \vdots \\ B(x) \text{ type}\end{matrix} \quad \begin{matrix}[x \in A] \\ \vdots \\ C(x) \text{ type}\end{matrix} \quad p \in \Sigma_{x\in A} B(x) + C(x)}{?a \in (\Sigma_{x\in A} B(x)) + (\Sigma_{x\in A} C(x))}$$

To begin, we bind the rule's premises — returned by the `goal` command — to the ML variable `prems`.

```
val prems = goal CTT.thy
    "[| A type;                          \
\       !!x. x:A ==> B(x) type;          \
\       !!x. x:A ==> C(x) type;          \
\       p: SUM x:A. B(x) + C(x)          \
\    |] ==>  ?a : (SUM x:A. B(x)) + (SUM x:A. C(x))";
  Level 0
  ?a : (SUM x:A. B(x)) + (SUM x:A. C(x))
   1. ?a : (SUM x:A. B(x)) + (SUM x:A. C(x))
```

```
val prems = ["A type  [A type]",
             "?x : A ==> B(?x) type  [!!x. x : A ==> B(x) type]",
             "?x : A ==> C(?x) type  [!!x. x : A ==> C(x) type]",
             "p : SUM x:A. B(x) + C(x)  [p : SUM x:A. B(x) + C(x)]"]
            : thm list
```

The last premise involves the sum type Σ. Since it is a premise rather than the assumption of a goal, it cannot be found by **eresolve_tac**. We could insert it (and the other atomic premise) by calling

```
cut_facts_tac prems 1;
```

A forward proof step is more straightforward here. Let us resolve the Σ-elimination rule with the premises using RL. This inference yields one result, which we supply to resolve_tac.

```
by (resolve_tac (prems RL [SumE]) 1);
  Level 1
  split(p,?c1) : (SUM x:A. B(x)) + (SUM x:A. C(x))
   1. !!x y.
           [| x : A; y : B(x) + C(x) |] ==>
           ?c1(x,y) : (SUM x:A. B(x)) + (SUM x:A. C(x))
```

The subgoal has two new parameters, x and y. In the main goal, $?a$ has been instantiated with a split term. The assumption $y \in B(x) + C(x)$ is eliminated next, causing a case split and creating the parameter xa. This inference also inserts **when** into the main goal.

```
by (eresolve_tac [PlusE] 1);
  Level 2
  split(p,%x y. when(y,?c2(x,y),?d2(x,y)))
  : (SUM x:A. B(x)) + (SUM x:A. C(x))
   1. !!x y xa.
           [| x : A; xa : B(x) |] ==>
           ?c2(x,y,xa) : (SUM x:A. B(x)) + (SUM x:A. C(x))
   2. !!x y ya.
           [| x : A; ya : C(x) |] ==>
           ?d2(x,y,ya) : (SUM x:A. B(x)) + (SUM x:A. C(x))
```

To complete the proof object for the main goal, we need to instantiate the terms $?c_2(x, y, xa)$ and $?d_2(x, y, xa)$. We attack subgoal 1 by a +-introduction rule; since the goal assumes $xa \in B(x)$, we take the left injection (inl).

```
by (resolve_tac [PlusI_inl] 1);
  Level 3
  split(p,%x y. when(y,%xa. inl(?a3(x,y,xa)),?d2(x,y)))
  : (SUM x:A. B(x)) + (SUM x:A. C(x))
   1. !!x y xa. [| x : A; xa : B(x) |] ==> ?a3(x,y,xa) : SUM x:A. B(x)
   2. !!x y xa. [| x : A; xa : B(x) |] ==> SUM x:A. C(x) type
   3. !!x y ya.
           [| x : A; ya : C(x) |] ==>
           ?d2(x,y,ya) : (SUM x:A. B(x)) + (SUM x:A. C(x))
```

A new subgoal 2 has appeared, to verify that $\sum_{x \in A} C(x)$ is a type. Continuing to work on subgoal 1, we apply the Σ-introduction rule. This instantiates the term

$?a_3(x, y, xa)$; the main goal now contains an ordered pair, whose components are two new unknowns.

```
by (resolve_tac [SumI] 1);
  Level 4
  split(p,%x y. when(y,%xa. inl(<?a4(x,y,xa),?b4(x,y,xa)>),?d2(x,y)))
  : (SUM x:A. B(x)) + (SUM x:A. C(x))
   1. !!x y xa. [| x : A; xa : B(x) |] ==> ?a4(x,y,xa) : A
   2. !!x y xa. [| x : A; xa : B(x) |] ==> ?b4(x,y,xa) : B(?a4(x,y,xa))
   3. !!x y xa. [| x : A; xa : B(x) |] ==> SUM x:A. C(x) type
   4. !!x y ya.
         [| x : A; ya : C(x) |] ==>
         ?d2(x,y,ya) : (SUM x:A. B(x)) + (SUM x:A. C(x))
```

The two new subgoals both hold by assumption. Observe how the unknowns $?a_4$ and $?b_4$ are instantiated throughout the proof state.

```
by (assume_tac 1);
  Level 5
  split(p,%x y. when(y,%xa. inl(<x,?b4(x,y,xa)>),?d2(x,y)))
  : (SUM x:A. B(x)) + (SUM x:A. C(x))
   1. !!x y xa. [| x : A; xa : B(x) |] ==> ?b4(x,y,xa) : B(x)
   2. !!x y xa. [| x : A; xa : B(x) |] ==> SUM x:A. C(x) type
   3. !!x y ya.
         [| x : A; ya : C(x) |] ==>
         ?d2(x,y,ya) : (SUM x:A. B(x)) + (SUM x:A. C(x))
by (assume_tac 1);
  Level 6
  split(p,%x y. when(y,%xa. inl(<x,xa>),?d2(x,y)))
  : (SUM x:A. B(x)) + (SUM x:A. C(x))
   1. !!x y xa. [| x : A; xa : B(x) |] ==> SUM x:A. C(x) type
   2. !!x y ya.
         [| x : A; ya : C(x) |] ==>
         ?d2(x,y,ya) : (SUM x:A. B(x)) + (SUM x:A. C(x))
```

Subgoal 1 is an example of a well-formedness subgoal [9]. Such subgoals are usually trivial; this one yields to `typechk_tac`, given the current list of premises.

```
by (typechk_tac prems);
  Level 7
  split(p,%x y. when(y,%xa. inl(<x,xa>),?d2(x,y)))
  : (SUM x:A. B(x)) + (SUM x:A. C(x))
   1. !!x y ya.
         [| x : A; ya : C(x) |] ==>
         ?d2(x,y,ya) : (SUM x:A. B(x)) + (SUM x:A. C(x))
```

This subgoal is the other case from the +-elimination above, and can be proved similarly. Quicker is to apply `pc_tac`. The main goal finally gets a fully instantiated proof object.

```
by (pc_tac prems 1);
  Level 8
  split(p,%x y. when(y,%xa. inl(<x,xa>),%y. inr(<x,y>)))
  : (SUM x:A. B(x)) + (SUM x:A. C(x))
  No subgoals!
```

Calling `pc_tac` after the first Σ-elimination above also proves this theorem.

20.11 Example: deriving a currying functional

In simply-typed languages such as ML, a currying functional has the type

$$(A \times B \to C) \to (A \to (B \to C)).$$

Let us generalize this to the dependent types Σ and Π. The functional takes a function f that maps $z : \Sigma(A, B)$ to $C(z)$; the resulting function maps $x \in A$ and $y \in B(x)$ to $C(\langle x, y\rangle)$.

Formally, there are three typing premises. A is a type; B is an A-indexed family of types; C is a family of types indexed by $\Sigma(A, B)$. The goal is expressed using `PROD f` to ensure that the parameter corresponding to the functional's argument is really called f; Isabelle echoes the type using `-->` because there is no explicit dependence upon f.

```
val prems = goal CTT.thy
    "[| A type; !!x. x:A ==> B(x) type;                                  \
\                 !!z. z: (SUM x:A. B(x)) ==> C(z) type                  \
\    |] ==> ?a : PROD f: (PROD z : (SUM x:A . B(x)) . C(z)).             \
\                     (PROD x:A . PROD y:B(x) . C(<x,y>))";
  Level 0
  ?a : (PROD z:SUM x:A. B(x). C(z)) -->
       (PROD x:A. PROD y:B(x). C(<x,y>))
   1. ?a : (PROD z:SUM x:A. B(x). C(z)) -->
           (PROD x:A. PROD y:B(x). C(<x,y>))
  val prems = ["A type  [A type]",
               "?x : A ==> B(?x) type  [!!x. x : A ==> B(x) type]",
               "?z : SUM x:A. B(x) ==> C(?z) type
                [!!z. z : SUM x:A. B(x) ==> C(z) type]"] : thm list
```

This is a chance to demonstrate `intr_tac`. Here, the tactic repeatedly applies Π-introduction and proves the rather tiresome typing conditions.

Note that $?a$ becomes instantiated to three nested λ-abstractions. It would be easier to read if the bound variable names agreed with the parameters in the subgoal. Isabelle attempts to give parameters the same names as corresponding bound variables in the goal, but this does not always work. In any event, the goal is logically correct.

```
by (intr_tac prems);
  Level 1
  lam x xa xb. ?b7(x,xa,xb)
  : (PROD z:SUM x:A. B(x). C(z)) --> (PROD x:A. PROD y:B(x). C(<x,y>))
   1. !!f x y.
         [| f : PROD z:SUM x:A. B(x). C(z); x : A; y : B(x) |] ==>
         ?b7(f,x,y) : C(<x,y>)
```

Using Π-elimination, we solve subgoal 1 by applying the function f.

```
by (eresolve_tac [ProdE] 1);
  Level 2
  lam x xa xb. x ` <xa,xb>
  : (PROD z:SUM x:A. B(x). C(z)) --> (PROD x:A. PROD y:B(x). C(<x,y>))
   1. !!f x y. [| x : A; y : B(x) |] ==> <x,y> : SUM x:A. B(x)
```

Finally, we verify that the argument's type is suitable for the function application. This is straightforward using introduction rules.

```
by (intr_tac prems);
  Level 3
  lam x xa xb. x ` <xa,xb>
  : (PROD z:SUM x:A. B(x). C(z)) --> (PROD x:A. PROD y:B(x). C(<x,y>))
  No subgoals!
```

Calling `pc_tac` would have proved this theorem in one step; it can also prove an example by Martin-Löf, related to $\vee$-elimination [33, page 58].

20.12 Example: proving the Axiom of Choice

Suppose we have a function $h \in \prod_{x \in A} \sum_{y \in B(x)} C(x, y)$, which takes $x \in A$ to some $y \in B(x)$ paired with some $z \in C(x, y)$. Interpreting propositions as types, this asserts that for all $x \in A$ there exists $y \in B(x)$ such that $C(x, y)$. The Axiom of Choice asserts that we can construct a function $f \in \prod_{x \in A} B(x)$ such that $C(x, f`x)$ for all $x \in A$, where the latter property is witnessed by a function $g \in \prod_{x \in A} C(x, f`x)$.

In principle, the Axiom of Choice is simple to derive in Constructive Type Theory. The following definitions work:

$$f \equiv \mathtt{fst} \circ h$$
$$g \equiv \mathtt{snd} \circ h$$

But a completely formal proof is hard to find. The rules can be applied in countless ways, yielding many higher-order unifiers. The proof can get bogged down in the details. But with a careful selection of derived rules (recall Fig. 20.10) and the type checking tactics, we can prove the theorem in nine steps.

```
val prems = goal CTT.thy
    "[| A type;  !!x. x:A ==> B(x) type;                        \
\        !!x y.[| x:A;  y:B(x) |] ==> C(x,y) type               \
\     |] ==> ?a : PROD h: (PROD x:A. SUM y:B(x). C(x,y)).      \
\                      (SUM f: (PROD x:A. B(x)). PROD x:A. C(x, f`x))";
  Level 0
  ?a : (PROD x:A. SUM y:B(x). C(x,y)) -->
       (SUM f:PROD x:A. B(x). PROD x:A. C(x,f ` x))
   1. ?a : (PROD x:A. SUM y:B(x). C(x,y)) -->
           (SUM f:PROD x:A. B(x). PROD x:A. C(x,f ` x))
  val prems = ["A type  [A type]",
               "?x : A ==> B(?x) type  [!!x. x : A ==> B(x) type]",
               "[| ?x : A; ?y : B(?x) |] ==> C(?x, ?y) type
                 [!!x y. [| x : A; y : B(x) |] ==> C(x, y) type]"]
             : thm list
```

First, `intr_tac` applies introduction rules and performs routine type checking. This instantiates $?a$ to a construction involving a λ-abstraction and an ordered pair. The pair's components are themselves λ-abstractions and there is a subgoal for each.

```
by (intr_tac prems);
  Level 1
  lam x. <lam xa. ?b7(x,xa),lam xa. ?b8(x,xa)>
  : (PROD x:A. SUM y:B(x). C(x,y)) -->
    (SUM f:PROD x:A. B(x). PROD x:A. C(x,f ` x))

   1. !!h x.
         [| h : PROD x:A. SUM y:B(x). C(x,y); x : A |] ==>
         ?b7(h,x) : B(x)

   2. !!h x.
         [| h : PROD x:A. SUM y:B(x). C(x,y); x : A |] ==>
         ?b8(h,x) : C(x,(lam x. ?b7(h,x)) ` x)
```

Subgoal 1 asks to find the choice function itself, taking $x \in A$ to some $?b_7(h,x) \in B(x)$. Subgoal 2 asks, given $x \in A$, for a proof object $?b_8(h,x)$ to witness that the choice function's argument and result lie in the relation C. This latter task will take up most of the proof.

```
by (eresolve_tac [ProdE RS SumE_fst] 1);
  Level 2
  lam x. <lam xa. fst(x ` xa),lam xa. ?b8(x,xa)>
  : (PROD x:A. SUM y:B(x). C(x,y)) -->
    (SUM f:PROD x:A. B(x). PROD x:A. C(x,f ` x))

   1. !!h x. x : A ==> x : A
   2. !!h x.
         [| h : PROD x:A. SUM y:B(x). C(x,y); x : A |] ==>
         ?b8(h,x) : C(x,(lam x. fst(h ` x)) ` x)
```

Above, we have composed `fst` with the function h. Unification has deduced that the function must be applied to $x \in A$. We have our choice function.

```
by (assume_tac 1);
  Level 3
  lam x. <lam xa. fst(x ` xa),lam xa. ?b8(x,xa)>
  : (PROD x:A. SUM y:B(x). C(x,y)) -->
    (SUM f:PROD x:A. B(x). PROD x:A. C(x,f ` x))
   1. !!h x.
         [| h : PROD x:A. SUM y:B(x). C(x,y); x : A |] ==>
         ?b8(h,x) : C(x,(lam x. fst(h ` x)) ` x)
```

Before we can compose `snd` with h, the arguments of C must be simplified. The derived rule `replace_type` lets us replace a type by any equivalent type, shown below as the schematic term $?A_{13}(h,x)$:

```
by (resolve_tac [replace_type] 1);
  Level 4
  lam x. <lam xa. fst(x ` xa),lam xa. ?b8(x,xa)>
  : (PROD x:A. SUM y:B(x). C(x,y)) -->
    (SUM f:PROD x:A. B(x). PROD x:A. C(x,f ` x))

   1. !!h x.
         [| h : PROD x:A. SUM y:B(x). C(x,y); x : A |] ==>
         C(x,(lam x. fst(h ` x)) ` x) = ?A13(h,x)
```

```
2. !!h x.
       [| h : PROD x:A. SUM y:B(x). C(x,y); x : A |] ==>
       ?b8(h,x) : ?A13(h,x)
```

The derived rule subst_eqtyparg lets us simplify a type's argument (by currying, $C(x)$ is a unary type operator):

```
by (resolve_tac [subst_eqtyparg] 1);
 Level 5
 lam x. <lam xa. fst(x ` xa),lam xa. ?b8(x,xa)>
 : (PROD x:A. SUM y:B(x). C(x,y)) -->
   (SUM f:PROD x:A. B(x). PROD x:A. C(x,f ` x))
  1. !!h x.
        [| h : PROD x:A. SUM y:B(x). C(x,y); x : A |] ==>
        (lam x. fst(h ` x)) ` x = ?c14(h,x) : ?A14(h,x)
  2. !!h x z.
        [| h : PROD x:A. SUM y:B(x). C(x,y); x : A;
           z : ?A14(h,x) |] ==>
        C(x,z) type
  3. !!h x.
        [| h : PROD x:A. SUM y:B(x). C(x,y); x : A |] ==>
        ?b8(h,x) : C(x,?c14(h,x))
```

Subgoal 1 requires simply β-contraction, which is the rule ProdC. The term $?c_{14}(h, x)$ in the last subgoal receives the contracted result.

```
by (resolve_tac [ProdC] 1);
 Level 6
 lam x. <lam xa. fst(x ` xa),lam xa. ?b8(x,xa)>
 : (PROD x:A. SUM y:B(x). C(x,y)) -->
   (SUM f:PROD x:A. B(x). PROD x:A. C(x,f ` x))
  1. !!h x.
        [| h : PROD x:A. SUM y:B(x). C(x,y); x : A |] ==>
        x : ?A15(h,x)
  2. !!h x xa.
        [| h : PROD x:A. SUM y:B(x). C(x,y); x : A;
           xa : ?A15(h,x) |] ==>
        fst(h ` xa) : ?B15(h,x,xa)
  3. !!h x z.
        [| h : PROD x:A. SUM y:B(x). C(x,y); x : A;
           z : ?B15(h,x,x) |] ==>
        C(x,z) type
  4. !!h x.
        [| h : PROD x:A. SUM y:B(x). C(x,y); x : A |] ==>
        ?b8(h,x) : C(x,fst(h ` x))
```

Routine type checking goals proliferate in Constructive Type Theory, but typechk_tac quickly solves them. Note the inclusion of SumE_fst along with the premises.

```
by (typechk_tac (SumE_fst::prems));
 Level 7
 lam x. <lam xa. fst(x ` xa),lam xa. ?b8(x,xa)>
 : (PROD x:A. SUM y:B(x). C(x,y)) -->
   (SUM f:PROD x:A. B(x). PROD x:A. C(x,f ` x))
```

```
1. !!h x.
        [| h : PROD x:A. SUM y:B(x). C(x,y); x : A |] ==>
        ?b8(h,x) : C(x,fst(h ` x))
```

We are finally ready to compose snd with h.

```
by (eresolve_tac [ProdE RS SumE_snd] 1);
  Level 8
  lam x. <lam xa. fst(x ` xa),lam xa. snd(x ` xa)>
  : (PROD x:A. SUM y:B(x). C(x,y)) -->
    (SUM f:PROD x:A. B(x). PROD x:A. C(x,f ` x))
   1. !!h x. x : A ==> x : A
   2. !!h x. x : A ==> B(x) type
   3. !!h x xa. [| x : A; xa : B(x) |] ==> C(x,xa) type
```

The proof object has reached its final form. We call typechk_tac to finish the type checking.

```
by (typechk_tac prems);
  Level 9
  lam x. <lam xa. fst(x ` xa),lam xa. snd(x ` xa)>
  : (PROD x:A. SUM y:B(x). C(x,y)) -->
    (SUM f:PROD x:A. B(x). PROD x:A. C(x,f ` x))
  No subgoals!
```

It might be instructive to compare this proof with Martin-Löf's forward proof of the Axiom of Choice [33, page 50].

A. Syntax of Isabelle Theories

Chapter 9.1 explains the meanings of these constructs. The syntax obeys the following conventions:

- `Typewriter font` denotes terminal symbols.
- *id*, *tid*, *nat*, *string* and *text* are the lexical classes of identifiers, type identifiers, natural numbers, ML strings (with their quotation marks) and arbitrary ML text. The first three are fully defined in Chap. 10.

Comments in theories take the form `(*` *text* `*)`, where *text* should not contain the character sequence `*)`.

theoryDef

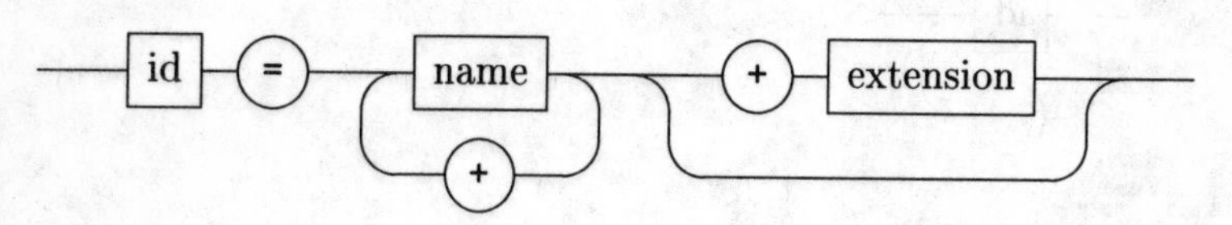

name

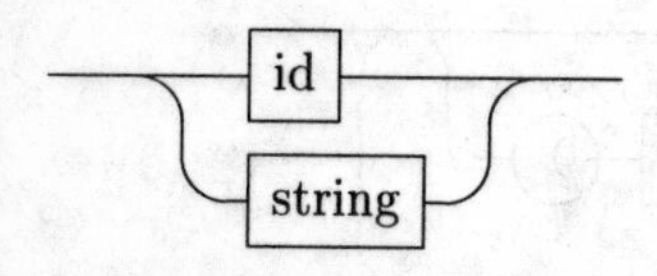

extension

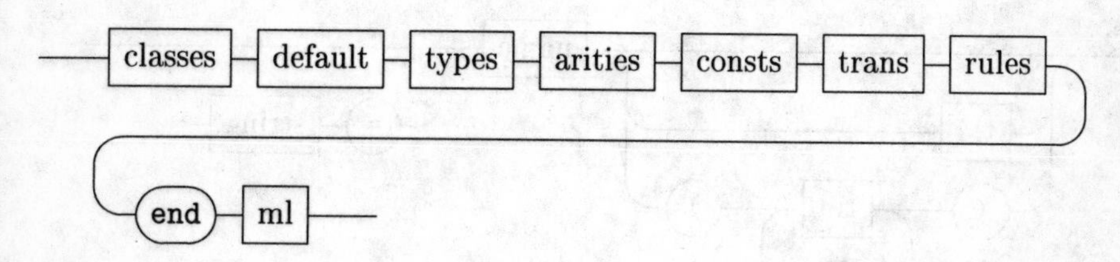

classes

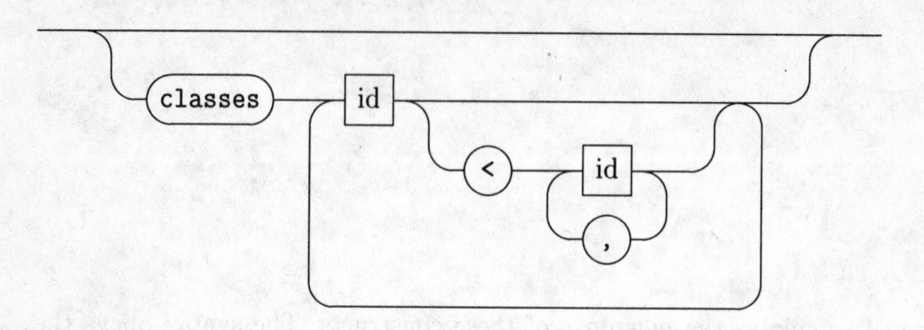

default

sort

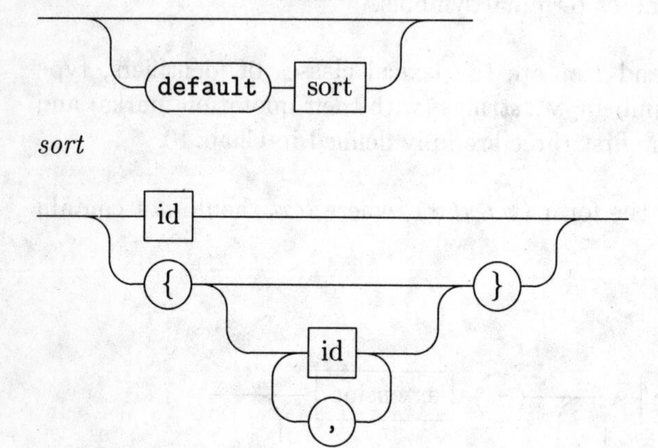

types

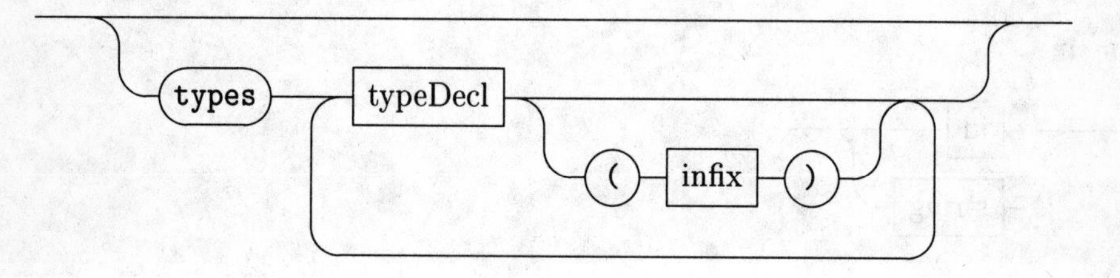

typeDecl

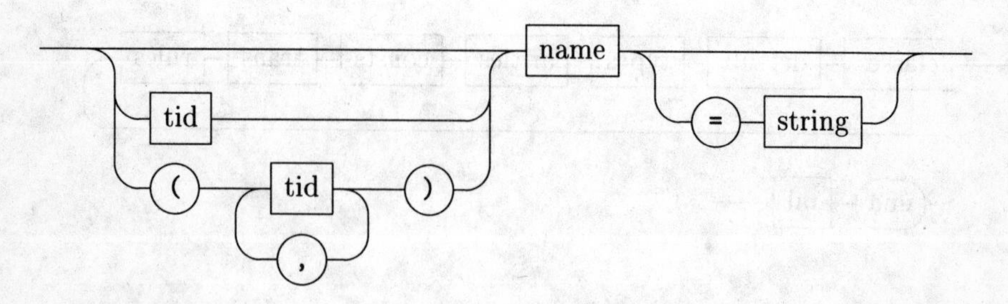

infix

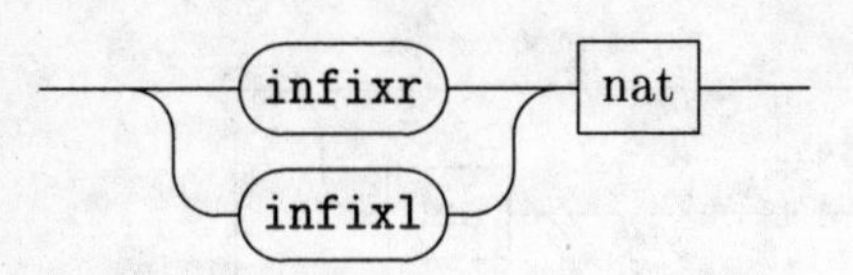

arities

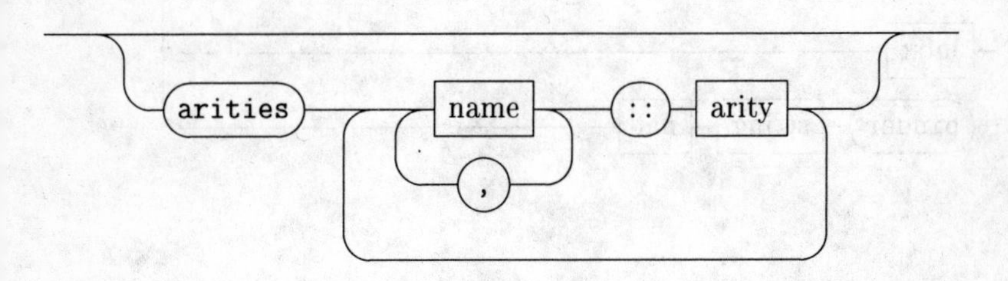

arity

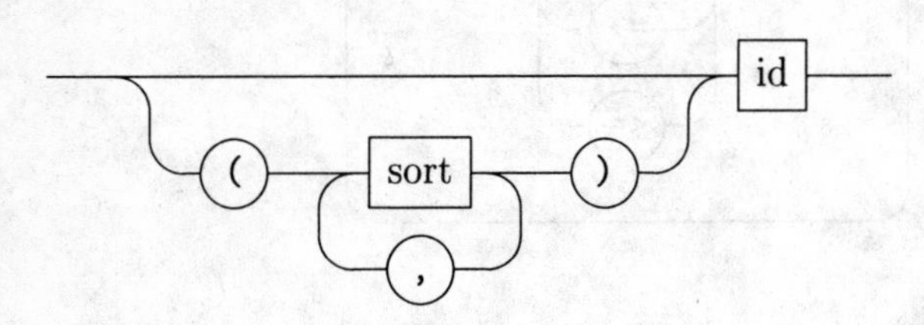

consts

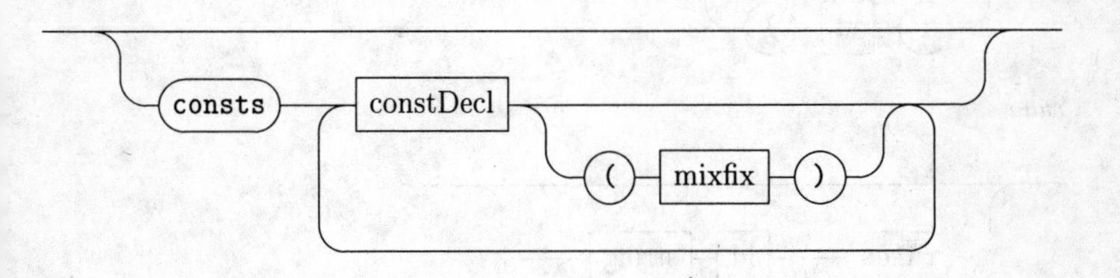

constDecl

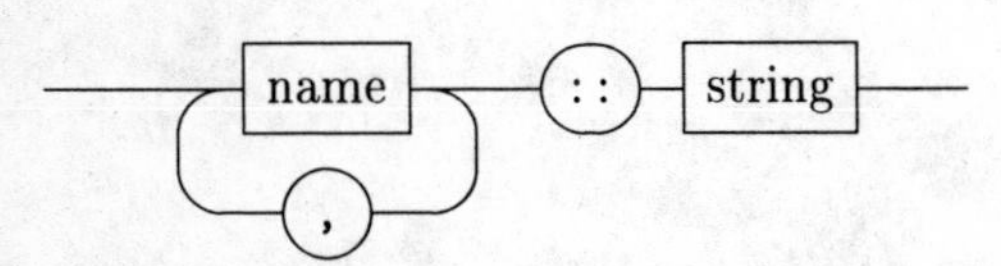

mixfix

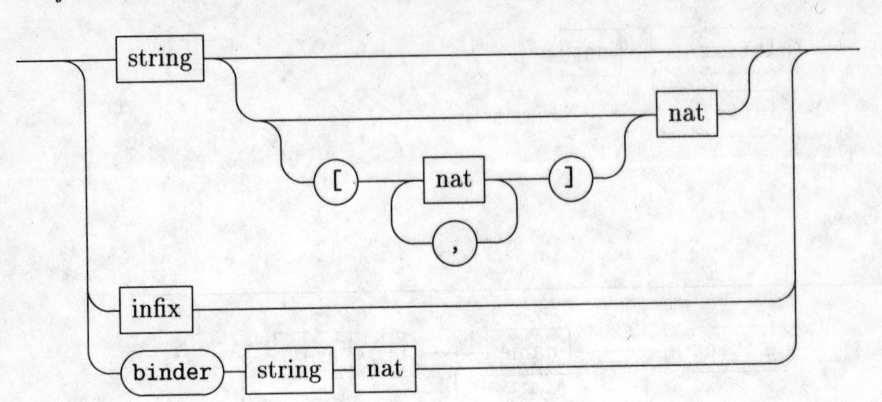

trans

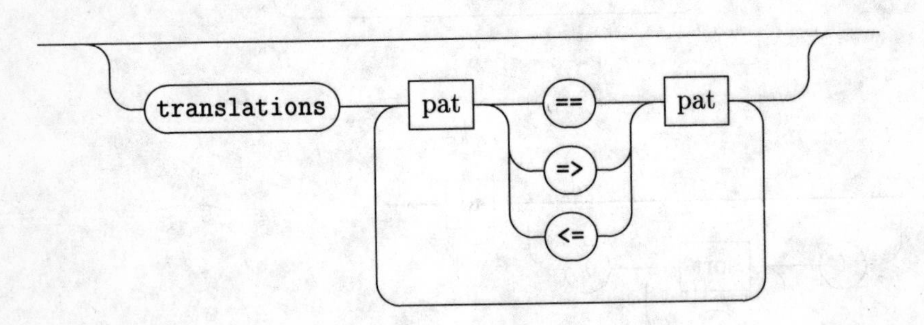

pat

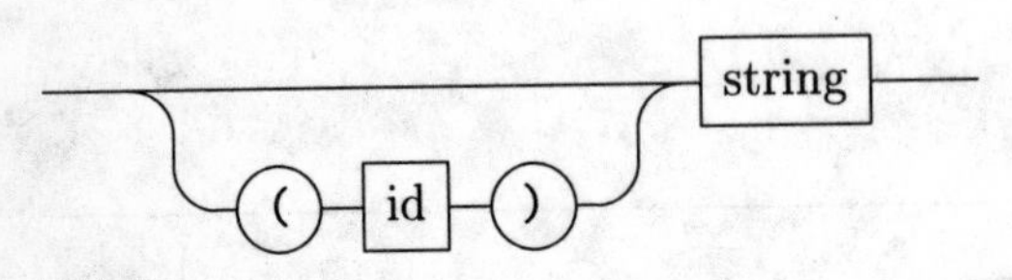

rules

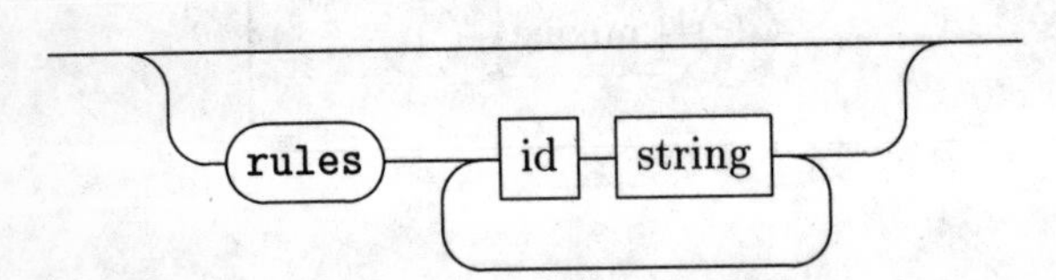

ml

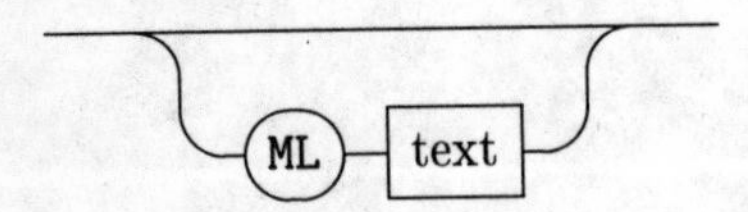

References

1. Andrews, P. B., *An Introduction to Mathematical Logic and Type Theory: To Truth Through Proof*, Academic Press, 1986

2. Basin, D., Kaufmann, M., The Boyer-Moore prover and Nuprl: An experimental comparison, In *Logical Frameworks*, G. Huet, G. Plotkin, Eds. Cambridge Univ. Press, 1991, pp. 89–119

3. Boyer, R., Lusk, E., McCune, W., Overbeek, R., Stickel, M., Wos, L., Set theory in first-order logic: Clauses for Gödel's axioms, *J. Auto. Reas.* **2**, 3 (1986), 287–327

4. Boyer, R. S., Moore, J. S., *A Computational Logic Handbook*, Academic Press, 1988

5. Camilleri, J., Melham, T. F., Reasoning with inductively defined relations in the HOL theorem prover, Tech. Rep. 265, Comp. Lab., Univ. Cambridge, Aug. 1992

6. Charniak, E., Riesbeck, C. K., McDermott, D. V., *Artificial Intelligence Programming*, Lawrence Erlbaum Associates, 1980

7. Church, A., A formulation of the simple theory of types, *J. Symb. Logic* **5** (1940), 56–68

8. Coen, M. D., *Interactive Program Derivation*, PhD thesis, University of Cambridge, 1992, Computer Laboratory Technical Report 272

9. Constable et al., R. L., *Implementing Mathematics with the Nuprl Proof Development System*, Prentice-Hall, 1986

10. Davey, B. A., Priestley, H. A., *Introduction to Lattices and Order*, Cambridge Univ. Press, 1990

11. Dawson, W. M., *A Generic Logic Environment*, PhD thesis, Imperial College, London, 1990

12. de Bruijn, N. G., Lambda calculus notation with nameless dummies, a tool for automatic formula manipulation, with application to the Church-Rosser Theorem, *Indag. Math.* **34** (1972), 381–392

13. Devlin, K. J., *Fundamentals of Contemporary Set Theory*, Springer, 1979

14. Dowek et al., G., The Coq proof assistant user's guide, Technical Report 134, INRIA-Rocquencourt, 1991

15. Dummett, M., *Elements of Intuitionism*, Oxford University Press, 1977

16. Dyckhoff, R., Contraction-free sequent calculi for intuitionistic logic, *J. Symb. Logic* **57**, 3 (1992), 795–807

17. Felty, A., A logic program for transforming sequent proofs to natural deduction proofs, In *Extensions of Logic Programming* (1991), P. Schroeder-Heister, Ed., Springer, pp. 157–178, LNAI 475

18. Felty, A., Implementing tactics and tacticals in a higher-order logic programming language, *J. Auto. Reas.* **11**, 1 (1993), 43–82

19. Frost, J., A case study of co-induction in Isabelle HOL, Tech. Rep. 308, Comp. Lab., Univ. Cambridge, Aug. 1993

20. Futatsugi, K., Goguen, J., Jouannaud, J.-P., Meseguer, J., Principles of OBJ2, In *Princ. Prog. Lang.* (1985), pp. 52–66

21. Gallier, J. H., *Logic for Computer Science: Foundations of Automatic Theorem Proving*, Harper & Row, 1986

22. Gordon, M. J. C., Melham, T. F., *Introduction to HOL: A Theorem Proving Environment for Higher Order Logic*, Cambridge Univ. Press, 1993

23. Halmos, P. R., *Naive Set Theory*, Van Nostrand, 1960

24. Harper, R., Honsell, F., Plotkin, G., A framework for defining logics, *J. ACM* **40**, 1 (1993), 143–184

25. Hudak, P., Fasel, J. H., A gentle introduction to Haskell, *SIGPLAN* **27**, 5 (May 1992)

26. Hudak, P., Jones, S. P., Wadler, P., Report on the programming language Haskell: A non-strict, purely functional language, *SIGPLAN* **27**, 5 (May 1992), Version 1.2

27. Huet, G. P., A unification algorithm for typed λ-calculus, *Theoretical Comput. Sci.* **1** (1975), 27–57

28. Huet, G. P., Lang, B., Proving and applying program transformations expressed with second-order patterns, *Acta Inf.* **11** (1978), 31–55

29. Jones, C. B., Jones, K. D., Lindsay, P. A., Moore, R., *Mural: A Formal Development Support System*, Springer, 1991

30. Magnusson, L., Nordström, B., The ALF proof editor and its proof engine, In *Types for Proofs and Programs: International Workshop TYPES '93* (published 1994), Springer, pp. 213–237, LNCS 806

31. Manna, Z., Waldinger, R., Deductive synthesis of the unification algorithm, *Sci. Comput. Programming* **1**, 1 (1981), 5–48

32. Martin, U., Nipkow, T., Ordered rewriting and confluence, In *10th Conf. Auto. Deduct.* (1990), M. E. Stickel, Ed., Springer, pp. 366–380, LNCS 449

33. Martin-Löf, P., *Intuitionistic type theory*, Bibliopolis, 1984

34. Melham, T. F., Automating recursive type definitions in higher order logic, In *Current Trends in Hardware Verification and Automated Theorem Proving*, G. Birtwistle, P. A. Subrahmanyam, Eds. Springer, 1989, pp. 341–386

35. Miller, D., Unification under a mixed prefix, *J. Symb. Comput.* **14**, 4 (1992), 321–358

36. Milner, R., Tofte, M., Co-induction in relational semantics, *Theoretical Comput. Sci.* **87** (1991), 209–220

37. Nipkow, T., Prehofer, C., Type checking type classes, In *20th Princ. Prog. Lang.* (1993), ACM Press, pp. 409–418, Revised version to appear in *J. Func. Prog.*

38. Noël, P., Experimenting with Isabelle in ZF set theory, *J. Auto. Reas.* **10**, 1 (1993), 15–58

39. Nordström, B., Petersson, K., Smith, J., *Programming in Martin-Löf's Type Theory. An Introduction*, Oxford University Press, 1990

40. Paulin-Mohring, C., Inductive definitions in the system Coq: Rules and properties, Research Report 92-49, LIP, Ecole Normale Supérieure de Lyon, Dec. 1992

41. Paulson, L. C., Verifying the unification algorithm in LCF, *Sci. Comput. Programming* **5** (1985), 143–170

42. Paulson, L. C., *Logic and Computation: Interactive proof with Cambridge LCF*, Cambridge Univ. Press, 1987

43. Paulson, L. C., The foundation of a generic theorem prover, *J. Auto. Reas.* **5**, 3 (1989), 363–397

44. Paulson, L. C., A formulation of the simple theory of types (for Isabelle), In *COLOG-88: International Conference on Computer Logic* (Tallinn, 1990), P. Martin-Löf, G. Mints, Eds., Estonian Academy of Sciences, Springer, LNCS 417

45. Paulson, L. C., Isabelle: The next 700 theorem provers, In *Logic and Computer Science*, P. Odifreddi, Ed. Academic Press, 1990, pp. 361–386

46. Paulson, L. C., *ML for the Working Programmer*, Cambridge Univ. Press, 1991

47. Paulson, L. C., Co-induction and co-recursion in higher-order logic, Tech. Rep. 304, Comp. Lab., Univ. Cambridge, July 1993

48. Paulson, L. C., A fixedpoint approach to implementing (co)inductive definitions, Tech. Rep. 320, Comp. Lab., Univ. Cambridge, Dec. 1993

49. Paulson, L. C., Set theory for verification: I. From foundations to functions, *J. Auto. Reas.* **11**, 3 (1993), 353–389

50. Paulson, L. C., Set theory for verification: II. Induction and recursion, Tech. Rep. 312, Comp. Lab., Univ. Cambridge, 1993

51. Paulson, L. C., A concrete final coalgebra theorem for ZF set theory, Tech. rep., Comp. Lab., Univ. Cambridge, 1994

52. Pelletier, F. J., Seventy-five problems for testing automatic theorem provers, *J. Auto. Reas.* **2** (1986), 191–216, Errata, JAR 4 (1988), 235–236

53. Plaisted, D. A., A sequent-style model elimination strategy and a positive refinement, *J. Auto. Reas.* **6**, 4 (1990), 389–402

54. Quaife, A., Automated deduction in von Neumann-Bernays-Gödel set theory, *J. Auto. Reas.* **8**, 1 (1992), 91–147

55. Sawamura, H., Minami, T., Ohashi, K., Proof methods based on sheet of thought in EUODHILOS, Research Report IIAS-RR-92-6E, International Institute for Advanced Study of Social Information Science, Fujitsu Laboratories, 1992

56. Suppes, P., *Axiomatic Set Theory*, Dover, 1972

57. Takeuti, G., *Proof Theory*, 2nd ed., North Holland, 1987

58. Thompson, S., *Type Theory and Functional Programming*, Addison-Wesley, 1991

59. Whitehead, A. N., Russell, B., *Principia Mathematica*, Cambridge Univ. Press, 1962, Paperback edition to *56, abridged from the 2nd edition (1927)

60. Wos, L., Automated reasoning and Bledsoe's dream for the field, In *Automated Reasoning: Essays in Honor of Woody Bledsoe*, R. S. Boyer, Ed. Kluwer Academic Publishers, 1991, pp. 297–342

Index

Printing: Weihert-Druck GmbH, Darmstadt
Binding: Theo Gansert Buchbinderei GmbH, Weinheim

Springer-Verlag and the Environment

We at Springer-Verlag firmly believe that an international science publisher has a special obligation to the environment, and our corporate policies consistently reflect this conviction.

We also expect our business partners – paper mills, printers, packaging manufacturers, etc. – to commit themselves to using environmentally friendly materials and production processes.

The paper in this book is made from low- or no-chlorine pulp and is acid free, in conformance with international standards for paper permanency.

Lecture Notes in Computer Science

For information about Vols. 1–751
please contact your bookseller or Springer-Verlag

Vol. 752: T. W. Finin, C. K. Nicholas, Y. Yesha (Eds.), Information and Knowledge Management. Proceedings, 1992. VII, 142 pages. 1993.

Vol. 753: L. J. Bass, J. Gornostaev, C. Unger (Eds.), Human-Computer Interaction. Proceedings, 1993. X, 388 pages. 1993.

Vol. 754: H. D. Pfeiffer, T. E. Nagle (Eds.), Conceptual Structures: Theory and Implementation. Proceedings, 1992. IX, 327 pages. 1993. (Subseries LNAI).

Vol. 755: B. Möller, H. Partsch, S. Schuman (Eds.), Formal Program Development. Proceedings. VII, 371 pages. 1993.

Vol. 756: J. Pieprzyk, B. Sadeghiyan, Design of Hashing Algorithms. XV, 194 pages. 1993.

Vol. 757: U. Banerjee, D. Gelernter, A. Nicolau, D. Padua (Eds.), Languages and Compilers for Parallel Computing. Proceedings, 1992. X, 576 pages. 1993.

Vol. 758: M. Teillaud, Towards Dynamic Randomized Algorithms in Computational Geometry. IX, 157 pages. 1993.

Vol. 759: N. R. Adam, B. K. Bhargava (Eds.), Advanced Database Systems. XV, 451 pages. 1993.

Vol. 760: S. Ceri, K. Tanaka, S. Tsur (Eds.), Deductive and Object-Oriented Databases. Proceedings, 1993. XII, 488 pages. 1993.

Vol. 761: R. K. Shyamasundar (Ed.), Foundations of Software Technology and Theoretical Computer Science. Proceedings, 1993. XIV, 456 pages. 1993.

Vol. 762: K. W. Ng, P. Raghavan, N. V. Balasubramanian, F. Y. L. Chin (Eds.), Algorithms and Computation. Proceedings, 1993. XIII, 542 pages. 1993.

Vol. 763: F. Pichler, R. Moreno Díaz (Eds.), Computer Aided Systems Theory – EUROCAST '93. Proceedings, 1993. IX, 451 pages. 1994.

Vol. 764: G. Wagner, Vivid Logic. XII, 148 pages. 1994. (Subseries LNAI).

Vol. 765: T. Helleseth (Ed.), Advances in Cryptology – EUROCRYPT '93. Proceedings, 1993. X, 467 pages. 1994.

Vol. 766: P. R. Van Loocke, The Dynamics of Concepts. XI, 340 pages. 1994. (Subseries LNAI).

Vol. 767: M. Gogolla, An Extended Entity-Relationship Model. X, 136 pages. 1994.

Vol. 768: U. Banerjee, D. Gelernter, A. Nicolau, D. Padua (Eds.), Languages and Compilers for Parallel Computing. Proceedings, 1993. XI, 655 pages. 1994.

Vol. 769: J. L. Nazareth, The Newton-Cauchy Framework. XII, 101 pages. 1994.

Vol. 770: P. Haddawy (Representing Plans Under Uncertainty. X, 129 pages. 1994. (Subseries LNAI).

Vol. 771: G. Tomas, C. W. Ueberhuber, Visualization of Scientific Parallel Programs. XI, 310 pages. 1994.

Vol. 772: B. C. Warboys (Ed.),Software Process Technology. Proceedings, 1994. IX, 275 pages. 1994.

Vol. 773: D. R. Stinson (Ed.), Advances in Cryptology – CRYPTO '93. Proceedings, 1993. X, 492 pages. 1994.

Vol. 774: M. Banâtre, P. A. Lee (Eds.), Hardware and Software Architectures for Fault Tolerance. XIII, 311 pages. 1994.

Vol. 775: P. Enjalbert, E. W. Mayr, K. W. Wagner (Eds.), STACS 94. Proceedings, 1994. XIV, 782 pages. 1994.

Vol. 776: H. J. Schneider, H. Ehrig (Eds.), Graph Transformations in Computer Science. Proceedings, 1993. VIII, 395 pages. 1994.

Vol. 777: K. von Luck, H. Marburger (Eds.), Management and Processing of Complex Data Structures. Proceedings, 1994. VII, 220 pages. 1994.

Vol. 778: M. Bonuccelli, P. Crescenzi, R. Petreschi (Eds.), Algorithms and Complexity. Proceedings, 1994. VIII, 222 pages. 1994.

Vol. 779: M. Jarke, J. Bubenko, K. Jeffery (Eds.), Advances in Database Technology — EDBT '94. Proceedings, 1994. XII, 406 pages. 1994.

Vol. 780: J. J. Joyce, C.-J. H. Seger (Eds.), Higher Order Logic Theorem Proving and Its Applications. Proceedings, 1993. X, 518 pages. 1994.

Vol. 781: G. Cohen, S. Litsyn, A. Lobstein, G. Zémor (Eds.), Algebraic Coding. Proceedings, 1993. XII, 326 pages. 1994.

Vol. 782: J. Gutknecht (Ed.), Programming Languages and System Architectures. Proceedings, 1994. X, 344 pages. 1994.

Vol. 783: C. G. Günther (Ed.), Mobile Communications. Proceedings, 1994. XVI, 564 pages. 1994.

Vol. 784: F. Bergadano, L. De Raedt (Eds.), Machine Learning: ECML-94. Proceedings, 1994. XI, 439 pages. 1994. (Subseries LNAI).

Vol. 785: H. Ehrig, F. Orejas (Eds.), Recent Trends in Data Type Specification. Proceedings, 1992. VIII, 350 pages. 1994.

Vol. 786: P. A. Fritzson (Ed.), Compiler Construction. Proceedings, 1994. XI, 451 pages. 1994.

Vol. 787: S. Tison (Ed.), Trees in Algebra and Programming – CAAP '94. Proceedings, 1994. X, 351 pages. 1994.

Vol. 788: D. Sannella (Ed.), Programming Languages and Systems – ESOP '94. Proceedings, 1994. VIII, 516 pages. 1994.

Vol. 789: M. Hagiya, J. C. Mitchell (Eds.), Theoretical Aspects of Computer Software. Proceedings, 1994. XI, 887 pages. 1994.

Vol. 790: J. van Leeuwen (Ed.), Graph-Theoretic Concepts in Computer Science. Proceedings, 1993. IX, 431 pages. 1994.

Vol. 791: R. Guerraoui, O. Nierstrasz, M. Riveill (Eds.), Object-Based Distributed Programming. Proceedings, 1993. VII, 262 pages. 1994.

Vol. 792: N. D. Jones, M. Hagiya, M. Sato (Eds.), Logic, Language and Computation. XII, 269 pages. 1994.

Vol. 793: T. A. Gulliver, N. P. Secord (Eds.), Information Theory and Applications. Proceedings, 1993. XI, 394 pages. 1994.

Vol. 794: G. Haring, G. Kotsis (Eds.), Computer Performance Evaluation. Proceedings, 1994. X, 464 pages. 1994.

Vol. 795: W. A. Hunt, Jr., FM8501: A Verified Microprocessor. XIII, 333 pages. 1994.

Vol. 796: W. Gentzsch, U. Harms (Eds.), High-Performance Computing and Networking. Proceedings, 1994, Vol. I. XXI, 453 pages. 1994.

Vol. 797: W. Gentzsch, U. Harms (Eds.), High-Performance Computing and Networking. Proceedings, 1994, Vol. II. XXII, 519 pages. 1994.

Vol. 798: R. Dyckhoff (Ed.), Extensions of Logic Programming. Proceedings, 1993. VIII, 362 pages. 1994.

Vol. 799: M. P. Singh, Multiagent Systems. XXIII, 168 pages. 1994. (Subseries LNAI).

Vol. 800: J.-O. Eklundh (Ed.), Computer Vision – ECCV '94. Proceedings 1994, Vol. I. XVIII, 603 pages. 1994.

Vol. 801: J.-O. Eklundh (Ed.), Computer Vision – ECCV '94. Proceedings 1994, Vol. II. XV, 485 pages. 1994.

Vol. 802: S. Brookes, M. Main, A. Melton, M. Mislove, D. Schmidt (Eds.), Mathematical Foundations of Programming Semantics. Proceedings, 1993. IX, 647 pages. 1994.

Vol. 803: J. W. de Bakker, W.-P. de Roever, G. Rozenberg (Eds.), A Decade of Concurrency. Proceedings, 1993. VII, 683 pages. 1994.

Vol. 804: D. Hernández, Qualitative Representation of Spatial Knowledge. IX, 202 pages. 1994. (Subseries LNAI).

Vol. 805: M. Cosnard, A. Ferreira, J. Peters (Eds.), Parallel and Distributed Computing. Proceedings, 1994. X, 280 pages. 1994.

Vol. 806: H. Barendregt, T. Nipkow (Eds.), Types for Proofs and Programs. VIII, 383 pages. 1994.

Vol. 807: M. Crochemore, D. Gusfield (Eds.), Combinatorial Pattern Matching. Proceedings, 1994. VIII, 326 pages. 1994.

Vol. 808: M. Masuch, L. Pólos (Eds.), Knowledge Representation and Reasoning Under Uncertainty. VII, 237 pages. 1994. (Subseries LNAI).

Vol. 809: R. Anderson (Ed.), Fast Software Encryption. Proceedings, 1993. IX, 223 pages. 1994.

Vol. 810: G. Lakemeyer, B. Nebel (Eds.), Foundations of Knowledge Representation and Reasoning. VIII, 355 pages. 1994. (Subseries LNAI).

Vol. 811: G. Wijers, S. Brinkkemper, T. Wasserman (Eds.), Advanced Information Systems Engineering. Proceedings, 1994. XI, 420 pages. 1994.

Vol. 812: J. Karhumäki, H. Maurer, G. Rozenberg (Eds.), Results and Trends in Theoretical Computer Science. Proceedings, 1994. X, 445 pages. 1994.

Vol. 813: A. Nerode, Yu. N. Matiyasevich (Eds.), Logical Foundations of Computer Science. Proceedings, 1994. IX, 392 pages. 1994.

Vol. 814: A. Bundy (Ed.), Automated Deduction—CADE-12. Proceedings, 1994. XVI, 848 pages. 1994. (Subseries LNAI).

Vol. 815: R. Valette (Ed.), Application and Theory of Petri Nets 1994. Proceedings. IX, 587 pages. 1994.

Vol. 816: J. Heering, K. Meinke, B. Möller, T. Nipkow (Eds.), Higher-Order Algebra, Logic, and Term Rewriting. Proceedings, 1993. VII, 344 pages. 1994.

Vol. 817: C. Halatsis, D. Maritsas, G. Philokyprou, S. Theodoridis (Eds.), PARLE '94. Parallel Architectures and Languages Europe. Proceedings, 1994. XV, 837 pages. 1994.

Vol. 818: D. L. Dill (Ed.), Computer Aided Verification. Proceedings, 1994. IX, 480 pages. 1994.

Vol. 819: W. Litwin, T. Risch (Eds.), Applications of Databases. Proceedings, 1994. XII, 471 pages. 1994.

Vol. 820: S. Abiteboul, E. Shamir (Eds.), Automata, Languages and Programming. Proceedings, 1994. XIII, 644 pages. 1994.

Vol. 821: M. Tokoro, R. Pareschi (Eds.), Object-Oriented Programming. Proceedings, 1994. XI, 535 pages. 1994.

Vol. 822: F. Pfenning (Ed.), Logic Programming and Automated Reasoning. Proceedings, 1994. X, 345 pages. 1994. (Subseries LNAI).

Vol. 823: R. A. Elmasri, V. Kouramajian, B. Thalheim (Eds.), Entity-Relationship Approach — ER '93. Proceedings, 1993. X, 531 pages. 1994.

Vol. 824: E. M. Schmidt, S. Skyum (Eds.), Algorithm Theory - SWAT '94. Proceedings. IX, 383 pages. 1994.

Vol. 825: J. L. Mundy, A. Zisserman, D. Forsyth (Eds.), Applications of Invariance in Computer Vision. Proceedings, 1993. IX, 510 pages.

Vol. 826: D. S. Bowers (Ed.), Directions in Databases. Proceedings, 1994. X, 234 pages. 1994.

Vol. 827: D. M. Gabbay, H. J. Ohlbach (Eds.), Temporal Logic. Proceedings, 1994. XI, 546 pages. 1994. (Subseries LNAI).

Vol. 828: L. C. Paulson, Isabelle. XVII, 321 pages. 1994.

Vol. 829: A. Chmora, S. B. Wicker (Eds.), Error Control, Cryptology, and Speech Compression. Proceedings, 1993. VIII, 121 pages. 1994.

Vol. 831: V. Bouchitté, M. Morvan (Eds.), Orders, Algorithms, and Applications. Proceedings, 1994. IX, 204 pages. 1994.